VOLUME THREE HUNDRED AND SIXTY SIX

INTERNATIONAL REVIEW OF CELL AND MOLECULAR BIOLOGY

Cellular and Molecular Aspects of Myeloproliferative Neoplasms - Part B

INTERNATIONAL REVIEW OF CELL AND MOLECULAR BIOLOGY

VOLUME THREE HUNDRED AND SIXTY SIX

INTERNATIONAL REVIEW OF CELL AND MOLECULAR BIOLOGY

Cellular and Molecular Aspects of Myeloproliferative Neoplasms - Part B

Edited by

NICCOLÒ BARTALUCCI
University of Florence, Florence, Italy

LORENZO GALLUZZI
Weill Cornell Medical College, New York, NY, United States

ELSEVIER

ACADEMIC PRESS
An imprint of Elsevier

Academic Press is an imprint of Elsevier
50 Hampshire Street, 5th Floor, Cambridge, MA 02139, United States
525 B Street, Suite 1650, San Diego, CA 92101, United States
The Boulevard, Langford Lane, Kidlington, Oxford OX5 1GB, United Kingdom
125 London Wall, London, EC2Y 5AS, United Kingdom

First edition 2022

Notices

Knowledge and best practice in this field are constantly changing. As new research and experience broaden our understanding, changes in research methods, professional practices, or medical treatment may become necessary.

Practitioners and researchers must always rely on their own experience and knowledge in evaluating and using any information, methods, compounds, or experiments described herein. In using such information or methods they should be mindful of their own safety and the safety of others, including parties for whom they have a professional responsibility.

To the fullest extent of the law, neither the Publisher nor the authors, contributors, or editors, assume any liability for any injury and/or damage to persons or property as a matter of products liability, negligence or otherwise, or from any use or operation of any methods, products, instructions, or ideas contained in the material herein.

ISBN: 978-0-323-89941-3
ISSN: 1937-6448

Publisher: Zoe Kruze
Acquisitions Editor: Ashlie M. Jackman
Developmental Editor: Tara Nadera
Production Project Manager: Denny Mansingh
Cover Designer: Alan Studholme

Typeset by STRAIVE, India

Contents

Contributors

Sarah Bailly
Ludwig Institute for Cancer Research; SIGN Unit, de Duve Institute, Université Catholique de Louvain; Department of Hematology, Cliniques Universitaires Saint Luc, Université Catholique de Louvain, Brussels, Belgium

Camelia Benlabiod
INSERM; Université Paris-Saclay, UMR 1287, Gustave Roussy; Gustave Roussy, UMR 1287, Villejuif, France

Prithviraj Bose
Leukemia Department, The University of Texas MD Anderson Cancer Center, Houston, TX, United States

Evan M. Braunstein
Hematology Division, Department of Medicine, Johns Hopkins University School of Medicine, Baltimore, Maryland, United States

Helen T. Chifotides
Leukemia Department, The University of Texas MD Anderson Cancer Center, Houston, TX, United States

Stefan N. Constantinescu
Ludwig Institute for Cancer Research; SIGN Unit, de Duve Institute, Université Catholique de Louvain; WELBIO (Walloon Excellence in Life Sciences and Biotechnology), Brussels, Belgium; Ludwig Institute for Cancer Research, Nuffield Department of Medicine, Oxford University, Oxford, United Kingdom

Tracy Dagher
INSERM; Université Paris-Saclay, UMR 1287, Gustave Roussy; Gustave Roussy, UMR 1287, Villejuif, France

Carmen C. Diaconu
Department of Cellular and Molecular Pathology, Stefan S. Nicolau Institute of Virology, Bucharest, Romania

Violaine Havelange
SIGN Unit, de Duve Institute, Université Catholique de Louvain; Department of Hematology, Cliniques Universitaires Saint Luc, Université Catholique de Louvain, Brussels, Belgium

Gabriel Levy
Ludwig Institute for Cancer Research; SIGN Unit, de Duve Institute; Department of Pediatric Hematology and Oncology, Cliniques Universitaires Saint-Luc, Université Catholique de Louvain, Brussels, Belgium

Cristina Mambet
Department of Cellular and Molecular Pathology, Stefan S. Nicolau Institute of Virology; Department of Hematology, Carol Davila University of Medicine and Pharmacy, Bucharest, Romania

Caroline Marty
INSERM; Université Paris-Saclay, UMR 1287, Gustave Roussy; Gustave Roussy, UMR 1287, Villejuif, France

Alison R. Moliterno
Hematology Division, Department of Medicine, Johns Hopkins University School of Medicine, Baltimore, Maryland, United States

Alice Moncada
Laboratory Medicine Department, Molecular and Clinical Pathology Sector, Azienda USL Toscana Sudest, Ospedale San Donato, Arezzo, Italy

Alessandro Pancrazzi
Laboratory Medicine Department, Molecular and Clinical Pathology Sector, Azienda USL Toscana Sudest, Ospedale San Donato, Arezzo, Italy

Sergiu Pasca
Leukemia Department, The University of Texas MD Anderson Cancer Center, Houston, TX, United States

Christian Pecquet
Ludwig Institute for Cancer Research; SIGN Unit, de Duve Institute, Université Catholique de Louvain; WELBIO (Walloon Excellence in Life Sciences and Biotechnology), Brussels, Belgium

Fiorella Schischlik
Cancer Data Science Laboratory, Center for Cancer Research, National Cancer Institute, Bethesda, MD, United States

Srdan Verstovsek
Leukemia Department, The University of Texas MD Anderson Cancer Center, Houston, TX, United States

Jean-Luc Villeval
INSERM; Université Paris-Saclay, UMR 1287, Gustave Roussy; Gustave Roussy, UMR 1287, Villejuif, France

Philadelphia-negative myeloproliferative neoplasms: From origins to new perspectives

Niccolò Bartalucci[a,b,*] and Lorenzo Galluzzi[c,d,e,f,g,*]
[a]Department of Experimental and Clinical Medicine, University of Florence, Florence, Italy
[b]DENOThe Excellence Center, University of Florence, Florence, Italy
[c]Department of Radiation Oncology, Weill Cornell Medical College, New York, NY, United States
[d]Sandra and Edward Meyer Cancer Center, New York, NY, United States
[e]Caryl and Israel Englander Institute for Precision Medicine, New York, NY, United States
[f]Department of Dermatology, Yale School of Medicine, New Haven, CT, United States
[g]Université de Paris, Paris, France
[*]Corresponding authors: e-mail address: niccolo.bartalucci@unifi.it; deadoc80@gmail.com

Myeloproliferative neoplasms (MPNs) are hematological disorders that originate from the clonal expansion of multipotent myeloid progenitors and are characterized by an overproduction of differentiated hematopoietic cells in the bone marrow and extramedullary tissues. The concept of myeloproliferation was originally introduced by William Dameshek in 1951 in the editorial "Some speculations on the myeloproliferative syndromes" (Dameshek, 1951). The pathological conditions described in the visionary work of Dameshek were characterized by the excessive production of mature blood cells and the proliferation of hematopoietic precursors in the bone marrow, culminating with various disorders such as chronic granulocytic leukemia, polycythemia vera, megakaryocytic/erythrocytic leukemia, and myeloid metaplasia of the spleen. The concept of myeloproliferation has been further refined by a large amount of work published over the past decades that concurred to unravel the mutational landscape of MPNs (Tefferi, 2016). While some of the original definitions from Dameshek changed over the years, the term "myeloproliferative neoplasms" was officially introduced by the World Health Organization in 2008 to highlight the clonal nature of such hematologic disorders (Tefferi and Vardiman, 2008). Nowadays, MPNs encompass various myeloid disorders such as chronic myeloid leukemia (CML), chronic neutrophilic leukemia, polycythemia vera (PV), essential thrombocythemia (ET), primary myelofibrosis (PMF), chronic eosinophilic leukemia and myeloid/lymphoid neoplasms with eosinophilia and genetic rearrangements of platelet-derived growth factor receptor alpha (*PDGFRA*), platelet-derived growth factor receptor beta (*PDGFRB*), fibroblast growth factor receptor 1 (*FGFR1*), or Janus kinase (*JAK2*) as well as a group of unclassifiable overlapping MPNs (Arber et al., 2016).

In modern clinical practice, MPNs are classified as Philadelphia positive (Ph^+) or negative (Ph^-) based on the presence of a reciprocal chromosomal translocation between the q arms of chromosomes 9 and 22 involving the BCR activator of RhoGEF and GTPase (*BCR*) and ABL proto-oncogene 1, nonreceptor tyrosine kinase (*ABL1*) that is reflected by the appearance of a minute chromosome commonly known as Philadelphia (Nowell, 1962). The resulting *BCR-ABL1* fusion gene (Rowley, 1973) codes for a chimeric tyrosine kinase responsible for the clinical phenotype of most cases of CML. Conversely, Ph^- MPNs represent a continuum of disorders with overlapping features and specific clinical manifestations including PV, ET, and PMF (Passamonti and Maffioli, 2016). Ph^- MPNs are characterized by the abnormal proliferation of different myeloid lineages (Van Egeren et al., 2021). They can not only transition from one to another (Masarova and Verstovsek, 2019) but also progress to acute leukemia (Dunbar et al., 2020; Mannelli, 2021). Despite the specificities of the distinct pathological entities, Ph^- MPNs share common clinical signs such as an increased propensity for thromboembolic manifestations, a subverted bone marrow architecture leading to a massive deposition of fibers, and the tendency to leukemic transformation (Spivak, 2017).

The dramatic progress in the understanding of the cellular and molecular basis of deregulated myeloproliferation achieved in the recent years provided intriguing insights into the mechanisms driving the origin and the progression of this disease (Bartalucci et al., 2020; Lanikova et al., 2019; Morotti et al., 2017). Specifically, the acquisition of somatic mutations leading to the clonal expansion of hematologic precursors is considered the hallmark of all myeloid malignancies. However, MPN-initiating hits are heterogeneous, encompassing large chromosomal aberrations, hot spot mutations, and epigenetics modifications that subvert the regulation of key signaling pathways involved in cell growth, survival, differentiation, and cell death (Galluzzi et al., 2018; Vainchenker et al., 2016). The vast majority of Ph^- MPNs are associated with somatic driver mutations that promote the development of malignant stem cells that are characterized by a survival advantage over normal hematopoietic precursors and dysregulated lineage commitment (Grabek et al., 2020; Vainchenker and Kralovics, 2017). The driver mutations of classical Ph^- MPNs involve *JAK2* (Baxter et al., 2005; James et al., 2005; Kralovics et al., 2005; Levine et al., 2005), MPL proto-oncogene, thrombopoietin receptor (*MPL*) (Pikman et al., 2006), or calreticulin (*CALR*) (Klampfl et al., 2013; Nangalia et al., 2013). Specifically, almost all cases of PV and about half of patients with ET and PMF harbor the *JAK2* V617F mutation in exon 14 (O'Sullivan and Mead, 2019). Other common alternations include mutations in *MPL* exon 10 (*MPL* W515), *CALR* exon 9, and *JAK2* exon 12 (O'Sullivan and Mead, 2019). These mutations generally result in the constitutive activation of intracellular pathways that ultimately break the fine-tuned balance of stimuli ensuring physiological myelopoiesis. Accordingly, hyperactivation of the signal transducer and activator of transcription (STAT) signaling downstream of *JAK*, *MPL*, or *CALR* mutations is considered the main etiological mechanism supporting Ph^{-} MPNs (Constantinescu et al., 2021). That said, approximately 10–15% of patients with clinical manifestations of MPN lack specific molecular markers and hence conventionally referred to as "triple-negative" (Barbui et al., 2018; Tefferi et al., 2014). According to the recent reports, overall survival is significantly longer for patients with *CALR* type 1 mutations (a 52 bp deletion) but otherwise similar in individuals bearing *JAK2*, *MPL*, and *CALR* nontype 1 mutations (Tefferi et al., 2018a).

The therapeutic options for patients with PV and ET are mainly related to age and the risk of thrombosis (Barbui et al., 2012, 2014). Specifically, low-risk PV and ET are managed to optimize the cardiovascular risk with low-dose aspirin, while phlebotomy is often adopted for PV to maintain a hematocrit lower than 45% (Marchioli et al., 2013). Cytoreductive therapy with hydroxyurea is the first option for high-risk patients (Cortelazzo et al., 1995; Fruchtman et al., 1997). Alternatively, pegylated interferon α can be used in younger subjects (Kiladjian et al., 2008). PMF patients with uncontrolled splenomegaly or constitutional symptoms should be evaluated for bone marrow transplantation or alternatively treated with the JAK1/2 inhibitor ruxolitinib (Verstovsek et al., 2017a,b) or the JAK2/FLT3 inhibitor fedratinib (Harrison et al., 2017), which represent the only targeted anticancer agents approved for this indication. Inhibition of the JAK/STAT axis generally results in clinical benefits including improvements in splenomegaly and symptoms not only in patients with PMF but also in subjects with secondary (post-ET/PV) myelofibrosis and chemotherapy-resistant PV (Brkic and Meyer, 2021; Vannucchi, 2015; Verstovsek et al., 2012). However, prolonged pan-JAK2 inhibition offers poor long-term benefits and may cause anemia (Greenfield et al., 2018; Schieber et al., 2019). This is probably due to additional molecular determinants that, together with genetic predisposition (O'Sullivan and Mead, 2019) and the detrimental effects of MPN-associated chronic inflammation of the bone marrow (Barone et al., 2020; Curto-Garcia et al., 2020; Fisher et al., 2019; Hoermann et al., 2015; Korn and Mendez-Ferrer, 2017), sculpt disease phenotype and influence progression. Furthermore, the signal transduction cascades originating from mutant *JAK2* (Dawson et al., 2009; Kleppe et al., 2015; Lai et al., 2019; Stetka et al., 2019), *MPL* (Cabagnols et al., 2016; Defour et al., 2016; Levy et al., 2020), and *CALR* (Arshad and Cresswell, 2018; Balliu et al., 2021; Di Buduo et al., 2020; Fucikova et al., 2021; Salati et al., 2019) remain to be fully elucidated. This is especially true for CALR, which has multiple cell-intrinsic and cell-extrinsic functions that may be differentially affected by exon 9 mutations (Fucikova et al., 2020a,b, 2021; Holmström et al., 2018a; Kepp et al., 2020; Liu et al., 2020). The dysregulation of ancillary signaling circuits observed in Ph$^-$ MPNs may also offer new biomarkers of the disease and molecular targets for therapy (Bartalucci et al., 2017; Coltro et al., 2020; Greenfield et al., 2020; Stivala et al., 2019; Subotički et al., 2021). Inhibiting some of such targets as a stand-alone strategy or in combination with JAK2 inhibition proved efficacious in both experimental (Bartalucci et al., 2013; Lu et al., 2014) and clinical studies (Guglielmelli et al., 2011; Mascarenhas et al., 2019a).

The advent of technologies enabling the in-depth screening of genomic and transcriptomic MPN profiles contributed to the identification of a complex mutational landscape modeled by the accumulation of somatic mutations as the result of an increasing genomic instability paralleling the disease evolution (Karantanos and Moliterno, 2018; Tefferi et al., 2016a, b; Vitale et al., 2021). In this context, various nondriver mutations in disease-modifying genes affect several cell functions (Lee, 2021; Milosevic and Kralovics, 2013), including epigenetic regulation, tumor suppression, regulation of transcription, and splicing, and seem to play a pivotal role in increasing the risk of progression to secondary myelofibrosis or leukemic transformation (Guglielmelli et al., 2014; Tefferi et al., 2016a). The study of these mutational hits and their order of acquisition allowed for the reconstruction of the complex clonal hierarchy of MPNs (Mead and Mullally, 2017; Ortmann et al., 2015). Furthermore, the integration of molecular findings with clinical data leads to the refinement of risk stratification (Guglielmelli et al., 2018; Tefferi et al., 2018b,c, 2020).

Despite the profound insights that were obtained on the factors shaping MPN stem cells and disrupting normal hematopoiesis, currently available therapeutic strategies remain inadequate to exhaust the neoplastic clone. In line with this notion, hematopoietic stem cell transplantation is the only curative option today (Davidson and Gupta, 2021). To challenge such an unmet clinical need, a wide variety of therapeutic strategies and novel drugs are under investigation as first-line, second-line, or ruxolitinib "add-on" interventions. Specifically, the pan-JAK inhibitors pacritinib (Mesa et al., 2017a) and momelotinib (Mesa et al., 2017b) are currently being evaluated in phase III clinical trials as first-line therapy in patients with PMF. Moreover, most of the recently discovered biomarkers of MPNs are being targeted to explore therapeutic efficacy. For instance, this is the case of agents targeting MDM2 proto-oncogene (MDM2), the negative regulator of tumor protein P53 (TP53) that is overexpressed in MPN cells expressing $JAK2^{V617F}$ (Lu et al., 2014), as well as BH3 (Mascarenhas et al., 2019b) and SMAC mimetics (Pemmaraju et al., 2019). These latter molecules are conceived to target the apoptotic machinery to specifically drive the demise of malignant over normal cells as only the former exhibit "mitochondrial priming" (Bhatt et al., 2020; Fairlie et al., 2020; Rasmussen and Gama, 2020; Sánchez-Rivera et al., 2021).

Previous studies have also demonstrated the ability of histone deacetylase inhibitors (HDACIs) to restrict the proliferation and induce apoptosis in MPN cells by favoring the downregulation of nuclear factor, erythroid 2 (NFE2) and MYB proto-oncogene transcription factor (MYB) (Amaru Calzada et al., 2012, 2013; Hobbs et al., 2018; Wang et al., 2009). Largely based on these preclinical findings, HDACIs have entered clinical development for the treatment of MPNs (Finazzi et al., 2013; Rambaldi et al., 2020). However, their activity seems to be restricted to PMF and not extend to ET and PV (Bose and Verstovsek, 2016). Also phosphoinositide-3-kinase (PI3K) signaling via AKT serine/threonine kinase 1 (AKT1) and the mechanistic target of rapamycin (MTOR) has attracted attention as a potential target for novel MPN treatments, largely reflecting the key role of this pathway in oncogenesis (Rosa et al., 2020) and immunosurveillance (Kepp et al., 2021), as well as the discovery that MPN cells with mutant *JAK2* exhibit robust MTOR activation (Grimwade et al., 2009; Vicari et al., 2012). Lately, such an interest has converged on the δ subunit of PI3K, the primary isoform expressed in $CD34^+$ PMF cells (Meadows et al., 2013). Accordingly, the PI3Kδ inhibitor parsaclisib is currently being tested as a combinatorial partner for ruxolitinib in a phase II clinical study enrolling patients with PMF (Shin et al., 2020). Along similar lines, cyclin-dependent kinase 4/6 (CDK4/6) inhibition (Petroni et al., 2020, 2021a,b) combined with ruxolitinib mediated promising therapeutic effects in preclinical mouse models of PMF (Dutta et al., 2021). However, the clinical development of this combinatorial regimen for the treatment of MPNs is still in its infancy. Moreover, the recent identification of new antigenic epitopes appears to pave the way for the development of various immunotherapeutic approaches (Holmström et al., 2018b; Schischlik et al., 2019), constituting a considerable step forward to the establishment of personalized immunotherapy as for a number of other tumors (Alcantara et al., 2020). For instance, a subset of T cells from patients with MPN bearing *CALR* mutations are responsive to the C-terminal domain of CALR, but their proliferation is abrogated by coinhibitory receptors such as cytotoxic T lymphocyte-associated protein 4 (CTLA4) and programmed cell death 1 (PDCD1, best known as PD-1) (Cimen Bozkus et al., 2019). These findings prompt the use of immune checkpoint inhibitors specific for CTLA4 or PD-1 to boost existing tumor-targeting T-cell immunity in patients with MPN bearing *CALR* mutations (Shulgin et al., 2020).

Since the original definition of myeloproliferation by Dameshek, a growing knowledge of the molecular landscape of Ph$^-$ MPN and its interaction with (1) the hematopoietic microenvironment and (2) the genetic predisposition of the patient not only laid the foundation for current therapeutic strategies but also constitute the beacon driving the future investigation toward improved disease management. In "Cellular and molecular aspects of Philadelphia-negative myeloproliferative neoplasms" (a thematic issue of the *International Review of Cell and Molecular Biology* series), world-leading experts in the field offer an overview of the major molecular and cellular aspects of Ph$^-$ MPNs, focusing on the prominent features of pathophysiological relevance. The book (which consists of two parts, part A and part B) is equally addressed to expert investigators, who may wish to expand their knowledge on the biology of Ph-MPN, and to newcomers to this exciting and rapidly expanding area of research.

Acknowledgments

The L.G. lab is supported by a Breakthrough Level 2 grant from the US DoD BRCP (#BC180476P1), by the 2019 Laura Ziskin Prize in Translational Research (#ZP-6177, PI: Formenti) from SU2C, by a Mantle Cell Lymphoma Research Initiative (MCL-RI, PI: Chen-Kiang) grant from the Leukemia and Lymphoma Society (LLS), by a startup grant from the Department of Radiation Oncology at Weill Cornell Medicine (New York, USA), by a Rapid Response Grant from the Functional Genomics Initiative (New York, USA), by industrial collaborations with Lytix (Oslo, Norway) and Phosplatin (New York, USA), and by donations from Phosplatin (New York, USA), the Luke Heller TECPR2 Foundation (Boston, USA), Onxeo (Paris, France), and Sotio a.s. (Prague, Czech Republic).

Conflict of Interest

L.G. reports research funding from Lytix and Phosplatin; and consulting/advisory honoraria from Boehringer Ingelheim, AstraZeneca, OmniSEQ, The Longevity Labs, Inzen, Onxeo, and the Luke Heller TECPR2 Foundation.

References

Alcantara, M., et al., 2020. Current clinical evidence and potential solutions to increase benefit of CAR T-cell therapy for patients with solid tumors. Oncoimmunology 9 (1), 1777064.

Amaru Calzada, A., et al., 2012. The HDAC inhibitor Givinostat modulates the hematopoietic transcription factors NFE2 and C-MYB in JAK2(V617F) myeloproliferative neoplasm cells. Exp. Hematol. 40 (8). 634–45.e10.

Amaru Calzada, A., et al., 2013. Givinostat and hydroxyurea synergize in vitro to induce apoptosis of cells from JAK2(V617F) myeloproliferative neoplasm patients. Exp. Hematol. 41 (3). 253–60.e2.

Arber, D.A., et al., 2016. The 2016 revision to the World Health Organization classification of myeloid neoplasms and acute leukemia. Blood 127 (20), 2391–2405.

Arshad, N., Cresswell, P., 2018. Tumor-associated calreticulin variants functionally compromise the peptide loading complex and impair its recruitment of MHC-I. J. Biol. Chem. 293 (25), 9555–9569.

Balliu, M., et al., 2021. Activated IL-6 signaling contributes to the pathogenesis of, and is a novel therapeutic target for, CALR-mutated MPNs. Blood Adv. 5 (8), 2184–2195.

Barbui, T., et al., 2012. Development and validation of an International Prognostic Score of thrombosis in World Health Organization-essential thrombocythemia (IPSET-thrombosis). Blood 120 (26), 5128–5133 (quiz 5252).

Barbui, T., et al., 2014. In contemporary patients with polycythemia vera, rates of thrombosis and risk factors delineate a new clinical epidemiology. Blood 124 (19), 3021–3023.

Barbui, T., et al., 2018. The 2016 WHO classification and diagnostic criteria for myeloproliferative neoplasms: document summary and in-depth discussion. Blood Cancer J. 8 (2), 15.

Barone, M., et al., 2020. The role of circulating monocytes and JAK inhibition in the infectious-driven inflammatory response of myelofibrosis. Oncoimmunology 9 (1), 1782575.

Bartalucci, N., et al., 2013. Co-targeting the PI3K/mTOR and JAK2 signalling pathways produces synergistic activity against myeloproliferative neoplasms. J. Cell. Mol. Med. 17 (11), 1385–1396.

Bartalucci, N., et al., 2017. Inhibitors of the PI3K/mTOR pathway prevent STAT5 phosphorylation in JAK2V617F mutated cells through PP2A/CIP2A axis. Oncotarget 8 (57), 96710–96724.

Bartalucci, N., et al., 2020. Polycythemia vera: the current status of preclinical models and therapeutic targets. Expert Opin. Ther. Targets 24 (7), 615–628.

Baxter, E.J., et al., 2005. Acquired mutation of the tyrosine kinase JAK2 in human myeloproliferative disorders. Lancet 365 (9464), 1054–1061.

Bhatt, S., et al., 2020. Reduced mitochondrial apoptotic priming drives resistance to BH3 mimetics in acute myeloid leukemia. Cancer Cell 38 (6). 872–890.e6.

Bose, P., Verstovsek, S., 2016. Investigational histone deacetylase inhibitors (HDACi) in myeloproliferative neoplasms. Expert Opin. Investig. Drugs 25 (12), 1393–1403.

Brkic, S., Meyer, S.C., 2021. Challenges and perspectives for therapeutic targeting of myeloproliferative neoplasms. Hemasphere 5 (1), e516.

Cabagnols, X., et al., 2016. Presence of atypical thrombopoietin receptor (MPL) mutations in triple-negative essential thrombocythemia patients. Blood 127 (3), 333–342.
Cimen Bozkus, C., et al., 2019. Immune checkpoint blockade enhances shared neoantigen-induced T-cell immunity directed against mutated calreticulin in myeloproliferative neoplasms. Cancer Discov. 9 (9), 1192–1207.
Coltro, G., et al., 2020. RAS/CBL mutations predict resistance to JAK inhibitors in myelofibrosis and are associated with poor prognostic features. Blood Adv. 4 (15), 3677–3687.
Constantinescu, S.N., et al., 2021. Functional consequences of mutations in myeloproliferative neoplasms. Hemasphere 5 (6), e578.
Cortelazzo, S., et al., 1995. Hydroxyurea for patients with essential thrombocythemia and a high risk of thrombosis. N. Engl. J. Med. 332 (17), 1132–1136.
Curto-Garcia, N., et al., 2020. Bone marrow niche dysregulation in myeloproliferative neoplasms. Haematologica 105 (5), 1189–1200.
Dameshek, W., 1951. Some speculations on the myeloproliferative syndromes. Blood 6 (4), 372–375.
Davidson, M.B., Gupta, V., 2021. Application of stem cell therapy in myelofibrosis. Hematol. Oncol. Clin. North Am. 35 (2), 391–407.
Dawson, M.A., et al., 2009. JAK2 phosphorylates histone H3Y41 and excludes HP1alpha from chromatin. Nature 461 (7265), 819–822.
Defour, J.P., et al., 2016. Oncogenic activation of MPL/thrombopoietin receptor by 17 mutations at W515: implications for myeloproliferative neoplasms. Leukemia 30 (5), 1214–1216.
Di Buduo, C.A., et al., 2020. Defective interaction of mutant calreticulin and SOCE in megakaryocytes from patients with myeloproliferative neoplasms. Blood 135 (2), 133–144.
Dunbar, A.J., et al., 2020. Leukemia secondary to myeloproliferative neoplasms. Blood 136 (1), 61–70.
Dutta, A., et al., 2021. CDK6 is a therapeutic target in myelofibrosis. Cancer Res. 81 (16), 4332–4345.
Fairlie, W.D., et al., 2020. Crosstalk between apoptosis and autophagy signaling pathways. Int. Rev. Cell Mol. Biol. 352, 115–158.
Finazzi, G., et al., 2013. A phase II study of Givinostat in combination with hydroxycarbamide in patients with polycythaemia vera unresponsive to hydroxycarbamide monotherapy. Br. J. Haematol. 161 (5), 688–694.
Fisher, D.A.C., et al., 2019. Cytokine production in myelofibrosis exhibits differential responsiveness to JAK-STAT, MAP kinase, and NFκB signaling. Leukemia 33 (8), 1978–1995.
Fruchtman, S.M., et al., 1997. From efficacy to safety: a Polycythemia Vera Study group report on hydroxyurea in patients with polycythemia vera. Semin. Hematol. 34 (1), 17–23.
Fucikova, J., et al., 2020a. Calreticulin arms NK cells against leukemia. Oncoimmunology 9 (1), 1671763.
Fucikova, J., et al., 2020b. Detection of immunogenic cell death and its relevance for cancer therapy. Cell Death Dis. 11 (11), 1013.
Fucikova, J., et al., 2021. Calreticulin and cancer. Cell Res. 31 (1), 5–16.
Galluzzi, L., et al., 2018. Molecular mechanisms of cell death: recommendations of the Nomenclature Committee on Cell Death 2018. Cell Death Differ. 25 (3), 486–541.
Grabek, J., et al., 2020. MPN: the molecular drivers of disease initiation, progression and transformation and their effect on treatment. Cells 9 (8).
Greenfield, G., et al., 2018. The ruxolitinib effect: understanding how molecular pathogenesis and epigenetic dysregulation impact therapeutic efficacy in myeloproliferative neoplasms. J. Transl. Med. 16 (1), 360.

Greenfield, G., et al., 2020. Modification of the histone landscape with JAK inhibition in myeloproliferative neoplasms. Cancers (Basel) 12 (9).

Grimwade, L.F., et al., 2009. Phospho-STAT5 and phospho-Akt expression in chronic myeloproliferative neoplasms. Br. J. Haematol. 147 (4), 495–506.

Guglielmelli, P., et al., 2011. Safety and efficacy of everolimus, a mTOR inhibitor, as single agent in a phase 1/2 study in patients with myelofibrosis. Blood 118 (8), 2069–2076.

Guglielmelli, P., et al., 2014. The number of prognostically detrimental mutations and prognosis in primary myelofibrosis: an international study of 797 patients. Leukemia 28 (9), 1804–1810.

Guglielmelli, P., et al., 2018. MIPSS70: mutation-enhanced International Prognostic Score system for transplantation-age patients with primary myelofibrosis. J. Clin. Oncol. 36 (4), 310–318.

Harrison, C.N., et al., 2017. Janus kinase-2 inhibitor fedratinib in patients with myelofibrosis previously treated with ruxolitinib (JAKARTA-2): a single-arm, open-label, non-randomised, phase 2, multicentre study. Lancet Haematol. 4 (7), e317–e324.

Hobbs, G.S., et al., 2018. Hsp90 inhibition disrupts JAK-STAT signaling and leads to reductions in splenomegaly in patients with myeloproliferative neoplasms. Haematologica 103 (1), e5–e9.

Hoermann, G., et al., 2015. Cytokine regulation of microenvironmental cells in myeloproliferative neoplasms. Mediators Inflamm. 2015, 869242.

Holmström, M.O., et al., 2018a. The calreticulin (CALR) exon 9 mutations are promising targets for cancer immune therapy. Leukemia 32 (2), 429–437.

Holmström, M.O., et al., 2018b. Spontaneous T-cell responses against the immune check point programmed-death-ligand 1 (PD-L1) in patients with chronic myeloproliferative neoplasms correlate with disease stage and clinical response. Oncoimmunology 7 (6), e1433521.

James, C., et al., 2005. A unique clonal JAK2 mutation leading to constitutive signalling causes polycythaemia vera. Nature 434 (7037), 1144–1148.

Karantanos, T., Moliterno, A.R., 2018. The roles of JAK2 in DNA damage and repair in the myeloproliferative neoplasms: opportunities for targeted therapy. Blood Rev. 32 (5), 426–432.

Kepp, O., et al., 2020. Surface-exposed and soluble calreticulin: conflicting biomarkers for cancer prognosis. Oncoimmunology 9 (1), 1792037.

Kepp, O., et al., 2021. ATP and cancer immunosurveillance. EMBO J. 40 (13), e108130.

Kiladjian, J.J., et al., 2008. Pegylated interferon-alfa-2a induces complete hematologic and molecular responses with low toxicity in polycythemia vera. Blood 112 (8), 3065–3072.

Klampfl, T., et al., 2013. Somatic mutations of calreticulin in myeloproliferative neoplasms. N. Engl. J. Med. 369 (25), 2379–2390.

Kleppe, M., et al., 2015. JAK-STAT pathway activation in malignant and nonmalignant cells contributes to MPN pathogenesis and therapeutic response. Cancer Discov. 5 (3), 316–331.

Korn, C., Mendez-Ferrer, S., 2017. Myeloid malignancies and the microenvironment. Blood 129 (7), 811–822.

Kralovics, R., et al., 2005. A gain-of-function mutation of JAK2 in myeloproliferative disorders. N. Engl. J. Med. 352 (17), 1779–1790.

Lai, H.Y., et al., 2019. Defective negative regulation of toll-like receptor signaling leads to excessive TNF-α in myeloproliferative neoplasm. Blood Adv. 3 (2), 122–131.

Lanikova, L., et al., 2019. Experimental modeling of myeloproliferative neoplasms. Genes (Basel) 10 (10).

Lee, S.E., 2021. Disease modifying agents of myeloproliferative neoplasms: a review. Blood Res. 56 (Suppl. 1), S26–S33.

Levine, R.L., et al., 2005. Activating mutation in the tyrosine kinase JAK2 in polycythemia vera, essential thrombocythemia, and myeloid metaplasia with myelofibrosis. Cancer Cell 7 (4), 387–397.

Levy, G., et al., 2020. MPL mutations in essential thrombocythemia uncover a common path of activation with eltrombopag dependent on W491. Blood 135 (12), 948–953.

Liu, P., et al., 2020. Immunosuppression by mutated calreticulin released from malignant cells. Mol. Cell 77 (4). 748–760.e9.

Lu, M., et al., 2014. The orally bioavailable MDM2 antagonist RG7112 and pegylated interferon α 2a target JAK2V617F-positive progenitor and stem cells. Blood 124 (5), 771–779.

Mannelli, F., 2021. Acute myeloid leukemia evolving from myeloproliferative neoplasms: many sides of a challenging disease. J. Clin. Med. 10 (3).

Marchioli, R., et al., 2013. Cardiovascular events and intensity of treatment in polycythemia vera. N. Engl. J. Med. 368 (1), 22–33.

Masarova, L., Verstovsek, S., 2019. The evolving understanding of prognosis in post-essential thrombocythemia myelofibrosis and post-polycythemia vera myelofibrosis vs primary myelofibrosis. Clin. Adv. Hematol. Oncol. 17 (5), 299–307.

Mascarenhas, J., et al., 2019a. Oral idasanutlin in patients with polycythemia vera. Blood 134 (6), 525–533.

Mascarenhas, J., et al., 2019b. MANIFEST, a phase 2 study of CPI-0610, a bromodomain and extraterminal domain inhibitor (BETi), as monotherapy or "add-on" to ruxolitinib, in patients with refractory or intolerant advanced myelofibrosis. Blood 134 (Suppl. 1), 670.

Mead, A.J., Mullally, A., 2017. Myeloproliferative neoplasm stem cells. Blood 129 (12), 1607–1616.

Meadows, S.A., et al., 2013. PI3Kδ inhibitor idelalisib inhibits AKT signaling in myelofibrosis patients on chronic JAK inhibitor therapy. Blood 122 (21), 4065.

Mesa, R.A., et al., 2017a. Pacritinib versus best available therapy for the treatment of myelofibrosis irrespective of baseline cytopenias (PERSIST-1): an international, randomised, phase 3 trial. Lancet Haematol. 4 (5), e225–e236.

Mesa, R.A., et al., 2017b. SIMPLIFY-1: a phase III randomized trial of momelotinib versus ruxolitinib in janus kinase inhibitor-naïve patients with myelofibrosis. J. Clin. Oncol. 35 (34), 3844–3850.

Milosevic, J.D., Kralovics, R., 2013. Genetic and epigenetic alterations of myeloproliferative disorders. Int. J. Hematol. 97 (2), 183–197.

Morotti, A., et al., 2017. Modeling myeloproliferative neoplasms: from mutations to mouse models and back again. Blood Rev. 31 (3), 139–150.

Nangalia, J., et al., 2013. Somatic CALR mutations in myeloproliferative neoplasms with nonmutated JAK2. N. Engl. J. Med. 369 (25), 2391–2405.

Nowell, P.C., 1962. The minute chromosome (Phl) in chronic granulocytic leukemia. Blut 8, 65–66.

Ortmann, C.A., et al., 2015. Effect of mutation order on myeloproliferative neoplasms. N. Engl. J. Med. 372 (7), 601–612.

O'Sullivan, J., Mead, A.J., 2019. Heterogeneity in myeloproliferative neoplasms: causes and consequences. Adv. Biol. Regul. 71, 55–68.

Passamonti, F., Maffioli, M., 2016. Update from the latest WHO classification of MPNs: a user's manual. Hematology Am. Soc. Hematol. Educ. Program 2016 (1), 534–542.

Pemmaraju, N., et al., 2019. Final results of phase 2 clinical trial of LCL161, a novel oral SMAC mimetic/IAP antagonist, for patients with intermediate to high risk myelofibrosis. Blood 134 (Suppl. 1), 555.

Petroni, G., et al., 2020. Immunomodulation by anticancer cell cycle inhibitors. Nat. Rev. Immunol. 20 (11), 669–679.

Petroni, G., et al., 2021a. Radiotherapy delivered before CDK4/6 inhibitors mediates superior therapeutic effects in ER(+) breast cancer. Clin. Cancer Res. 27 (7), 1855–1863.
Petroni, G., et al., 2021b. Immunomodulation by targeted anticancer agents. Cancer Cell 39 (3), 310–345.
Pikman, Y., et al., 2006. MPLW515L is a novel somatic activating mutation in myelofibrosis with myeloid metaplasia. PLoS Med. 3 (7), e270.
Rambaldi, A., et al., 2020. Safety and efficacy of the maximum tolerated dose of givinostat in polycythemia vera: a two-part phase Ib/II study. Leukemia 34 (8), 2234–2237.
Rasmussen, M.L., Gama, V., 2020. A connection in life and death: the BCL-2 family coordinates mitochondrial network dynamics and stem cell fate. Int. Rev. Cell Mol. Biol. 353, 255–284.
Rosa, N., et al., 2020. Type 3 IP(3) receptors: the chameleon in cancer. Int. Rev. Cell Mol. Biol. 351, 101–148.
Rowley, J.D., 1973. Letter: a new consistent chromosomal abnormality in chronic myelogenous leukaemia identified by quinacrine fluorescence and Giemsa staining. Nature 243 (5405), 290–293.
Salati, S., et al., 2019. Calreticulin Ins5 and Del52 mutations impair unfolded protein and oxidative stress responses in K562 cells expressing CALR mutants. Sci. Rep. 9 (1), 10558.
Sánchez-Rivera, F.J., et al., 2021. Mitochondrial apoptotic priming is a key determinant of cell fate upon p53 restoration. Proc. Natl. Acad. Sci. U. S. A. 118 (23).
Schieber, M., et al., 2019. Myelofibrosis in 2019: moving beyond JAK2 inhibition. Blood Cancer J. 9 (9), 74.
Schischlik, F., et al., 2019. Mutational landscape of the transcriptome offers putative targets for immunotherapy of myeloproliferative neoplasms. Blood 134 (2), 199–210.
Shin, N., et al., 2020. Parsaclisib is a next-generation phosphoinositide 3-kinase δ inhibitor with reduced hepatotoxicity and potent antitumor and immunomodulatory activities in models of B-cell malignancy. J. Pharmacol. Exp. Ther. 374 (1), 211–222.
Shulgin, B., et al., 2020. Dose dependence of treatment-related adverse events for immune checkpoint inhibitor therapies: a model-based meta-analysis. Oncoimmunology 9 (1), 1748982.
Spivak, J.L., 2017. Myeloproliferative neoplasms. N. Engl. J. Med. 376 (22), 2168–2181.
Stetka, J., et al., 2019. Addiction to DUSP1 protects JAK2V617F-driven polycythemia vera progenitors against inflammatory stress and DNA damage, allowing chronic proliferation. Oncogene 38 (28), 5627–5642.
Stivala, S., et al., 2019. Targeting compensatory MEK/ERK activation increases JAK inhibitor efficacy in myeloproliferative neoplasms. J. Clin. Invest. 129 (4), 1596–1611.
Subotički, T., et al., 2021. VEGF regulation of angiogenic factors via inflammatory signaling in myeloproliferative neoplasms. Int. J. Mol. Sci. 22 (13).
Tefferi, A., 2016. Myeloproliferative neoplasms: a decade of discoveries and treatment advances. Am. J. Hematol. 91 (1), 50–58.
Tefferi, A., Vardiman, J.W., 2008. Classification and diagnosis of myeloproliferative neoplasms: the 2008 World Health Organization criteria and point-of-care diagnostic algorithms. Leukemia 22 (1), 14–22.
Tefferi, A., et al., 2014. CALR vs JAK2 vs MPL-mutated or triple-negative myelofibrosis: clinical, cytogenetic and molecular comparisons. Leukemia 28 (7), 1472–1477.
Tefferi, A., et al., 2016a. Targeted deep sequencing in polycythemia vera and essential thrombocythemia. Blood Adv. 1 (1), 21–30.
Tefferi, A., et al., 2016b. Targeted deep sequencing in primary myelofibrosis. Blood Adv. 1 (2), 105–111.
Tefferi, A., et al., 2018a. Driver mutations and prognosis in primary myelofibrosis: Mayo-Careggi MPN alliance study of 1,095 patients. Am. J. Hematol. 93 (3), 348–355.

Tefferi, A., et al., 2018b. MIPSS70+ version 2.0: mutation and karyotype-enhanced international prognostic scoring system for primary myelofibrosis. J. Clin. Oncol. 36 (17), 1769–1770.

Tefferi, A., et al., 2018c. GIPSS: genetically inspired prognostic scoring system for primary myelofibrosis. Leukemia 32 (7), 1631–1642.

Tefferi, A., et al., 2020. Mutation-enhanced international prognostic systems for essential thrombocythaemia and polycythaemia vera. Br. J. Haematol. 189 (2), 291–302.

Vainchenker, W., Kralovics, R., 2017. Genetic basis and molecular pathophysiology of classical myeloproliferative neoplasms. Blood 129 (6), 667–679.

Vainchenker, W., et al., 2016. Recent advances in understanding myelofibrosis and essential thrombocythemia. F1000Res 5.

Van Egeren, D., et al., 2021. Reconstructing the lineage histories and differentiation trajectories of individual cancer cells in myeloproliferative neoplasms. Cell Stem Cell 28 (3). 514–523.e9.

Vannucchi, A.M., 2015. Ruxolitinib versus standard therapy for the treatment of polycythemia vera. N. Engl. J. Med. 372 (17), 1670–1671.

Verstovsek, S., et al., 2012. A double-blind, placebo-controlled trial of ruxolitinib for myelofibrosis. N. Engl. J. Med. 366 (9), 799–807.

Verstovsek, S., et al., 2017a. A phase 1 study of the janus kinase 2 (JAK2)(V617F) inhibitor, gandotinib (LY2784544), in patients with primary myelofibrosis, polycythemia vera, and essential thrombocythemia. Leuk. Res. 61, 89–95.

Verstovsek, S., et al., 2017b. A phase 2 study of momelotinib, a potent JAK1 and JAK2 inhibitor, in patients with polycythemia vera or essential thrombocythemia. Leuk. Res. 60, 11–17.

Vicari, L., et al., 2012. Increased phospho-mTOR expression in megakaryocytic cells derived from CD34+ progenitors of essential thrombocythaemia and myelofibrosis patients. Br. J. Haematol. 159 (2), 237–240.

Vitale, I., et al., 2021. Intratumoral heterogeneity in cancer progression and response to immunotherapy. Nat. Med. 27 (2), 212–224.

Wang, Y., et al., 2009. Cotreatment with panobinostat and JAK2 inhibitor TG101209 attenuates JAK2V617F levels and signaling and exerts synergistic cytotoxic effects against human myeloproliferative neoplastic cells. Blood 114 (24), 5024–5033.

CHAPTER ONE

The roles of sex and genetics in the MPN

Alison R. Moliterno*,† and Evan M. Braunstein†
Hematology Division, Department of Medicine, Johns Hopkins University School of Medicine, Baltimore, Maryland, United States
*Corresponding author: e-mail address: amoliter@jhmi.edu

Contents

Abstract

The Philadelphia chromosome negative myeloproliferative neoplasms(MPNs), polycythemia vera (PV), essential thrombocythemia (ET), and primary myelofibrosis (PMF) are acquired hematopoietic stem cell disorders driven by activating mutations of intracellular signal transduction pathways that control the production of circulating blood cells. The MPN are characterized clinically by marked variation in degrees of vascular risk, familial clustering, and evolution to myelofibrosis and acute leukemia. MPN disease presentations and outcomes are highly variable, and are markedly influenced by both sex and germline genetic variation. This chapter will focus on the evidence of sex and germline genetic background as modifiers of MPN development and outcomes. Large population genome wide association studies in both clonal hematopoiesis and MPN have revealed novel mechanisms, including inflammatory pathways and genomic instability, which further our understanding of how sex and genetic

† E.M.B. and A.R.M. conceived the research, developed the manuscript and created the figures.

International Review of Cell and Molecular Biology, Volume 366
ISSN 1937-6448
https://doi.org/10.1016/bs.ircmb.2021.08.001

background mediate MPN risk. Recent advances in our understanding of clonal hematopoiesis and MPN development in various contexts informs the mechanisms by which sex, inflammation, exposures and genetics influence MPN incidence and outcomes, and provide opportunities to develop new strategies for prognostics and therapeutics in the MPN.

Key Points

- Sex is a major modifier of incidence and evolution of the MPN
- Male sex is associated with both qualitative and quantitative mutation burdens that drive higher risk MPN presentation and outcomes
- Sex and genetic variation interact with exposure, inflammation and age to influence incidence and progression of both clonal hematopoiesis and MPN

1. Introduction

The Philadelphia chromosome negative myeloproliferative neoplasms (MPN), polycythemia vera (PV), essential thrombocythemia (ET), and primary myelofibrosis (PMF) are acquired hematopoietic stem cell disorders driven by activating mutations of intracellular signal transduction pathways that control the production of circulating blood cells (Stein et al., 2015). The MPN are characterized clinically by the overproduction of platelets, red cells and neutrophils to varying degrees, high rates of vascular events, frequent familial clustering, and evolution to myelofibrosis or transformation to acute leukemia. MPN disease presentation and outcomes are highly variable, and are markedly influenced by both sex and other inherited genetic factors. The genomic revolution has not only redefined the pathogenesis of the MPNs, but has elucidated the interplay of genetic variation and sex on clonal disease. This chapter will focus on the evidence of sex as a factor central to the MPN, and the mounting evidence of germline influence on the MPN. We will incorporate recent advances in our understanding of clonal hematopoiesis and inflammation to postulate mechanisms by which sex and inheritance interact in the MPN.

2. Sex as a factor in MPN epidemiology

Aging is a major risk factor for cancer development, while male sex is an additional independent risk factor for most cancers (Dunford et al., 2017; Haupt et al., 2021). Men shoulder increased cancer risks, even after adjusting for socioeconomic class, global location, environmental exposure and tobacco exposure (Edgren et al., 2012; Radkiewicz et al., 2017).

Men carry an increase in hematologic cancer risk, in both incidence and outcomes. Men have a higher incidence of AML (De-Morgan et al., 2021; Siegel et al., 2021), MDS (Wang et al., 2019), and CML (Hehlmann et al., 2020) compared to women. Male predominance is even more prominent in specific myeloid malignancies, including chronic myelomonocytic leukemia (CMMOL) (Coltro et al., 2020; Pleyer et al., 2021), FIP1L1-PDGFRA-myeloid neoplasm (Khoury et al., 2016), blastic plasmacytoid dendritic-cell neoplasm (BPDCN) (Pemmaraju et al., 2019) and the most recently reported entity, VEXAS syndrome (Table 1) (Beck et al., 2020).

Within the classical MPN, essential thrombocytosis (ET), polycythemia vera (PV) and primary myelofibrosis (MF), there are even more complex differences related to sex. Whether ascertained through large population based surveys or from large MPN cohorts, MPN is more prevalent in women (Hultcrantz et al., 2012, 2018; Karantanos et al., 2020). Much of

Table 1 Mutation prevalence in male-biased myeloid malignancy cohorts.

Myeloid malignancy	% male	Prevalent acquired somatic mutations
MPN- All (Hultcrantz et al., 2018)	45%	*JAK2V617F, CALR, MPL*
JAK2V617F CH (Cordua et al., 2019)	54%	*JAK2V617F*
MDS (Wang et al., 2019; Zheng et al., 2019)	55%	*TET2, SF3B1,* ***ASXL1, TP53,*** *DNMT3A, SRSF2,* ***RUNX1, STAG2, U2AF1, SETBP1,*** *BCOR, NF1*
MPN-MF (Karantanos et al., 2020)	63%	*JAK2, CALR,* ***ASXL1,*** *TET2,* ***SRSF2, U2AF1,*** *SF3B1, DNMT3A,* ***EZH2, IDH1/2,*** *CBL*
CALR CH (Cordua et al., 2019)	66%	*CALR*
CMMOL (Coltro et al., 2020)	68%	***SRSF2, ASLX1,*** *TET2, DNMT3A,* ***IDH1/2, EZH2,*** *UTX, JAK2, KRAS,* ***NRAS,*** *CBL,* ***RUNX1***
BPDCN (Pemmaraju et al., 2019; Renosi et al., 2021)	83%	*TET2,* ***ASXL1,*** *ZRSR2*
VEXAS (Beck et al., 2020)	100%	*UBA1*

X-linked genes are underlined. High-risk MPN and MDS genes are in bold typeface.

female MPN predominance can be attributed to JAK2V617F-positive ET, in which females overwhelm males 2:1 in every reported survey. Women may equal men in PV, whereas men outnumber women in MF (Table 1). However, due to the higher incidence of ET and PV compared to that of MF, and due to longer survival of ET and PV compared to that of MF, the prevalence of women in MPN is increased compared to men (Karantanos et al., 2020).

Within PV, age influences gender distribution. While median age at diagnosis is broad within PV, examination of younger age at diagnosis compared to older age at diagnosis revealed a striking predominance of females in the younger group (Boddu et al., 2018; Stein et al., 2011a, 2013). Emerging pediatric and young adult surveys confirm this interaction of age, sex, and JAK2V617F. There is a female predominance in pediatric ET (60%), whether JAK2V617F-positive or negative, and in pediatric JAK2V617F-positive PV, females again outnumber males (Boddu et al., 2018).

Within the classic MPN drivers JAK2V617F and CALR, there are strong sex-specific risks in presentation and outcomes. Regardless of driver, females at presentation cluster in lower risk disease, ET or PV, compared to males who present more frequently in PV or PMF, even when adjusted for age at presentation (Karantanos et al., 2020). Factors that determine higher risk disease presentation include clonal burden at the stem cell level, and acquisition of additional cytogenetic lesions or somatic mutations, both of which are more prevalent in men compared to women (Moliterno et al., 2008). For instance, in 815 patients with ET, PV, or PMF evaluated between 2005 and 2019, the association of sex with presenting phenotype, JAK2 V617F burden, progression, and survival showed significant sex biases. Men presented more often with PMF vs ET (relative risk, 3.2; $P < 0.001$) and PV (relative risk, 2.1; $P < 0.001$), had higher rates of transformation to secondary MF (hazard ratio 1.55; $P = 0.013$) and acute myeloid leukemia (hazard ration 3.67; $P < 0.001$), and worse survival (hazard ration, 1.63; $P = 0.001$) independent of age, phenotype at diagnosis, and MPN-specific mutation. Men had higher JAK2 V617F allele burdens in their CD34+ cells, acquired more somatic mutations apart from the MPN-specific mutations, and had an increased frequency of high-risk mutations (*ASXL, EZH2, SRSF2, U2AF1,* or *IDH2)* independent of age, phenotype, and driver mutation (Karantanos et al., 2020). Even in low risk MPN such as ET, male sex is an independent predictor of poorer overall survival (Grinfeld et al., 2018; Tefferi et al., 2017).

In myelodysplasia/myeloproliferative (MDS/MPN) syndromes, a higher risk myeloid malignancy compared to classical MPN, men predominated over women (63%) and had worse overall survival (hazard ratio 2.09, 95% confidence interval 1.16–3.75; $P=0.013$) independent of subtype, Revised International Prognostic Scoring System score and age at diagnosis, compared to women. Men had higher number of somatic mutations independent of subtype, sample source and blast percentage, and more somatic mutations were associated with a higher incidence of transformation to AML. Similar to the findings in classical MPN, in MDS/MPN, men had more mutations in high-risk genes (*ASXL1, EZH2, RUNX1, SETBP1, NRAS, STAG2*), and 13-times higher odds of harboring an *EZH2* mutation, both findings attributed to worse overall survival (Karantanos et al., 2021). Whether increased risk in men is solely attributed to the quantity of the male mutation burden, or to specific high-risk mutations, is not yet defined. In the MPN hematopoietic stem cell context, however, mutation burden and quality influence clonal dominance of disease clones over unaffected clones, and clonal dominance and expansion is a fundamental risk factor for MPN disease progression and transformation (Rumi et al., 2014a; Stein et al., 2011b,c) (Fig. 1).

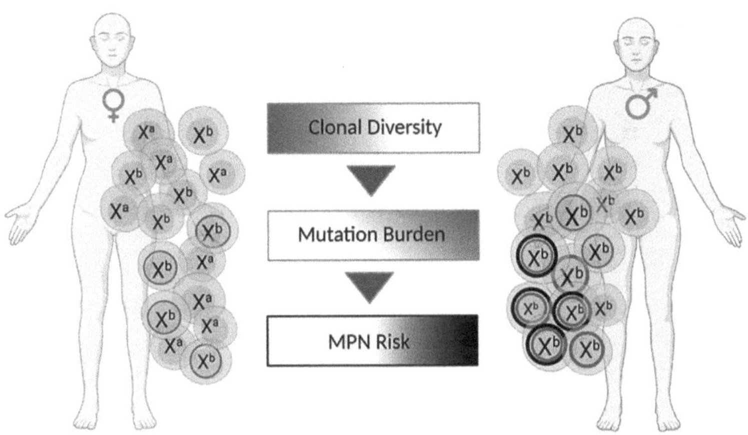

Fig. 1 Sex differences in the MPN incidence and evolution. Women (left) have more hematopoietic stem cell diversity and protection from X chromosome lesions due to random inactivation of paternal (Xa) or maternal (Xb) alleles, whereas men have only a maternal (Xb) allele. Males are have higher rates of CH (red circles), higher rates of high risk somatic mutations (black circles), higher numbers of mutations, and higher rates of mutation of X linked genes *STAG2, UTX, ZRSR2, BCOR* and *UBA1*. The extra somatic burden that men carry increases their rate of CH, CH progression to MPN, and mutant stem cell dominance. This mutation burden and dynamic in men translates clinically to a higher risk at presentation and worse outcomes in MPN compared to women.

3. Biologic basis of sex differences in somatic mutation burden

Male and female sex are defined by the X- and Y-chromosomes. Sex chromosome expression confers different gene expression by specific mechanisms in all tissues, and even in the hematopoietic stem cell context (Spivak et al., 2014). In females, normal development does not allow for expression of the entirety of the X chromosome, and thus at early stage of embryogenesis, X-chromosome inactivation (XCI) events occur for most of the genes on the X chromosome. XCI occurs randomly, and thus females are a mosaic for X chromosome genes, adding to clonal diversity in hematopoietic stem cell populations. Mosaicism in X chromosome expression in HSCs introduces genetic diversity which may protect females from deleterious genes inherited from paternal or maternal source; in males, the X chromosome is only from one maternal X, and thus HSC diversity and protection from a poorly functioning X-linked gene is not the case (Fig. 1) (Haupt et al., 2021). In females, an additional mechanism controls X gene expression. Some genes escape X-inactivation and are expressed from both the active and inactive X chromosome, with recent estimates that nearly 15% of X-linked genes escape XCI (Carrel and Willard, 2005). Relevant to blood cancer risk, genes that escape inactivation include *ATRX* and *KDM6A* (*UTX*), and these play important roles as mutational targets that drive blood cancers (Table 2). (Bailey et al., 2018; Herbaux et al., 2012, 2015; Jankowska et al., 2011; van Haaften et al., 2009). Moreover, recent studies show that some genes that escape X inactivation are less likely to have acquired somatic mutation in females, including both *ATRX* and *KDM6A,* and that this protection from mutation of X-linked tumor suppressors may explain to some extent the reduced cancer incidence in females across a diverse set of tissues, including blood (Dunford et al., 2017).

The X chromosome harbors not only tumor suppressors, but also genes important to chromatin remodeling and the cohesion complex, genes that are strongly implicated in HSC function and genomic protection. These X-linked genes, *BCOR*, *PHF6*, *STAG2* and *ZRSR2*, when mutated in the male context, may have a far stronger impact in driving an HSC malignant phenotype compared to the female context, given the lack of mosaicism at this locus (Tables 1 and 2). In males, an oncogenic allele is expressed from the single copy of the X chromosome. In females, a mutation in an oncogene carried on the inactive X chromosome will have no deleterious effect if this gene does not escape X-chromosome inactivation,

Table 2 X chromosome genes relevant to blood cancers.

X chromosome gene; locus	Gene function	Loss of function implications in blood cancer context
ATRX; Xq21.1	Core chromatin remodeler, genomic stability	Increased DNA damage; TP53 interaction
UTX (KDM6A); Xp11.3	Histone lysine demethylase, catalyzes the demethylation of histone H3	UTX interacts with MLL3/4 lysine methyl transferases.
ZRSR2; Xp22.2	Splicing factor	Aberrant splicing in MDS
UBA1; Xp11.3	Catalyzes ubiquitin conjugation to mark cellular proteins for degradation	VEXAS syndrome; adult onset severe auto-inflammatory syndrome, MDS association
DDX3X; Xp11.4	RNA helicase	Loss of function mutations in CLL, in males predominantly
STAG2; Xq25	Subunit of the cohesin complex, regulates the separation of chromatids during cell division	Loss of function associated with aneuploidy, genomic instability
BCOR; Xp11.4	Polycomb repressor complex	HSC aberrant gene expression

similar to the situation in X-linked mutations in tumor suppressor genes described above. Indeed, *STAG2* and *ZRSR2* mutations are enriched in male dominated phenotypes of CMML, MDS, and AML (Table 1) (Kon et al., 2013; Madan et al., 2020). In an analysis of targeted resequencing of 68 genes in 664 patients with AML treated through the German Cooperative Group, mutations in *RUNX1*, *SRSF2*, *ASXL1*, *STAG2* and *BCOR* were more common in men compared to women (Metzeler et al., 2016). More recently, an analysis of mutations from more than 11,000 AML and 2700 MDS cases combined from the COSMIC database and the German Münchner Leukämielabor GmbH again showed male-biased mutation bias. Male mutation bias was not only in in X-linked genes, *BCOR, PHF6, STAG2* and *ZRSR2,* but also in autosomal genes *ASXL1, RUNX1, TET2, SRSF2*, and *U2AF1* (De-Morgan et al., 2021). These genes in the MPN context are not drivers, but are associated with clonal evolution and increased risk for myelofibrosis and AML transformation. Thus their more frequent occurrence in men in general may drive worse outcomes for men specifically in MPN (Fig. 1, Table 1) (Grinfeld et al., 2018; Guglielmelli et al., 2014; Karantanos et al., 2020).

4. Genetic predisposition to CH and the MPN

Acquired HSC somatic mutations are not unique to individuals with blood cancers, but occur in the normal population, and when low in variant allele fraction (VAF) are termed clonal hematopoiesis (CH). While CH lesions are associated with increased cardiovascular and cancer risk, they can be quite latent, suggesting the presence of many different modifiers that may exert varying levels of risk (Acuna-Hidalgo et al., 2017; Bolton et al., 2020b; Genovese et al., 2014; Jaiswal et al., 2014; McKerrell et al., 2015; Zink et al., 2017). More recently, by examining clonal mutation architecture in individuals with MPN, acquisition of JAK2V617F has been shown to occur decades before clinical MPN diagnosis, suggesting that prolonged latency occurs, and further implies that specific genetic, sex or other factors contribute to clonal evolution and expansion from CH to a myeloid neoplasm (Van Egeren et al., 2021). Blood counts are influenced by gender, race, age, obesity, smoking, inflammation, iron metabolism and a myriad of other factors, and it is likely that these same factors may influence acquisition of CH (Fig. 2) (Furuncuoglu et al., 2016; Jayasuriya et al., 2019; Segal and Moliterno, 2006; Xavier-Ferrucio et al., 2019).

The study of CH offers a window to examine the genetic factors relevant to the MPN, as JAK2V617F is one of the more common CH lesions, after *DNMT3A*, *TET2*, and *ASXL1* (Table 2). CH was more prevalent in the aged, and among persons 60 years of age or older, men had an increased likelihood of having a detectable mutation as compared with women (odds ratio, 1.3; 95% CI: 1.1–1.5; $P=0.005$ by logistic regression) (Jaiswal et al., 2014). Several surveys that examined only JAK2V617F CH in unselected populations in the United States and Europe also found older age, tobacco exposure and male sex as independent risk factors for acquiring JAK2V617F (Cordua et al., 2019; Hinds et al., 2016; Nielsen et al., 2011, 2013). Cordua et al. found that both JAK2V617F CH and CALR CH were more common in aging males, and that male predominance occurred whether CH levels were below 1%, greater than 1%, and in the small proportion of the population that had an MPN diagnosis (Table 1). By this analysis, one can infer that male sex influences both incidence of JAK2V617F or CALR CH, and also influences the expansion over time to levels of VAF associated with an MPN diagnosis (Fig. 2) (Cordua et al., 2019). In 2011, Nielsen et al., 2011 showed a similar male predominance in the JAK2V617F CH cohort ascertained from 10,000 individuals

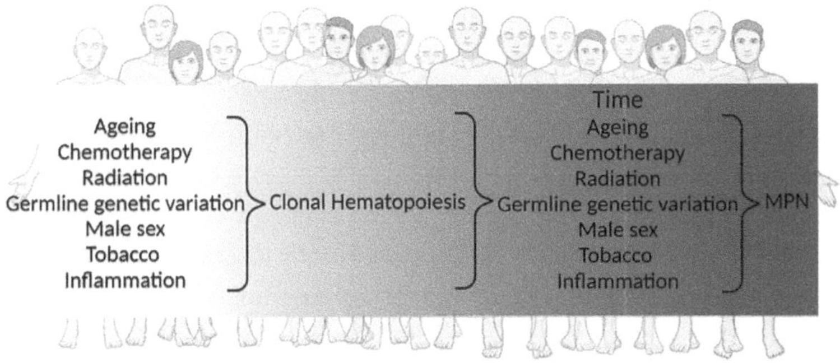

Fig. 2 Risk factors for CH and MPN in the general population. Risk factors for developing CH in the general population include aging, exposure to tobacco, chemotherapy or radiation therapy, male sex, and inflammation. Heritable factors that vary across the general population also influence CH risk, whether the heritable factors are common low penetrance alleles such as the JAK2 haplotype, or rare but high penetrance alleles such the 14q32 duplication (Pegliasco et al., 2021). Latency and progression of CH to MPN is governed by the same risk factors as for CH, in addition to time.

from the Copenhagen City Heart Study. Multifactor adjusted hazard ratios for any cancer, hematologic cancer and myeloproliferative cancer were 3.7 (1.7–8.0), 58 (13–261) and 161 (12–2197), respectively. However, hazard ratios were significantly higher for men compared to women. Women with CH in this study tended to have lower JAK2V617F VAF, and less commonly an MPN clinical diagnosis. Taken together, MPN-specific drivers CALR and JAK2V617F seem to arise more commonly in males compared to females, and in males associate with higher mortality risk and development of a blood cancer. This excess risk for males may be mediated by a more rapid rise in clonal burden, or may be pushed along by other mutational events, or other factors (Figs. 1 and 2). In this sense, risks specific to men in CH echo the risks specific to men in the MPN.

5. Heritable risk factors for CH and MPN

5.1 Familial MPN and myeloid malignancy predisposition syndromes

The (MPN) are usually sporadic diseases characterized by increased JAK/STAT pathway signaling due to a somatic driver mutation in either the *JAK2*, *CALR*, or *MPL* gene. A heritable component, however, is implied by the fact that approximately 10% of MPN cases display familial clustering, and there is a five- to sevenfold increased risk of developing an MPN among first-degree relatives of MPN patients (Landgren et al., 2008; Sud et al., 2019). In contrast to other myeloid malignancies, investigation of large pedigrees with familial clustering of MPN has failed to identify highly penetrant predisposition genes relevant to the broad MPN population. Rather, highly penetrant risk alleles have been confined to rare families, or to unique cancer predisposition syndromes. For example, variants in *RBBP6* were found to co-segregate with disease in three pedigrees with familial MPN, albeit with incomplete penetrance (Harutyunyan et al., 2016). A germline duplication on chromosome 14q32 implicating *ATG2B/GSKIP* was linked to predisposition to MPN as well as other myeloid malignancies in multiple unrelated families, though the specific causative gene(s) in the 14q32 amplification and their mechanisms remains unclear (Babushok et al., 2018; Saliba et al., 2015). An *ERBB2* germline variant that co-segregated in a family with both MPN and hereditary cancer was identified through an unbiased whole exome approach. Subsequently, in 1604 individuals with sporadic MPN, rare *ERBB2* germline variants were associated with an increased risk of MPN compared to controls (8.9% vs 4.1%, OR 2.4, 95% CI: 1.4–4.0, $P=0.0028$), or to non-MPN blood neoplasms (8.9% vs 2.7%, OR 3.5, 95% CI: 1.4–8.3, $P=0.0053$) (Braunstein et al., 2021). Thus, predisposition genes identified from rare, penetrant familial variants may exert a risk of variable degree in the broader, sporadic MPN population.

Recent studies of rare predisposition syndromes associated with MPN have yielded fascinating insights related CH and MPN risk. In a re-examination of a number of West Indian families with a 14q32 duplication and predisposition to MPN, investigators found a very high prevalence of CH in young, asymptomatic 14q32 duplication carriers. In addition, the most common CH lesions in the 14q32 duplication carrier cohort (whether asymptomatic or with MPN) involved *TET2 and JAK2*, while *DNMT3A*, generally the most common CH lesion in general population surveys, was absent (Table 3) (Pegliasco et al., 2021). In another predisposition context, MPN was found to be part of the spectrum of hematologic malignancies in a large cohort with germline *DDX41* variants. *DDX41* germline variant carriers intriguingly had a long latency to hematologic malignancy presentation, a strong male predominance, a high frequency of *TP53*, *ASXL1* and *JAK2* CH, and similar to the case in the 14q32 context, a low prevalence of DNMT3A CH (Quesada et al., 2019). Lim et al. (2021) examined the prevalence of germline *POT1* variants in 2744 individuals with blood cancers. In individuals with myeloid neoplasms, particularly

Table 3 CH in selected populations, lesion frequency and clinical associations.

Study	Population features	CH lesion frequency	CH associations
Jaiswal et al.(2017)	Cardiovascular cohorts	DNMT3A, TET2, ASXL1, JAK2 TP53, CBL, KDM6A (UTX), NF1	Thrombosis risk highest with JAK2, TET2, and ASXL1 CH
Bolton et al. (2020a)	Cancer cohort	DNMT3A, TET2, PPM1D, ASXL1, TP53, CHEK2, ATM, JAK2, SF3B1, SRSF2	Cancer treatment associated with more prevalent PPM1D, TP53, and CHEK2
Hecker et al. (2021)	Arthroplasty cohort	DNMT3A, TET2, ASXL1, TP53, JAK2, U2AF1, SF3B1, SRSF2, NFE2, ZRSR2	50% CH prevalence, high prevalence of autoimmune disease in CH
Quesada et al. (2019)	DDX41 germline variant carriers	TP53, ASXL1, JAK2, EZH2, TET2, SRSF2, U2AF1, BCOR, DNMT3A, ZRSR2	Male predominance; older age with CH; low DNMT3A prevalence
Pegliasco et al. (2021)	14q32 dup germline variant carriers	TET2, JAK2, SRSF2, SF3B1, IDH1, NRAS, CALR, MPL, CBL, ASXL1	High prevalence of MPN in carriers; absence of DNMT3A; young age with CH;
Lim et al. (2021)	POT1 germline variant carriers	TET2, JAK2, DNMT3A, SRSF2, NRAS, FLT3, SF3B1, STAG2, EZH2, TP53	High prevalence of MPN in POT1 carriers

Lesion frequency listed in order from most prevalent to least prevalent. X-linked genes are underlined.

MPN, germline *POT1* variants occurred more commonly compared to a cancer free population (OR 6.7, 95% CI: 1.7–20.2, $P=0.012$), similar to the case of *ERBB2* germline variants. Moreover, sex seemed to exert an influence in the myeloid malignancies with *POT1* variants: men made up the majority of cases overall and had higher rates of acute myeloid leukemia compared to women. Intriguingly, lesions in *POT1* germline variant associated myeloid disease were most commonly *TET2* and *JAK2*, followed by *DNMT3A* (Table 3). In contrast to the presumed long telomere syndromes engendered by *POT1* variants, in the short telomere syndromes CH lesions were highly prevalent in carriers without myeloid disease (30%), and the lesions in both CH and MDS were most commonly

TP53, TET2, U2AF1 and *ASXL1* (Schratz et al., 2020). Taken together, these studies indicate that specific germline lesions may drive CH prevalence, latency, sex predominance, and even the types of CH lesions. Examination of CH in other cancer predisposition syndromes, such as ataxia-telangiectasia (ATM), BRCA1/2, and neurofibromatosis (NF1), will likely uncover important insights into CH and hematologic malignancy risk.

5.2 GWAS identified risk alleles in the general population

Genome-wide association studies (GWAS) offer insight into low penetrance, heritable alleles important to the development of disease. The most well studied germline predisposition variant for MPN is a common haplotype of the *JAK2* gene, referred to as 46/1 or GGCC, which is associated with a two- to threefold increased risk of acquiring the JAK2V617F mutation (Hinds et al., 2016; Jones et al., 2009; Kilpivaara et al., 2009). In a very large GWAS study, Hinds et al. found loci in the *JAK2, TERT, SH2B3, TET2, ATM, CHEK2, PINT*, and *GFI1B* genes that were associated with both the development of MPN and JAK2V617F CH (Hinds et al., 2016). In a smaller GWAS of MPN, Tapper identified loci in *MECOM, TERT* and *HBSIL-MYB* (Tapper et al., 2015). These 10 loci plus an addition novel seven loci (including *HMGA1, MKLN1, FOXO1, RUNX1*) were identified by a recent very large MPN GWAS analysis, which concluded that the 17 loci explain approximately 18.4% of the total heritable relative risk for MPN acquisition in the general population (Bao et al., 2020). It should be noted that both large GWAS studies showed identical associations for either MPN or for JAK2V617F CH (Bao et al., 2020; Hinds et al., 2016). These more common but low penetrance alleles may have an additive or multiplicative effect, such that a propensity risk score can be calculated; this propensity score no doubt is another factor that influences MPN risk in the general population and in specific predisposition syndrome populations described above.

6. Inflammation as a risk for CH and MPN

The importance of inflammation as a risk factor for developing MPN acquisition and evolution has been mathematically modeled, and has been an emerging research focus in both MPN and CH (Andersen et al., 2017). JAK2V617F is uniquely connected to inflammation, both proximally and distally: inflammatory states may predispose to acquiring JAK2V617F, and JAK2V617F signaling drives inflammatory pathways (Hasselbalch, 2012;

Pedersen et al., 2020). More recently, large genotyping population studies allow a Mendelian randomization approach to understand causal relationships between processes such as inflammation and MPN risk. Mendelian randomization takes advantage of genetic variants as proxies for modifiable risk factors to infer causality. A large Danish study performed genotyping for the interleukin-6 receptor (IL6R) variant (rs2228145), which is associated with impaired interleukin-6 signaling and reduced inflammation (Garbers et al., 2014). Among the 107,969 individuals from the Copenhagen General Population Study, 352 had or developed MPN; a subset of 49,143 individuals from the cohort were examined specifically for JAK2V617F CH, and 62 were positive. The IL6R loss of function variant was significantly underrepresented in both the MPN cohort, and in the JAK2V617F-positive CH cohort, implicating chronic inflammation as a risk factor for developing CH or MPN (Pedersen et al., 2020). The investigators demonstrated that the risk of acquiring any MPN was 40% lower among carriers of the loss of function IL-6 receptor variant, even when controlled for age, sex and smoking status. This relative risk reduction was even more prominent when examined in the JAK2V617F-positve CH sub cohort, further substantiating the concept that acquisition of JAK2V617F-positive CH and JAK2V617F-positive MPN subtend the same risk factors (Hinds et al., 2016; Nielsen et al., 2013; Pedersen et al., 2020). Interestingly, the protection was more prominent for PV and MF, consistent with the increased levels of inflammation that have been described in MF, and consistent with the higher prevalence and burden of JAK2V617F in PV (Hasselbalch, 2012). More recently investigators examined the same IL6R loss of function variant and its association with CH in 35,416 individuals from the UK Biobank without prevalent cardiovascular disease (CVD). The investigators tested the association of *DNMT3A* and *TET2* CH status with incident CVD events (myocardial infarction, coronary revascularization, stroke, or death), and whether CH risk was modified by the IL6R loss of function variant. While CVD risk was particularly associated with larger CH VAF, the risk was mitigated in the setting of the loss of function IL6R variant. Specifically, the IL6R loss of function variant attenuated CVD event risk among participants with large CH clones (hazard ratio, 0.46 [95% CI: 0.29–0.73], $P<0.001$) but not in individuals without CH (hazard ratio, 0.95 [95% CI: 0.89–1.01) (Bick et al., 2020). Taken together, both GWAS and Mendelian randomization approaches suggest that genetic background strongly influences both the acquisition and evolution of CH and MPN.

Chronic inflammation as a risk factor for developing CH is further supported by a recent study that examined the prevalence of CH in 200 individuals undergoing total hip replacement due to osteoarthritis (Hecker et al., 2021). The prevalence of CH, as assessed by resequencing of 68 myeloid genes, was a striking 50%. The most commonly mutated genes were similar to the large unselected population CH studies, with *DNMT3A* the most prevalent, followed by *TET2* and *ASXL1* (Table 3). Remarkably, the investigators observed an association of CH with autoimmune diseases, with a multivariate adjusted odds ratio of 6.6 (95% confidence interval (1.7, 30); $P = 0.0081$). The most common autoimmune syndromes were polymyalgia rheumatica and psoriatic arthritis, followed by autoimmune thyroid disease, rheumatoid arthritis, chronic inflammatory demyelinating polyneuropathy, and pemphigus. Indeed, in vivo murine models of *TET2* deficiency showed that an inflammatory insult enhanced *TET2* deficiency associated clonal expansion. This was associated with both increased secretion of inflammatory cytokines and increased HSC stress responses in the *TET2* deficiency context. In this model, the sensitivity to inflammation could be attenuated with anti-inflammatories (Cai et al., 2018). These findings underscore the association between CH and inflammatory diseases, and taken together with the Mendelian randomization studies and in vivo modeling, imply that inflammation may be causal to the development of CH. How an individual's background genetics modulates their response to inflammation will be important to factor into understanding their risk of subsequent CH and MPN.

7. Mutation burden and clinical risk

Mutation burden, relating to both the quality and quantity, is a key factor for risk of progression to hematologic malignancy in both the CH (Genovese et al., 2014) and MPN contexts (Grinfeld et al., 2018). In the decade and a half after the discovery of JAK2V617F, the importance of JAK2V617F risk both as a categorical and quantitative risk factor emerged. Not long after the 2005 JAK2V617F discovery, many studies observed a higher risk of thrombotic events in JAK2 mutation positive ET as compared to JAK2 mutation negative ET (Campbell et al., 2005; Rumi et al., 2014a) and MF (Rumi et al., 2014b). In a parallel fashion, JAK2V617F CH was highly associated with vascular thrombotic risk when assessed in a large cardiovascular cohort (Table 3) (Jaiswal et al., 2017). Furthermore, the significance of the JAK2V617F VAF in PV emerged. The VAF was paradoxically lower in younger individuals with PV yet still highly associated with an increased risk of splanchnic venous thrombosis. JAK2V617F VAF increase over time in PV associated with higher risk disease, and higher thrombosis risk (Moliterno et al., 2020). JAK2V617F VAF is an important variable in assessing risk of MPN at presentation and at transformation: lower JAK2V617F VAF associates with ET, whereas increases over time, or presentation with high VAF in PV, are risk factors for MF transformation (Vannucchi et al., 2011).

High bone marrow burden of somatic mutations has been associated with more aggressive MDS, and higher CD34-positive cell JAK2V617F burden associates with more aggressive MPN (Stein et al., 2011c; Xu et al., 2014). Higher CD34-positive cell burdens in MPN occurred in men, and this imparted more risk to men even in lower risk disease such as ET and PV (Karantanos et al., 2020). In a parallel fashion, higher mutation burden in CH was associated with higher risk and faster evolution to hematologic malignancy (Genovese et al., 2014; Nielsen et al., 2014). More rapid expansion of neoplastic clones in the stem and progenitor compartments in men could provide an explanation for their worse outcomes and faster progression to high-risk phenotypes and AML. High clonal stem cell burdens in men may relate to their number of mutations, to the high-risk quality of their lesions, to higher rates of mosaic chromosomal alterations (particularly related to the development of JAK2V617F homozygosity), to less fit residual normal clones due to effects of aging and inflammation, or to enhanced sensitivity of males to effects of inflammation (Fig. 1) (Gao et al., 2021). In both men and women, germline genetics that modulate inflammatory response, genomic instability, and mutation acquisition and latency, play into CH and MPN risk (Fig. 2 and Table 3).

8. Conclusion

Sex and genetic variation are critical modifiers of the presentation and outcomes in the MPN, and deserve further study to identify the mechanisms driving these effects. Designing therapies that do not address sex and genetic influences relevant to the MPN is a missed opportunity to make therapies that are more effective. For example, understanding that men have more complex and higher clonal mutational burden, may direct therapies away from symptom management and more to earlier referral for stem cell transplant. With regard to women, lower clonal burdens and longer latency may make a stem cell directed therapy, such as interferon alpha 2a, a therapy that may promote an effective cure, or drive burden reduction to that of CH levels (Gisslinger et al., 2020). Moreover, recognizing sex and genetic variation that might influence response to specific therapies deserves ongoing investigation (DeZern et al., 2017; Jäger et al., 2021; Nazha et al., 2019). Examining the impact of genetic background in natural history and outcomes will also provide insights to improve risk stratification and outcomes (Khan et al., 2016). Finally, recognizing that the sex-based and genetic factors that drive MPN also drive CH elevates the responsibility of the research community to identify risks and design optimization strategies to mitigate CH in all populations, and particularly in cardiovascular, autoimmune, and cancer cohorts, where CH has already demonstrated to be an independent negative prognostic factor.

Funding

The authors acknowledge the support of the MPNRF (A.R.M.) and NHLBI 1K08HL138142 (E.M.B.).

Conflicts of interest

The authors declare no conflict of interest.

References

Acuna-Hidalgo, R., Sengul, H., Steehouwer, M., van de Vorst, M., Vermeulen, S.H., Lalm, K., Veltman, J.A., Gilissen, C., Hoischen, A., 2017. Ultra-sensitive sequencing identifies high prevalence of clonal hematopoiesis-associated mutations throughout adult life. Am. J. Hum. Genet. 101, 50–64.

Andersen, M., Sajid, Z., Pedersen, R.K., Gudmand-Hoeyer, J., Ellervik, C., Skov, V., Kjær, L., Pallisgaard, N., Kruse, T.A., Thomassen, M., Troelsen, J., Hasselbalch, H.C., Ottesen, J.T., 2017. Mathematical modelling as a proof of concept for MPNs as a human inflammation model for cancer development. PLoS One 12, e0183620.

Babushok, D.V., Stanley, N.L., Morrissette, J.J.D., Lieberman, D.B., Olson, T.S., Chou, S.T., Hexner, E.O., 2018. Germline duplication of ATG2B and GSKIP genes is not required for the familial myeloid malignancy syndrome associated with the duplication of chromosome 14q32. Leukemia 32, 2720–2723.

Bailey, M.H., Tokheim, C., Porta-Pardo, E., Sengupta, S., Bertrand, D., Weerasinghe, A., Colaprico, A., Wendl, M.C., Kim, J., Reardon, B., Ng, P.K., Jeong, K.J., Cao, S., Wang, Z., Gao, J., Gao, Q., Wang, F., Liu, E.M., Mularoni, L., Rubio-Perez, C., Nagarajan, N., Cortés-Ciriano, I., Zhou, D.C., Liang, W.W., Hess, J.M., Yellapantula, V.D., Tamborero, D., Gonzalez-Perez, A., Suphavilai, C., Ko, J.Y., Khurana, E., Park, P.J., Van Allen, E.M., Liang, H., Lawrence, M.S., Godzik, A., Lopez-Bigas, N., Stuart, J., Wheeler, D., Getz, G., Chen, K., Lazar, A.J., Mills, G.B., Karchin, R., Ding, L., 2018. Comprehensive characterization of cancer driver genes and mutations. Cell 173, 371-85.e18.

Bao, E.L., Nandakumar, S.K., Liao, X., Bick, A.G., Karjalainen, J., Tabaka, M., Gan, O.I., Havulinna, A.S., Kiiskinen, T.T.J., Lareau, C.A., de Lapuente Portilla, A.L., Li, B., Emdin, C., Codd, V., Nelson, C.P., Walker, C.J., Churchhouse, C., de la

Chapelle, A., Klein, D.E., Nilsson, B., Wilson, P.W.F., Cho, K., Pyarajan, S., Gaziano, J.M., Samani, N.J., Regev, A., Palotie, A., Neale, B.M., Dick, J.E., Natarajan, P., O'Donnell, C.J., Daly, M.J., Milyavsky, M., Kathiresan, S., Sankaran, V.G., 2020. Inherited myeloproliferative neoplasm risk affects haematopoietic stem cells. Nature 586, 769–775.

Beck, D.B., Ferrada, M.A., Sikora, K.A., Ombrello, A.K., Collins, J.C., Pei, W., Balanda, N., Ross, D.L., Ospina Cardona, D., Wu, Z., Patel, B., Manthiram, K., Groarke, E.M., Gutierrez-Rodrigues, F., Hoffmann, P., Rosenzweig, S., Nakabo, S., Dillon, L.W., Hourigan, C.S., Tsai, W.L., Gupta, S., Carmona-Rivera, C., Asmar, A.J., Xu, L., Oda, H., Goodspeed, W., Barron, K.S., Nehrebecky, M., Jones, A., Laird, R.S., Deuitch, N., Rowczenio, D., Rominger, E., Wells, K.V., Lee, C.R., Wang, W., Trick, M., Mullikin, J., Wigerblad, G., Brooks, S., Dell'Orso, S., Deng, Z., Chae, J.J., Dulau-Florea, A., Malicdan, M.C.V., Novacic, D., Colbert, R.A., Kaplan, M.J., Gadina, M., Savic, S., Lachmann, H.J., Abu-Asab, M., Solomon, B.D., Retterer, K., Gahl, W.A., Burgess, S.M., Aksentijevich, I., Young, N.S., Calvo, K.R., Werner, A., Kastner, D.L., Grayson, P.C., 2020. Somatic mutations in UBA1 and severe adult-onset autoinflammatory disease. N. Engl. J. Med. 383, 2628–2638.

Bick, A.G., Pirruccello, J.P., Griffin, G.K., Gupta, N., Gabriel, S., Saleheen, D., Libby, P., Kathiresan, S., Natarajan, P., 2020. Genetic interleukin 6 signaling deficiency attenuates cardiovascular risk in clonal hematopoiesis. Circulation 141, 124–131.

Boddu, P., Masarova, L., Verstovsek, S., Strati, P., Kantarjian, H., Cortes, J., Estrov, Z., Pierce, S., Pemmaraju, N., 2018. Patient characteristics and outcomes in adolescents and young adults with classical Philadelphia chromosome-negative myeloproliferative neoplasms. Ann. Hematol. 97, 109–121.

Bolton, K.L., Ptashkin, R.N., Gao, T., Braunstein, L., Devlin, S.M., Kelly, D., Patel, M., Berthon, A., Syed, A., Yabe, M., Coombs, C.C., Caltabellotta, N.M., Walsh, M., Offit, K., Stadler, Z., Mandelker, D., Schulman, J., Patel, A., Philip, J., Bernard, E., Gundem, G., Ossa, J.E.A., Levine, M., Martinez, J.S.M., Farnoud, N., Glodzik, D., Li, S., Robson, M.E., Lee, C., Pharoah, P.D.P., Stopsack, K.H., Spitzer, B., Mantha, S., Fagin, J., Boucai, L., Gibson, C.J., Ebert, B.L., Young, A.L., Druley, T., Takahashi, K., Gillis, N., Ball, M., Padron, E., Hyman, D.M., Baselga, J., Norton, L., Gardos, S., Klimek, V.M., Scher, H., Bajorin, D., Paraiso, E., Benayed, R., Arcila, M.E., Ladanyi, M., Solit, D.B., Berger, M.F., Tallman, M., Garcia-Closas, M., Chatterjee, N., Diaz Jr., L.A., Levine, R.L., Morton, L.M., Zehir, A., Papaemmanuil, E., 2020a. Cancer therapy shapes the fitness landscape of clonal hematopoiesis. Nat. Genet. 52, 1219–1226.

Bolton, K.L., Zehir, A., Ptashkin, R.N., Patel, M., Gupta, D., Sidlow, R., Papaemmanuil, E., Berger, M.F., Levine, R.L., 2020b. The clinical management of clonal hematopoiesis: creation of a clonal hematopoiesis clinic. Hematol. Oncol. Clin. North Am. 34, 357–367.

Braunstein, E.M., Chen, H., Juarez, F., Yang, F., Tao, L., Makhlin, I., Williams, D.M., Chaturvedi, S., Pallavajjala, A., Karantanos, T., Martin, R., Wohler, E., Sobreira, N., Gocke, C.D., Moliterno, A.R., 2021. Germline ERBB2/HER2 coding variants are associated with increased risk of myeloproliferative neoplasms. Cancers (Basel) 13, 3246.

Cai, Z., Kotzin, J.J., Ramdas, B., Chen, S., Nelanuthala, S., Palam, L.R., Pandey, R., Mali, R.S., Liu, Y., Kelley, M.R., Sandusky, G., Mohseni, M., Williams, A., Henao-Mejia, J., Kapur, R., 2018. Inhibition of inflammatory signaling in Tet2 mutant preleukemic cells mitigates stress-induced abnormalities and clonal hematopoiesis. Cell Stem Cell 23, 833-49.e5.

Campbell, P.J., Scott, L.M., Buck, G., Wheatley, K., East, C.L., Marsden, J.T., Duffy, A., Boyd, E.M., Bench, A.J., Scott, M.A., Vassiliou, G.S., Milligan, D.W., Smith, S.R.,

Erber, W.N., Bareford, D., Wilkins, B.S., Reilly, J.T., Harrison, C.N., Green, A.R., 2005. Definition of subtypes of essential thrombocythaemia and relation to polycythaemia vera based on JAK2 V617F mutation status: a prospective study. Lancet 366, 1945–1953.
Carrel, L., Willard, H.F., 2005. X-inactivation profile reveals extensive variability in X-linked gene expression in females. Nature 434, 400–404.
Coltro, G., Mangaonkar, A.A., Lasho, T.L., Finke, C.M., Pophali, P., Carr, R., Gangat, N., Binder, M., Pardanani, A., Fernandez-Zapico, M., Robertson, K.D., Bosi, A., Droin, N., Vannucchi, A.M., Tefferi, A., Hunter, A., Padron, E., Solary, E., Patnaik, M.M., 2020. Clinical, molecular, and prognostic correlates of number, type, and functional localization of TET2 mutations in chronic myelomonocytic leukemia (CMML)-a study of 1084 patients. Leukemia 34, 1407–1421.
Cordua, S., Kjaer, L., Skov, V., Pallisgaard, N., Hasselbalch, H.C., Ellervik, C., 2019. Prevalence and phenotypes of JAK2 V617F and calreticulin mutations in a Danish general population. Blood 134, 469–479.
De-Morgan, A., Meggendorfer, M., Haferlach, C., Shlush, L., 2021. Male predominance in AML is associated with specific preleukemic mutations. Leukemia 35, 867–870.
DeZern, A.E., Zeidan, A.M., Barnard, J., Hand, W., Al Ali, N., Brown, F., Zimmerman, C., Roboz, G.J., Garcia-Manero, G., Steensma, D.P., Komrokji, R.S., Sekeres, M.A., 2017. Differential response to hypomethylating agents based on sex: a report on behalf of the MDS Clinical Research Consortium (MDS CRC). Leuk. Lymphoma 58, 1325–1331.
Dunford, A., Weinstock, D.M., Savova, V., Schumacher, S.E., Cleary, J.P., Yoda, A., Sullivan, T.J., Hess, J.M., Gimelbrant, A.A., Beroukhim, R., Lawrence, M.S., Getz, G., Lane, A.A., 2017. Tumor-suppressor genes that escape from X-inactivation contribute to cancer sex bias. Nat. Genet. 49, 10–16.
Edgren, G., Liang, L., Adami, H.O., Chang, E.T., 2012. Enigmatic sex disparities in cancer incidence. Eur. J. Epidemiol. 27, 187–196.
Furuncuoglu, Y., Tulgar, S., Dogan, A.N., Cakar, S., Tulgar, Y.K., Cakiroglu, B., 2016. How obesity affects the neutrophil/lymphocyte and platelet/lymphocyte ratio, systemic immune-inflammatory index and platelet indices: a retrospective study. Eur. Rev. Med. Pharmacol. Sci. 20, 1300–1306.
Gao, T., Ptashkin, R., Bolton, K.L., Sirenko, M., Fong, C., Spitzer, B., Menghrajani, K., Ossa, J.E.A., Zhou, Y., Bernard, E., Levine, M., Martinez, J.S.M., Zhang, Y., Franch-Expósito, S., Patel, M., Braunstein, L.Z., Kelly, D., Yabe, M., Benayed, R., Caltabellotta, N.M., Philip, J., Paraiso, E., Mantha, S., Solit, D.B., Diaz Jr., L.A., Berger, M.F., Klimek, V., Levine, R.L., Zehir, A., Devlin, S.M., Papaemmanuil, E., 2021. Interplay between chromosomal alterations and gene mutations shapes the evolutionary trajectory of clonal hematopoiesis. Nat. Commun. 12, 338.
Garbers, C., Monhasery, N., Aparicio-Siegmund, S., Lokau, J., Baran, P., Nowell, M.A., Jones, S.A., Rose-John, S., Scheller, J., 2014. The interleukin-6 receptor Asp358Ala single nucleotide polymorphism rs2228145 confers increased proteolytic conversion rates by ADAM proteases. Biochim. Biophys. Acta 1842, 1485–1494.
Genovese, G., Kähler, A.K., Handsaker, R.E., Lindberg, J., Rose, S.A., Bakhoum, S.F., Chambert, K., Mick, E., Neale, B.M., Fromer, M., Purcell, S.M., Svantesson, O., Landén, M., Höglund, M., Lehmann, S., Gabriel, S.B., Moran, J.L., Lander, E.S., Sullivan, P.F., Sklar, P., Grönberg, H., Hultman, C.M., McCarroll, S.A., 2014. Clonal hematopoiesis and blood-cancer risk inferred from blood DNA sequence. N. Engl. J. Med. 371, 2477–2487.
Gisslinger, H., Klade, C., Georgiev, P., Krochmalczyk, D., Gercheva-Kyuchukova, L., Egyed, M., Rossiev, V., Dulicek, P., Illes, A., Pylypenko, H., Sivcheva, L., Mayer, J., Yablokova, V., Krejcy, K., Grohmann-Izay, B., Hasselbalch, H.C., Kralovics, R., Kiladjian, J.J., 2020. Ropeginterferon alfa-2b versus standard therapy

for polycythaemia vera (PROUD-PV and CONTINUATION-PV): a randomised, non-inferiority, phase 3 trial and its extension study. Lancet Haematol. 7, e196–e208.

Grinfeld, J., Nangalia, J., Baxter, E.J., Wedge, D.C., Angelopoulos, N., Cantrill, R., Godfrey, A.L., Papaemmanuil, E., Gundem, G., MacLean, C., Cook, J., O'Neil, L., O'Meara, S., Teague, J.W., Butler, A.P., Massie, C.E., Williams, N., Nice, F.L., Andersen, C.L., Hasselbalch, H.C., Guglielmelli, P., McMullin, M.F., Vannucchi, A.M., Harrison, C.N., Gerstung, M., Green, A.R., Campbell, P.J., 2018. Classification and personalized prognosis in myeloproliferative neoplasms. N. Engl. J. Med. 379, 1416–1430.

Guglielmelli, P., Lasho, T.L., Rotunno, G., Score, J., Mannarelli, C., Pancrazzi, A., Biamonte, F., Pardanani, A., Zoi, K., Reiter, A., Duncombe, A., Fanelli, T., Pietra, D., Rumi, E., Finke, C., Gangat, N., Ketterling, R.P., Knudson, R.A., Hanson, C.A., Bosi, A., Pereira, A., Manfredini, R., Cervantes, F., Barosi, G., Cazzola, M., Cross, N.C., Vannucchi, A.M., Tefferi, A., 2014. The number of prognostically detrimental mutations and prognosis in primary myelofibrosis: an international study of 797 patients. Leukemia 28, 1804–1810.

Harutyunyan, A.S., Giambruno, R., Krendl, C., Stukalov, A., Klampfl, T., Berg, T., Chen, D., Milosevic Feenstra, J.D., Jäger, R., Gisslinger, B., Gisslinger, H., Rumi, E., Passamonti, F., Pietra, D., Müller, A.C., Parapatics, K., Breitwieser, F.P., Herrmann, R., Colinge, J., Bennett, K.L., Superti-Furga, G., Cazzola, M., Hammond, E., Kralovics, R., 2016. Germline RBBP6 mutations in familial myeloproliferative neoplasms. Blood 127, 362–365.

Hasselbalch, H.C., 2012. Perspectives on chronic inflammation in essential thrombocythemia, polycythemia vera, and myelofibrosis: is chronic inflammation a trigger and driver of clonal evolution and development of accelerated atherosclerosis and second cancer? Blood 119, 3219–3225.

Haupt, S., Caramia, F., Klein, S.L., Rubin, J.B., Haupt, Y., 2021. Sex disparities matter in cancer development and therapy. Nat. Rev. Cancer 21, 393–407.

Hecker, J.S., Hartmann, L., Rivière, J., Buck, M.C., van der Garde, M., Rothenberg-Thurley, M., Fischer, L., Winter, S., Ksienzyk, B., Ziemann, F., Solovey, M., Rauner, M., Tsourdi, E., Sockel, K., Schneider, M., Kubasch, A.S., Nolde, M., Hausmann, D., Paulus, A.C., Lützner, J., Roth, A., Bassermann, F., Spiekermann, K., Marr, C., Hofbauer, L.C., Platzbecker, U., Metzeler, K.H., Götze, K.S., 2021. CHIP & HIPs: clonal hematopoiesis is common in hip arthroplasty patients and associates with autoimmune disease. Blood.

Hehlmann, R., Voskanyan, A., Lauseker, M., Pfirrmann, M., Kalmanti, L., Rinaldetti, S., Kohlbrenner, K., Haferlach, C., Schlegelberger, B., Fabarius, A., Seifarth, W., Spieß, B., Wuchter, P., Krause, S., Kolb, H.J., Neubauer, A., Hossfeld, D.K., Nerl, C., Gratwohl, A., Baerlocher, G.M., Burchert, A., Brümmendorf, T.H., Hasford, J., Hochhaus, A., Saußele, S., Baccarani, M., 2020. High-risk additional chromosomal abnormalities at low blast counts herald death by CML. Leukemia 34, 2074–2086.

Herbaux, C., Badens, C., Guidez, S., Lacoste, C., Maboudou, P., Rose, C., 2012. A new ATRX mutation in a patient with acquired α-thalassemia myelodysplastic syndrome. Hemoglobin 36, 581–585.

Herbaux, C., Duployez, N., Badens, C., Poret, N., Gardin, C., Decamp, M., Eclache, V., Daliphard, S., Murati, A., Cony-Makhoul, P., Cheze, S., Beve, B., Lacoste, C., Prebet, T., Hunault-Berger, M., Maloisel, F., Renneville, A., Figeac, M., Stamatoullas-Bastard, A., Bastard, C., Fenaux, P., Preudhomme, C., Rose, C., 2015. Incidence of ATRX mutations in myelodysplastic syndromes, the value of microcytosis. Am. J. Hematol. 90, 737–738.

Hinds, D.A., Barnholt, K.E., Mesa, R.A., Kiefer, A.K., Do, C.B., Eriksson, N., Mountain, J.L., Francke, U., Tung, J.Y., Nguyen, H.M., Zhang, H., Gojenola, L.,

Zehnder, J.L., Gotlib, J., 2016. Germ line variants predispose to both JAK2 V617F clonal hematopoiesis and myeloproliferative neoplasms. Blood 128, 1121–1128.

Hultcrantz, M., Kristinsson, S.Y., Andersson, T.M., Landgren, O., Eloranta, S., Derolf, A.R., Dickman, P.W., Bjorkholm, M., 2012. Patterns of survival among patients with myeloproliferative neoplasms diagnosed in Sweden from 1973 to 2008: a population-based study. J. Clin. Oncol. 30, 2995–3001.

Hultcrantz, M., Bjorkholm, M., Dickman, P.W., Landgren, O., Derolf, A.R., Kristinsson, S.Y., Andersson, T.M.L., 2018. Risk for arterial and venous thrombosis in patients with myeloproliferative neoplasms: a population-based cohort study. Ann. Intern. Med. 168, 317–325.

Jäger, R., Gisslinger, H., Fuchs, E., Bogner, E., Milosevic Feenstra, J.D., Weinzierl, J., Schischlik, F., Gisslinger, B., Schalling, M., Zörer, M., Krejcy, K., Klade, C., Kralovics, R., 2021. Germline genetic factors influence the outcome of interferon-α therapy in polycythemia vera. Blood 137, 387–391.

Jaiswal, S., Fontanillas, P., Flannick, J., Manning, A., Grauman, P.V., Mar, B.G., Lindsley, R.C., Mermel, C.H., Burtt, N., Chavez, A., Higgins, J.M., Moltchanov, V., Kuo, F.C., Kluk, M.J., Henderson, B., Kinnunen, L., Koistinen, H.A., Ladenvall, C., Getz, G., Correa, A., Banahan, B.F., Gabriel, S., Kathiresan, S., Stringham, H.M., McCarthy, M.I., Boehnke, M., Tuomilehto, J., Haiman, C., Groop, L., Atzmon, G., Wilson, J.G., Neuberg, D., Altshuler, D., Ebert, B.L., 2014. Age-related clonal hematopoiesis associated with adverse outcomes. N. Engl. J. Med. 371, 2488–2498.

Jaiswal, S., Natarajan, P., Silver, A.J., Gibson, C.J., Bick, A.G., Shvartz, E., McConkey, M., Gupta, N., Gabriel, S., Ardissino, D., Baber, U., Mehran, R., Fuster, V., Danesh, J., Frossard, P., Saleheen, D., Melander, O., Sukhova, G.K., Neuberg, D., Libby, P., Kathiresan, S., Ebert, B.L., 2017. Clonal hematopoiesis and risk of atherosclerotic cardiovascular disease. N. Engl. J. Med. 377, 111–121.

Jankowska, A.M., Makishima, H., Tiu, R.V., Szpurka, H., Huang, Y., Traina, F., Visconte, V., Sugimoto, Y., Prince, C., O'Keefe, C., Hsi, E.D., List, A., Sekeres, M.A., Rao, A., McDevitt, M.A., Maciejewski, J.P., 2011. Mutational spectrum analysis of chronic myelomonocytic leukemia includes genes associated with epigenetic regulation: UTX, EZH2, and DNMT3A. Blood 118, 3932–3941.

Jayasuriya, N.A., Kjaergaard, A.D., Pedersen, K.M., Sorensen, A.L., Bak, M., Larsen, M.K., Nordestgaard, B.G., Bojesen, S.E., Colak, Y., Skov, V., Kjaer, L., Tolstrup, J.S., Hasselbalch, H.C., Ellervik, C., 2019. Smoking, blood cells and myeloproliferative neoplasms: meta-analysis and Mendelian randomization of 2.3 million people. Br. J. Haematol. 189, 323–334.

Jones, A.V., Chase, A., Silver, R.T., Oscier, D., Zoi, K., Wang, Y.L., Cario, H., Pahl, H.L., Collins, A., Reiter, A., Grand, F., Cross, N.C., 2009. JAK2 haplotype is a major risk factor for the development of myeloproliferative neoplasms. Nat. Genet. 41, 446–449.

Karantanos, T., Chaturvedi, S., Braunstein, E.M., Spivak, J., Resar, L., Karanika, S., Williams, D.M., Rogers, O., Gocke, C.D., Moliterno, A.R., 2020. Sex determines the presentation and outcomes in MPN and is related to sex-specific differences in the mutational burden. Blood Adv. 4, 2567–2576.

Karantanos, T., Gondek, L.P., Varadhan, R., Moliterno, A.R., DeZern, A.E., Jones, R.J., Jain, T., 2021. Gender-related differences in the outcomes and genomic landscape of patients with myelodysplastic syndrome/myeloproliferative neoplasm overlap syndromes. Br. J. Haematol. 193, 1142–1150.

Khan, I., Shergill, A., Saraf, S.L., Chen, Y.F., Patel, P.R., Quigley, J.G., Peace, D., Gordeuk, V.R., Hoffman, R., Rondelli, D., 2016. Outcome disparities in Caucasian and non-Caucasian patients with myeloproliferative neoplasms. Clin. Lymphoma Myeloma Leuk. 16, 350–357.

Khoury, P., Desmond, R., Pabon, A., Holland-Thomas, N., Ware, J.M., Arthur, D.C., Kurlander, R., Fay, M.P., Maric, I., Klion, A.D., 2016. Clinical features predict responsiveness to imatinib in platelet-derived growth factor receptor-alpha-negative hypereosinophilic syndrome. Allergy 71, 803–810.

Kilpivaara, O., Mukherjee, S., Schram, A.M., Wadleigh, M., Mullally, A., Ebert, B.L., Bass, A., Marubayashi, S., Heguy, A., Garcia-Manero, G., Kantarjian, H., Offit, K., Stone, R.M., Gilliland, D.G., Klein, R.J., Levine, R.L., 2009. A germline JAK2 SNP is associated with predisposition to the development of JAK2(V617F)-positive myeloproliferative neoplasms. Nat. Genet. 41, 455–459.

Kon, A., Shih, L.Y., Minamino, M., Sanada, M., Shiraishi, Y., Nagata, Y., Yoshida, K., Okuno, Y., Bando, M., Nakato, R., Ishikawa, S., Sato-Otsubo, A., Nagae, G., Nishimoto, A., Haferlach, C., Nowak, D., Sato, Y., Alpermann, T., Nagasaki, M., Shimamura, T., Tanaka, H., Chiba, K., Yamamoto, R., Yamaguchi, T., Otsu, M., Obara, N., Sakata-Yanagimoto, M., Nakamaki, T., Ishiyama, K., Nolte, F., Hofmann, W.K., Miyawaki, S., Chiba, S., Mori, H., Nakauchi, H., Koeffler, H.P., Aburatani, H., Haferlach, T., Shirahige, K., Miyano, S., Ogawa, S., 2013. Recurrent mutations in multiple components of the cohesin complex in myeloid neoplasms. Nat. Genet. 45, 1232–1237.

Landgren, O., Goldin, L.R., Kristinsson, S.Y., Helgadottir, E.A., Samuelsson, J., Björkholm, M., 2008. Increased risks of polycythemia vera, essential thrombocythemia, and myelofibrosis among 24,577 first-degree relatives of 11,039 patients with myeloproliferative neoplasms in Sweden. Blood 112, 2199–2204.

Lim, T.L., Lieberman, D.B., Davis, A.R., Loren, A.W., Hausler, R., Bigdeli, A., Li, Y., Powers, J., Raper, A., Genetics, C.R., Carty, S.A., Nathanson, K.L., Bagg, A., Hexner, E.O., Maxwell, K.N., Morrissette, J.J.D., Babushok, D.V., 2021. Germline POT1 variants can predispose to myeloid and lymphoid neoplasms. Leukemia.

Madan, V., Li, J., Zhou, S., Teoh, W.W., Han, L., Meggendorfer, M., Malcovati, L., Cazzola, M., Ogawa, S., Haferlach, T., Yang, H., Koeffler, H.P., 2020. Distinct and convergent consequences of splice factor mutations in myelodysplastic syndromes. Am. J. Hematol. 95, 133–143.

McKerrell, T., Park, N., Moreno, T., Grove, C.S., Ponstingl, H., Stephens, J., Crawley, C., Craig, J., Scott, M.A., Hodkinson, C., Baxter, J., Rad, R., Forsyth, D.R., Quail, M.A., Zeggini, E., Ouwehand, W., Varela, I., Vassiliou, G.S., 2015. Leukemia-associated somatic mutations drive distinct patterns of age-related clonal hemopoiesis. Cell Rep. 10, 1239–1245.

Metzeler, K.H., Herold, T., Rothenberg-Thurley, M., Amler, S., Sauerland, M.C., Görlich, D., Schneider, S., Konstandin, N.P., Dufour, A., Bräundl, K., Ksienzyk, B., Zellmeier, E., Hartmann, L., Greif, P.A., Fiegl, M., Subklewe, M., Bohlander, S.K., Krug, U., Faldum, A., Berdel, W.E., Wörmann, B., Büchner, T., Hiddemann, W., Braess, J., Spiekermann, K., 2016. Spectrum and prognostic relevance of driver gene mutations in acute myeloid leukemia. Blood 128, 686–698.

Moliterno, A.R., Williams, D.M., Rogers, O., Isaacs, M.A., Spivak, J.L., 2008. Phenotypic variability within the JAK2 V617F-positive MPD: roles of progenitor cell and neutrophil allele burdens. Exp. Hematol. 36, 1480–1486.

Moliterno, A.R., Ginzburg, Y.Z., Hoffman, R., 2020. Clinical insights into the origins of thrombosis in myeloproliferative neoplasms. Blood 137, 1145–1153.

Nazha, A., Sekeres, M.A., Bejar, R., Rauh, M.J., Othus, M., Komrokji, R.S., Barnard, J., Hilton, C.B., Kerr, C.M., Steensma, D.P., DeZern, A., Roboz, G., Garcia-Manero, G., Erba, H., Ebert, B.L., Maciejewski, J.P., 2019. Genomic biomarkers to predict resistance to hypomethylating agents in patients with myelodysplastic syndromes using artificial intelligence. JCO Precis. Oncol. 3. https://doi.org/10.1200/po.19.00119.

Nielsen, C., Birgens, H.S., Nordestgaard, B.G., Kjaer, L., Bojesen, S.E., 2011. The JAK2 V617F somatic mutation, mortality and cancer risk in the general population. Haematologica 96, 450–453.

Nielsen, C., Birgens, H.S., Nordestgaard, B.G., Bojesen, S.E., 2013. Diagnostic value of JAK2 V617F somatic mutation for myeloproliferative cancer in 49 488 individuals from the general population. Br. J. Haematol. 160, 70–79.

Nielsen, C., Bojesen, S.E., Nordestgaard, B.G., Kofoed, K.F., Birgens, H.S., 2014. JAK2V617F somatic mutation in the general population: myeloproliferative neoplasm development and progression rate. Haematologica 99, 1448–1455.

Pedersen, K.M., Colak, Y., Ellervik, C., Hasselbalch, H.C., Bojesen, S.E., Nordestgaard, B.G., 2020. Loss-of-function polymorphism in IL6R reduces risk of JAK2V617F somatic mutation and myeloproliferative neoplasm: a Mendelian randomization study. EClinicalMedicine 21, 100280.

Pegliasco, J., Hirsch, P., Marzac, C., Isnard, F., Meniane, J.C., Deswarte, C., Pellet, P., Lemaitre, C., Leroy, G., Rabadan Moraes, G., Guermouche, H., Schmaltz-Panneau, B., Pasquier, F., Colas, C., Benusiglio, P.R., Bera, O., Bourhis, J.H., Brissot, E., Caron, O., Chraibi, S., Cony-Makhoul, P., Delaunay-Darivon, C., Lapusan, S., de Fontbrune, F.S., Fuseau, P., Najman, A., Vainchenker, W., Delhommeau, F., Micol, J.B., Plo, I., Bellanné-Chantelot, C., 2021. Germline ATG2B/GSKIP-containing 14q32 duplication predisposes to early clonal hematopoiesis leading to myeloid neoplasms. Leukemia.

Pemmaraju, N., Lane, A.A., Sweet, K.L., Stein, A.S., Vasu, S., Blum, W., Rizzieri, D.A., Wang, E.S., Duvic, M., Sloan, J.M., Spence, S., Shemesh, S., Brooks, C.L., Balser, J., Bergstein, I., Lancet, J.E., Kantarjian, H.M., Konopleva, M., 2019. Tagraxofusp in blastic plasmacytoid dendritic-cell neoplasm. N. Engl. J. Med. 380, 1628–1637.

Pleyer, L., Leisch, M., Kourakli, A., Padron, E., Maciejewski, J.P., Xicoy Cirici, B., Kaivers, J., Ungerstedt, J., Heibl, S., Patiou, P., Hunter, A.M., Mora, E., Geissler, K., Dimou, M., Jimenez Lorenzo, M.J., Melchardt, T., Egle, A., Viniou, A.N., Patel, B.J., Arnan, M., Valent, P., Roubakis, C., Bernal Del Castillo, T., Galanopoulos, A., Calabuig Muñoz, M., Bonadies, N., Medina de Almeida, A., Cermak, J., Jerez, A., Montoro, M.J., Cortés, A., Avendaño Pita, A., Lopez Andrade, B., Hellstroem-Lindberg, E., Germing, U., Sekeres, M.A., List, A.F., Symeonidis, A., Sanz, G.F., Larcher-Senn, J., Greil, R., 2021. Outcomes of patients with chronic myelomonocytic leukaemia treated with non-curative therapies: a retrospective cohort study. Lancet Haematol. 8, e135-e48.

Quesada, A.E., Routbort, M.J., DiNardo, C.D., Bueso-Ramos, C.E., Kanagal-Shamanna, R., Khoury, J.D., Thakral, B., Zuo, Z., Yin, C.C., Loghavi, S., Ok, C.Y., Wang, S.A., Tang, Z., Bannon, S.A., Benton, C.B., Garcia-Manero, G., Kantarjian, H., Luthra, R., Medeiros, L.J., Patel, K.P., 2019. DDX41 mutations in myeloid neoplasms are associated with male gender, TP53 mutations and high-risk disease. Am. J. Hematol. 94, 757–766.

Radkiewicz, C., Johansson, A.L.V., Dickman, P.W., Lambe, M., Edgren, G., 2017. Sex differences in cancer risk and survival: a Swedish cohort study. Eur. J. Cancer 84, 130–140.

Renosi, F., Roggy, A., Giguelay, A., Soret, L., Viailly, P.J., Cheok, M., Biichle, S., Angelot-Delettre, F., Asnafi, V., Macintyre, E., Geffroy, S., Callanan, M., Petrella, T., Deconinck, E., Daguindau, E., Harrivel, V., Bouyer, S., Salaun, V., Saussoy, P., Feuillard, J., Fuseau, P., Saas, P., Adotévi, O., Jardin, F., Ferrand, C., Preudhomme, C., Colinge, J., Roumier, C., Garnache-Ottou, F., 2021. Transcriptomic and genomic heterogeneity in blastic plasmacytoid dendritic cell neoplasms: from ontogeny to oncogenesis. Blood Adv. 5, 1540–1551.

Rumi, E., Pietra, D., Ferretti, V., Klampfl, T., Harutyunyan, A.S., Milosevic, J.D., Them, N.C., Berg, T., Elena, C., Casetti, I.C., Milanesi, C., Sant'antonio, E., Bellini, M., Fugazza, E., Renna, M.C., Boveri, E., Astori, C., Pascutto, C.,

Kralovics, R., Cazzola, M., 2014a. JAK2 or CALR mutation status defines subtypes of essential thrombocythemia with substantially different clinical course and outcomes. Blood 123, 1544–1551.

Rumi, E., Pietra, D., Pascutto, C., Guglielmelli, P., Martinez-Trillos, A., Casetti, I., Colomer, D., Pieri, L., Pratcorona, M., Rotunno, G., Sant'Antonio, E., Bellini, M., Cavalloni, C., Mannarelli, C., Milanesi, C., Boveri, E., Ferretti, V., Astori, C., Rosti, V., Cervantes, F., Barosi, G., Vannucchi, A.M., Cazzola, M., 2014b. Clinical effect of driver mutations of JAK2, CALR, or MPL in primary myelofibrosis. Blood 124, 1062–1069.

Saliba, J., Saint-Martin, C., Di Stefano, A., Lenglet, G., Marty, C., Keren, B., Pasquier, F., Valle, V.D., Secardin, L., Leroy, G., Mahfoudhi, E., Grosjean, S., Droin, N., Diop, M., Dessen, P., Charrier, S., Palazzo, A., Merlevede, J., Meniane, J.C., Delaunay-Darivon, C., Fuseau, P., Isnard, F., Casadevall, N., Solary, E., Debili, N., Bernard, O.A., Raslova, H., Najman, A., Vainchenker, W., Bellanné-Chantelot, C., Plo, I., 2015. Germline duplication of ATG2B and GSKIP predisposes to familial myeloid malignancies. Nat. Genet. 47, 1131–1140.

Schratz, K.E., Haley, L., Danoff, S.K., Blackford, A., DeZern, A., Gocke, C.D., Duffield, A.S., Armanios, M., 2020. Cancer spectrum and outcomes in the mendelian short telomere syndromes. Blood 135, 1946–1956.

Segal, J.B., Moliterno, A.R., 2006. Platelet counts differ by sex, ethnicity, and age in the United States. Ann. Epidemiol. 16, 123–130.

Siegel, R.L., Miller, K.D., Fuchs, H.E., Jemal, A., 2021. Cancer statistics, 2021. CA Cancer J. Clin. 71, 7–33.

Spivak, J.L., Considine, M., Williams, D.M., Talbot Jr., C.C., Rogers, O., Moliterno, A.R., Jie, C., Ochs, M.F., 2014. Two clinical phenotypes in polycythemia vera. N. Engl. J. Med. 371, 808–817.

Stein, B.L., Rademaker, A., Spivak, J.L., Moliterno, A.R., 2011a. Gender and vascular complications in the JAK2 V617F-positive myeloproliferative neoplasms. Thrombosis 2011, 874146.

Stein, B.L., Williams, D.M., O'Keefe, C., Rogers, O., Ingersoll, R.G., Spivak, J.L., Verma, A., Maciejewski, J.P., McDevitt, M.A., Moliterno, A.R., 2011b. Disruption of the ASXL1 gene is frequent in primary, post-essential thrombocytosis and post-polycythemia vera myelofibrosis, but not essential thrombocytosis or polycythemia vera: analysis of molecular genetics and clinical phenotypes. Haematologica 96, 1462–1469.

Stein, B.L., Williams, D.M., Rogers, O., Isaacs, M.A., Spivak, J.L., Moliterno, A.R., 2011c. Disease burden at the progenitor level is a feature of primary myelofibrosis: a multivariable analysis of 164 JAK2 V617F-positive myeloproliferative neoplasm patients. Exp. Hematol. 39, 95–101.

Stein, B.L., Saraf, S., Sobol, U., Halpern, A., Shammo, J., Rondelli, D., Michaelis, L., Odenike, O., Rademaker, A., Zakarija, A., McMahon, B., Spivak, J.L., Moliterno, A.R., 2013. Age-related differences in disease characteristics and clinical outcomes in polycythemia vera. Leuk. Lymphoma 54, 1989–1995.

Stein, B.L., Gotlib, J., Arcasoy, M., Nguyen, M.H., Shah, N., Moliterno, A., Jamieson, C., Pollyea, D.A., Scott, B., Wadleigh, M., Levine, R., Komrokji, R., Klisovic, R., Gundabolu, K., Kropf, P., Wetzler, M., Oh, S.T., Ribeiro, R., Paschal, R., Mohan, S., Podoltsev, N., Prchal, J., Talpaz, M., Snyder, D., Verstovsek, S., Mesa, R.A., 2015. Historical views, conventional approaches, and evolving management strategies for myeloproliferative neoplasms. J. Natl. Compr. Canc. Netw. 13, 424–434.

Sud, A., Chattopadhyay, S., Thomsen, H., Sundquist, K., Sundquist, J., Houlston, R.S., Hemminki, K., 2019. Analysis of 153 115 patients with hematological malignancies refines the spectrum of familial risk. Blood 134, 960–969.

Tapper, W., Jones, A.V., Kralovics, R., Harutyunyan, A.S., Zoi, K., Leung, W., Godfrey, A.L., Guglielmelli, P., Callaway, A., Ward, D., Aranaz, P., White, H.E.,

Waghorn, K., Lin, F., Chase, A., Baxter, E.J., Maclean, C., Nangalia, J., Chen, E., Evans, P., Short, M., Jack, A., Wallis, L., Oscier, D., Duncombe, A.S., Schuh, A., Mead, A.J., Griffiths, M., Ewing, J., Gale, R.E., Schnittger, S., Haferlach, T., Stegelmann, F., Dohner, K., Grallert, H., Strauch, K., Tanaka, T., Bandinelli, S., Giannopoulos, A., Pieri, L., Mannarelli, C., Gisslinger, H., Barosi, G., Cazzola, M., Reiter, A., Harrison, C., Campbell, P., Green, A.R., Vannucchi, A., Cross, N.C., 2015. Genetic variation at MECOM, TERT, JAK2 and HBS1L-MYB predisposes to myeloproliferative neoplasms. Nat. Commun. 6, 6691.

Tefferi, A., Betti, S., Barraco, D., Mudireddy, M., Shah, S., Hanson, C.A., Ketterling, R.P., Pardanani, A., Gangat, N., Coltro, G., Guglielmelli, P., Vannucchi, A.M., 2017. Gender and survival in essential thrombocythemia: a two-center study of 1,494 patients. Am. J. Hematol. 92, 1193–1197.

Van Egeren, D., Escabi, J., Nguyen, M., Liu, S., Reilly, C.R., Patel, S., Kamaz, B., Kalyva, M., DeAngelo, D.J., Galinsky, I., Wadleigh, M., Winer, E.S., Luskin, M.R., Stone, R.M., Garcia, J.S., Hobbs, G.S., Camargo, F.D., Michor, F., Mullally, A., Cortes-Ciriano, I., Hormoz, S., 2021. Reconstructing the lineage histories and differentiation trajectories of individual cancer cells in myeloproliferative neoplasms. Cell Stem Cell 28, 514-23.e9.

van Haaften, G., Dalgliesh, G.L., Davies, H., Chen, L., Bignell, G., Greenman, C., Edkins, S., Hardy, C., O'Meara, S., Teague, J., Butler, A., Hinton, J., Latimer, C., Andrews, J., Barthorpe, S., Beare, D., Buck, G., Campbell, P.J., Cole, J., Forbes, S., Jia, M., Jones, D., Kok, C.Y., Leroy, C., Lin, M.L., McBride, D.J., Maddison, M., Maquire, S., McLay, K., Menzies, A., Mironenko, T., Mulderrig, L., Mudie, L., Pleasance, E., Shepherd, R., Smith, R., Stebbings, L., Stephens, P., Tang, G., Tarpey, P.S., Turner, R., Turrell, K., Varian, J., West, S., Widaa, S., Wray, P., Collins, V.P., Ichimura, K., Law, S., Wong, J., Yuen, S.T., Leung, S.Y., Tonon, G., DePinho, R.A., Tai, Y.T., Anderson, K.C., Kahnoski, R.J., Massie, A., Khoo, S.K., Teh, B.T., Stratton, M.R., Futreal, P.A., 2009. Somatic mutations of the histone H3K27 demethylase gene UTX in human cancer. Nat. Genet. 41, 521–523.

Vannucchi, A.M., Pieri, L., Guglielmelli, P., 2011. JAK2 allele burden in the myeloproliferative neoplasms: effects on phenotype, prognosis and change with treatment. Ther Adv Hematol 2, 21–32.

Wang, F., Ni, J., Wu, L., Wang, Y., He, B., Yu, D., 2019. Gender disparity in the survival of patients with primary myelodysplastic syndrome. J. Cancer 10, 1325–1332.

Xavier-Ferrucio, J., Scanlon, V., Li, X., Zhang, P.X., Lozovatsky, L., Ayala-Lopez, N., Tebaldi, T., Halene, S., Cao, C., Fleming, M.D., Finberg, K.E., Krause, D.S., 2019. Low iron promotes megakaryocytic commitment of megakaryocytic-erythroid progenitors in humans and mice. Blood 134, 1547–1557.

Xu, L., Gu, Z.H., Li, Y., Zhang, J.L., Chang, C.K., Pan, C.M., Shi, J.Y., Shen, Y., Chen, B., Wang, Y.Y., Jiang, L., Lu, J., Xu, X., Tan, J.L., Chen, Y., Wang, S.Y., Li, X., Chen, Z., Chen, S.J., 2014. Genomic landscape of CD34+ hematopoietic cells in myelodysplastic syndrome and gene mutation profiles as prognostic markers. Proc. Natl. Acad. Sci. U. S. A. 111, 8589–8594.

Zheng, G., Chen, P., Pallavajjalla, A., Haley, L., Gondek, L., Dezern, A., Ling, H., De Marchi, F., Lin, M.T., Gocke, C., 2019. The diagnostic utility of targeted gene panel sequencing in discriminating etiologies of cytopenia. Am. J. Hematol. 94, 1141–1148.

Zink, F., Stacey, S.N., Norddahl, G.L., Frigge, M.L., Magnusson, O.T., Jonsdottir, I., Thorgeirsson, T.E., Sigurdsson, A., Gudjonsson, S.A., Gudmundsson, J., Jonasson, J.G., Tryggvadottir, L., Jonsson, T., Helgason, A., Gylfason, A., Sulem, P., Rafnar, T., Thorsteinsdottir, U., Gudbjartsson, D.F., Masson, G., Kong, A., Stefansson, K., 2017. Clonal hematopoiesis, with and without candidate driver mutations, is common in the elderly. Blood 130, 742–752.

CHAPTER TWO

Transcriptional configurations of myeloproliferative neoplasms

Fiorella Schischlik*
Cancer Data Science Laboratory, Center for Cancer Research, National Cancer Institute, Bethesda, MD, United States
*Corresponding author: e-mail address: fiorella.schischlik@nih.gov

Contents

Abstract

Myeloproliferative neoplasms (MPNs) is an umbrella term for several heterogenous diseases, which are characterized by their stem cell origin, clonal hematopoiesis and increase of blood cells of the myeloid lineage. The focus will be on *BCR-ABL1* negative MPNs, polycythemia vera (PV), primary myelofibrosis (PMF), essential thrombocythemia (ET). Seminal findings in the field of MPN were driven by genomic analysis, focusing on dissecting genomic changes MPN patients. This led to identification of major MPN driver genes, *JAK2*, *MPL* and *CALR*. Transcriptomic analysis promises to bridge the gap between genetic and phenotypic characterization of each patient's tumor and with the advent of single cell sequencing even for each MPN cancer cell. This review will focus on efforts to mine the bulk transcriptome of MPN patients, including analysis of fusion genes and splicing alterations which can be addressed with RNA-seq technologies. Furthermore, this paper aims to review recent endeavors to elucidate tumor heterogeneity in MPN hematopoietic stem and progenitor cells using single cell technologies. Finally, it will highlight current shortcoming and future applications to advance the field in MPN biology and improve patient diagnostics using RNA-based assays.

1. Introduction

Polycythemia vera (PV), primary myelofibrosis (PMF), essential thrombocythemia (ET) are three manifestations of a chronic hematologic

International Review of Cell and Molecular Biology, Volume 366
ISSN 1937-6448
https://doi.org/10.1016/bs.ircmb.2021.07.003

malignancy referred to as Myeloproliferative Neoplasms MPN. Key defining features are increased red blood cells in the peripheral blood in PV, increased platelet counts in ET and bone marrow fibrosis in PMF patients (Arber et al., 2016). Mutations in driver genes such as *JAK2*, *CALR* and *PMF* are major criteria but not required for an MPN diagnosis. Patients with no mutation in these genes are referred to as triple-negatives. Mutations in any of these driver genes are mutually exclusive and have been shown to lead to cytokine independent JAK-STAT signaling activation (Vainchenker and Kralovics, 2017).

RNA is a truly versatile molecule. Transcriptomic analysis combined with high-throughput technologies enable us to harness the diverse properties of this unique molecule (Cieślik and Chinnaiyan, 2018). The primary readouts in transcriptomic data can either be classified into "*quantitative readouts*" such as measures of gene and isoform expression and analysis of tumor microenvironment and "*qualitative readouts*" which aim to capture genotypic differences such as structural variants (eg. fusion genes and copy number alterations) and mutations (e.g. SNVs and SNPs).

Depending on the technology used, only a single readout can be addressed (e.g. gene expression for microarray and NanoString) or a combination of both readouts is feasible (RNA-seq for bulk and single cell). While quantitative readouts of gene abundance can be measured by a diverse variety of assays, qualitative readouts that detect genetic changes require protocols capable of achieving high depth and base pair resolution. The choice of readout and protocol will depend on the research question being asked, the available research budget and, not to underestimate, the availability of computational resources and skilled personnel to process and analyze the growing amount of data.

This review will focus on selected applications and aspects of transcriptomic analysis performed to advance the field of MPN. First, it will focus on the bulk transcriptome followed by recent efforts to phenotypically characterize cells of MPN patients on a single cell level. Finally, we will review a few applications using genetic readouts such as fusion discovery and analysis of splicing aberrations.

2. Studies on the bulk transcriptome of MPN patients

Numerous studies have analyzed bulk expression data of either sorted (CD34+, granulocytes or platelets) or unsorted cells using various technologies for measuring gene expression (e.g. microarray, NanoString or RNA-seq) in human and mice peripheral blood or bone marrow (Berkofsky-Fessler et al., 2010; Rice et al., 2011; Schubert et al., 2019; Skov et al., 2012; Vannucchi et al., 2013). Most studies focusing on solely on gene expression included only small sample sizes and were therefore likely under-powered to reliably capture the variability seen in patients.

A few notable exceptions is a microarray study by Rampal et al. which analyzed the granulocyte transcriptome of 97 ET, PV and PMF patients (Rampal et al., 2014). Using a cell line system with induced silencing of *JAK2*, they derived a "JAK-STAT" signature. This signature was shown to be significantly enriched in all MPN disease entities regardless of genotype. Most importantly, it was enriched when comparing *CALR* mutated patients with controls, emphasizing the importance of the JAK-STAT pathway in the disease pathology of MPN. In addition, a *CALR* signature, derived comparing *CALR* mutated vs control patients, was enriched when comparing *JAK2*-V617F-mutant MPN patients relative to controls. Zini et al., tried to derive a more refined CALR-signature by directly comparing CALR mutated vs *JAK2*-V617F mutated CD34+ derived cells in ET samples and thereby showing that patients with the same phenotype show distinct transcriptional profiles when stratified by mutation status (Zini et al., 2017). The above-mentioned signature was enriched for pathways related mTOR, MAPK/PI3K and MYC signaling. Importantly, although these signatures provide interesting insights into how different MPN driver mutations shape the transcriptional landscape, none of the signatures were validated in independent MPN cohorts.

Other groups have focused on efforts to associated clinical variables with changes in gene expression. For instance, Wong et al. derived a signature using bone marrow biopsies from 108 MPN patients using a targeted gene panel (NanoString) approach. They could confirm a highly inflammatory environment in the bone marrow of MPN patients graded as overt fibrotic (Wong et al., 2019). The authors did not clearly mention whether it is possible to distinguish pre-fibrotic PMF from ET. A similar study used platelet derived RNA followed by whole transcriptome sequencing in small training cohort of 25 MPN patients. They identified a 3-gene signature, that could predict the progression to marrow fibrosis and could therefore be valuable as surrogate markers of pathological marrow reticulin in MPN patients (Zhou et al., 2019). Both studies had training and test set to validate their signature, however, additional validation is needed in independent cohorts to verify if these signatures will further generalize and may therefore serve as biomarkers to guide patient treatment.

The largest RNA sequencing-based cohort analyzing granulocytes from 113 MPN patients (including controls, ET, PMF, PV and post-MPN AML) focused on the mutational aspects derived from RNA-seq (Schischlik et al., 2019). It included mutation analysis, fusion discovery, splicing aberrations detection and finally neoantigen discovery. Although technological challenging, using RNA-seq to identify neoantigens may harbor other advantages, such as the confirmation that the mutation in question is expressed and the possibility to identify several distinct mutation classes such as fusions, indels and splicing related aberrations (Olweus and Lund-Johansen, 2019).

3. From bulk to single cell transcriptomic sequencing of MPN patients

Recent technological advances have enabled to capture the entire transcriptome within a single cell (Hwang et al., 2018). While high cost still renders single cell sequencing of large patient cohorts prohibitive, seminal discoveries have been made to further understand tumor heterogeneity within each patient's tumor (Fan et al., 2020).

Two recent transcriptome single cell studies in the field of MPN harness the power of single cell sequencing to elucidate how driver mutations in *CALR* and *JAK2* lead to a proliferative (fitness) advantage of megakaryocytes compared to HSPCs through myeloid differentiation in patients with essential thrombocythemia and myelofibrosis (Nam et al., 2019; Psaila et al., 2020).

In Nam et al. the authors developed a new method to interrogate a specific genotype as well the whole transcriptome of a single cell. This enables a direct comparison of transcriptional changes within mutated and wild-type cells of MPN driver genes. The authors further compared genes differentially expressed between mutant (*CALR*) and wild-type megakaryocyte cells in an ET patient and identified several upregulated genes which are key player in protein quality control such as *BiP* and other *UPR* (unfolded protein response) genes. A similar observation has been reported in in vitro studies, showing an increase of UPR in cells with mutated *CALR* (Lau et al., 2015; Pronier et al., 2018). In contrast, the same comparison in HSP cells in an ET patient, identified upregulated genes related to the NF-kB pathways not previously described and may serve as putative therapeutic avenue to specifically target HSP cells in *CALR* mutated ET patients (Nam et al., 2019).

The study of Psaila et al. focused on patients with myelofibrosis. The authors were able to identify genes specifically enriched on the surface of myelofibrosis stem and progenitor cells. More specifically, G6B was found to be aberrantly expressed, however, the true cancer specificity, and therefore the utility for selective targeting with immunotherapeutic approaches (such as monoclonal antibody or chimeric antigen receptor (CAR)-T based therapies), remains to be confirmed.

While the study of Nam et al. was limited to the analysis of specific genotypes of interest, Rodriguez-Meira, developed a method that is able to detect up to 12 different mutations and perform scRNAseq simultaneously of a single cell (Rodriguez-Meira et al., 2019). In their study they used single cells from ET or PMF patients as a proof of principle to show transcriptional differences between in HSP cells which are either heterozygous or homozygous for *JAK2*-V617F and/or co-mutated with other genes relevant for MPN biology (eg. epigenetic or spliceosome related genes). Interestingly, their method allowed them to annotate WT HSP cells in normal donor and patients with myelofibrosis, which allowed them to identify a gene signature of inflammation including TNF-α and TGF-β signaling. This is in line with previous findings in isolated granulocytes in patients with myelofibrosis showing on a single cell level that malignant as well as nonmalignant cells secret inflammatory cytokines and reduction of cytokines in both population using JAK inhibitors (such as Ruxolitinib) is therefore necessary to achieve therapeutic benefit (Belver and Ferrando, 2015; Kleppe et al., 2015).

The above-mentioned studies exemplify the potential of single cell technologies to address fundamental question of MPN pathogenesis and treatment, bridging the gap between genetic mutations and cellular identities, as well as opening a new area of target discovery.

4. Analysis of splicing abnormalities with RNA-seq

Mutations in splicing related genes and the splicing machinery are frequently found in hematological cancers and typically involve *SF3B1*, *SRSF2* and *U2AF1* (Papaemmanuil et al., 2011; Yoshida et al., 2011). In MPN, mutations in the above-mentioned genes are enriched in PMF and post-MPN AML (3–14% in PMF for *SF3B1*, 9% in PMF and 19% post-MPN AML for *SRSF2*, 11% in PMF for *U2AF1*) (Lasho et al., 2012; Nangalia et al., 2013; Schischlik et al., 2019; Vannucchi et al., 2013; Zhang et al., 2012).

Mutations in these genes are encountered in hotspot regions, which suggest a gain or novel function of the altered protein. Hotspot mutations in *SF3B1* are located in amino acid (aa) residue K700 and K666, for *SRSF2* in position P95 and *U2AF1* S34 and Q157 aa residue (Abdel-Wahab and Levine, 2011; Dvinge et al., 2016). In MPN, *SF3B1* are often encountered as initiating event, occurring early in cancer development (Grinfeld et al., 2018; Nam et al., 2019).

Survival differences between splicing factor mutations can vary substantially, depending on the specific gene in questions and disease subtype (Harbour et al., 2013; Rossi et al., 2011; Zhang et al., 2012). For instance, in MPN, mutations in *SRSF2* are predictors of worse outcome (Zhang et al., 2012). However, in a study of 155 PMF patients, *SF3B1* mutations were not associated with worse survival. Notably, the number of *SF3B1* mutated patients was small ($n=10$) (Lasho et al., 2012).

While the occurrence of splicing factor mutations, can be valuable as a prognostic marker, recent studies analyzed and annotated the downstream effects of specific mutations on splicing in an effort to identify potential target genes of aberrant splicing. The identification of splicing alterations at nucleotide resolution has been greatly facilitated with the advent of RNA sequencing (Dvinge et al., 2016). For instance, an in silico study by DeBoever et al. could show that the hotspot *SF3B1* mutations affects 3′ splice splicing and proposed a model in which the mutated SF3B1 protein preferentially used an alternative (or cryptic) 3′ splice site which is located downstream of the splicing branch point (DeBoever et al., 2015). However, the exact mechanism why one splice site is chosen over another is not clear (Alsafadi et al., 2016; Darman et al., 2015).

The mentioned studies further explored and identified genes with aberrant 3′ splicing as a consequence of altered 3′ splice site recognition. This was analyzed in cell lines and patient samples with mutated *SF3B1* and mapped several hundred mis-spliced junctions in MDS, CLL, UV and breast cancer using differential junction analysis. A similar analysis performed by Schischlik et al. in PMF patients with *SF3B1* mutations. This study compared differentially spliced junction with 6 additional cancer types and could show that the junction expression varies significantly with cancer types (even within myeloid diseases) and only few aberrant junctions were expressed across all cancer types (Alsafadi et al., 2016; Bueno et al., 2016; Darman et al., 2015; Obeng et al., 2016; Schischlik et al., 2019). Interestingly, in several of these studies, despite the clonal nature of mutations in *SF3B1*, a high variability in the ratio of expression between the wildtype and the mutant allele is suggestive for a preference for one or the other allele.

The study by Schischlik et al. is the first to systematically mine PMF-specific splicing aberrations in a gene *SF3B1*, which forms part of the splicing machinery. It would be of interest to further investigate whether any of the identified mis-spliced genes are downstream targets of *SF3B1* and whether the co-mutated oncogenes *JAK2* or *CALR* lead to distinct splicing patterns. Furthermore, analysis of splicing aberration in *U2AF1* and *SRSF2* in MPN patients is of particular interest, given the known association of these genes with post-MPN AML. It remains to be seen whether any genes identified as downstream targets in conjunction with MPN driver mutation are responsible for disease progression.

5. Analysis of fusion genes in MPN and post-MPN AML patients

Diagnosis of myeloid malignancies still heavily relies on classic cytogenetic karyotyping (Arber et al., 2016; Hussein et al., 2009). This technology is able to identify large structural changes such as translocation and other larger chromosomal aberrations including deletions and inversions. A major limitation is the restricted resolution and prior knowledge of the genes of interest. However, their identification is of great value, given that these rearrangements help to directly classify hematological malignancies. In addition, gene fusions arising from chromosomal rearrangements are tumor-specific and may therefore act as independent biomarkers as well as targets for personalized precision medicine. The *BCR-ABL1* fusion gene in CML (chronic myeloid leukemia)—a subtype of MPN- targeted by the inhibitor Imatinib is a prominent example. Frequently, gene fusions in either hematologic but also solid tumors are driver mutations of the disease (Mertens et al., 2015; Mitelman et al., 2007; Yoshihara et al., 2014).

Traditionally, their discovery has either been a serendipitous finding or driven by high cost and effort using methods such as Fluorescence in situ hybridization (FISH) or reverse transcription-PCR (RT-PCR). With the emergence of RNA-sequencing, their direct identification and unbiased and cost-efficient systematic discovery has enabled several novel discoveries (Cieślik and Chinnaiyan, 2018; Gao et al., 2018). Although, their discovery has been vastly simplified, several challenges remain. These include technological challenges due to small overlap of detected fusions between tools (Wang et al., 2013) and high false positivity rate (Abate et al., 2012; Carrara et al., 2013; Kumar et al., 2016).

Similar to other mutation classes such as SNVs and Indels, the complexity to describe their functional consequence is amplified by the variety of possible functional outcomes. For instance, the formation of a fusion genes may result in enhanced overexpression of an oncogene (e.g. *IgH-MYC* fusions) or it results in the formation of a novel chimeric protein such as the previously mentioned *BCR-ABL1* or *NPM1-ALK*. In both cases, *BCR* and *NPM1* are considered "activator" genes, driving the increases expression of the tyrosine kinases *ABL1* and *ALK*.

Fusion genes involving the MPN oncogene *JAK2* are similar examples where the fusion of two genes leads to aberrant kinase activation. The *JAK2* gene is known to form fusions with other partner genes such as *BCR*, *PCM1*, *ETV6* and *RUNX1* (Bousquet and Brousset, 2006; Griesinger et al., 2005; Lacronique et al., 1997; Peeters et al., 1997). Interestingly, a *PCM1-JAK2* fusion was found in several hematological cancers such as AML, MDS/MPN and T-cell lymphomas and were associated with myeloproliferation and myelofibrosis (Hoeller et al., 2011).

Previous examples of fusion genes are classified as "gain of functions." However, a vastly understudied mechanism, involves the loss of function, where the function of both genes in inactivated (usually through shifts in open reading frame). Inactivation of one or both genes forming part of a fusion, may play important roles in tumor formation. As an example, the loss of a tumor suppressor *CHEK2*, was shown to be accompanied by the formation of a fusion gene *PPP2RA-CHEK2* (Jin et al., 2006).

The fusion landscape in MPN has long been dominated by isolated reports of fusion, particularly in PV and PMF (Murati et al., 2006; Storlazzi et al., 2014; Tanaka et al., 2010). Interestingly, rare co-occurrences of *BCR-ABL1* fusion with MPN driver mutations in *JAK2* and *CALR* were reported (Bornhäuser et al., 2007; Schischlik et al., 2019; Seghatoleslami et al., 2016; Yamada et al., 2014; Zhou et al., 2015). These patients often experience symptoms and have phenotypes resembling CML. To date, the largest unbiased study using RNA-seq based approaches to map the fusion landscape of MPN analyzed the 113 MPN patients including a few triple-negative and post-MPN AML cases (Schischlik et al., 2019). Although small in number ($n=9$), isolated novel fusions (*INO80D-GPR1* and *XPO5-RUNX2*) were found in post-MPN AML, however, no typical de novo AML fusion gene was detected, emphasizing the distinct disease etiology of these diseases (Milosevic et al., 2012). However, none of the identified fusions were recurrent, therefore the significance of these fusion event in MPN patients remains inconclusive.

6. Concluding remarks and future directions

The MPN field has traditionally focused on the characterization of MPN genomes which has led to important discoveries of MPN driver mutations (*JAK2*, *MPL*, *CALR*) and other co-mutate genes with prognostic value (Grabek et al., 2020; Schischlik and Kralovics, 2017). Genotyping of these mutations using genomic DNA is part of the 2016 World Health Organization criteria for MPN diagnosis (Arber et al., 2016). RNA-based approaches on the contrary, have so far been restricted to exploratory analysis with few potential clinical applications such as development of RNA-based signatures as markers of fibrosis and progression.

From an analyte perspective, it is reasonable to have DNA as the analyte of choice for molecular diagnostics in precision oncology, given its high stability and standardized protocols and assays for processing. Nevertheless, RNA-based approaches as a tool used in clinical settings, is emerging as a promising alternative and has shown to identify a large number of actionable genetic events (Mody et al., 2015; Robinson et al., 2017). In addition, single cell RNA-seq is steadily maturing to a reliable technology for assessing tumor heterogeneity and is being used for a variety of other clinical applications in the context of cancer biology (Fan et al., 2020; Haque et al., 2017). However, phenotyping large cohorts of patients using single cell technologies is still prohibitively expensive, therefore associating single cell transcriptomic readouts with important clinical variables such as survival and drug response is still best addressed with tumor bulk sequencing of large patient cohorts.

Within the past years, emerging new therapies in the field of cancer immunotherapy have opened new therapeutic avenues for MPN patients (Braun and Zeiser, 2020). First, was the discovery that *PD-L1* expression was elevated in T-cells, monocytes and megakaryocytes in mice with *JAK2*-V617F mutations compared to wild-type controls (Prestipino et al., 2018) and *PD-L1* expression was increased in primary cells isolated from patients with mutated *JAK2*-V617F. Summarizing, the authors hypothesize that platelets reduce T cell activation and could therefore be a mechanism of immune escape.

This has flourished the field to further prove efficacy to PD1/PD-L1 checkpoint therapies in MPN patients. Past experience with checkpoint therapy has shown, that the majority of patients will not respond to therapy and that genetic aberrations alone are insufficient to predict response to immunotherapy (Meng et al., 2015). Besides proving efficacy, it will be necessary to further characterize the immunogenomic landscape in MPN patients in order to better understand the determinates of response to cancer immunotherapy. A recent meta study, has performed such an analysis in more than 10,000 hematological cancers integrating data of over 8000 transcriptomes and other data modalities with the purpose to systematically describe immune infiltration, anti-tumor immune responses and antigen presentation and should be exemplary for what is needed in the MPN field (Dufva et al., 2020).

Second, tumor neoantigens arising through somatic mutations form ideal tumor specific targets for adoptive cell transfer and therapeutic cancer vaccination (Kitano et al., 2015; Schumacher and Schreiber, 2015). In MPN, several studies have identified T cells that respond to either mutant *JAK2* or *CALR,* emphasizing their growing potential for cellular based therapies in MPN (Holmström et al., 2016, 2017a,b; Kjaer et al., 2018). Additional, it was shown in a large cohort of 113 MPN patients, that more than 62% of patients carried neoantigens identified solely using RNA-seq, which is encouraging if this approach is aimed to benefit a larger fraction of MPN patients (Olweus and Lund-Johansen, 2019; Schischlik et al., 2019).

References

Abate, F., Acquaviva, A., Paciello, G., Foti, C., Ficarra, E., Ferrarini, A., Delledonne, M., et al., 2012. Bellerophontes: an RNA-seq data analysis framework for chimeric transcripts discovery based on accurate fusion model. Bioinformatics (Oxford, England) 28 (16), 2114–2121. https://doi.org/10.1093/bioinformatics/bts334.

Abdel-Wahab, O., Levine, R., 2011. The spliceosome as an indicted conspirator in myeloid malignancies. Cancer Cell 20 (4), 420–423. https://doi.org/10.1016/j.ccr.2011.10.004.

Alsafadi, S., Houy, A., Battistella, A., Popova, T., Wassef, M., Henry, E., Tirode, F., et al., 2016. Cancer-associated SF3B1 mutations affect alternative splicing by promoting alternative branchpoint usage. Nat. Commun. 7 (January), 10615. https://doi.org/10.1038/ncomms10615.

Arber, D.A., Orazi, A., Hasserjian, R., Thiele, J., Borowitz, M.J., Le Beau, M.M., Bloomfield, C.D., Cazzola, M., Vardiman, J.W., 2016. The 2016 revision to the World Health Organization classification of myeloid neoplasms and acute leukemia. Blood 127 (20), 2391–2405. https://doi.org/10.1182/blood-2016-03-643544.

Belver, L., Ferrando, A.A., 2015. Aberrant cytokine production by nonmalignant cells in the pathogenesis of myeloproliferative tumors and response to Jak inhibitor therapies. Cancer Discov. 5 (3), 316–376. https://doi.org/10.1158/2159-8290.CD-15-0095.

Berkofsky-Fessler, W., Buzzai, M., Kim, M.K.H., Fruchtman, S., Najfeld, V., Min, D.J., Fabricio F. Costa, et al., 2010. Transcriptional profiling of polycythemia vera identifies gene expression patterns both dependent and independent from the action of JAK2V617F. Clin. Cancer Res. 16 (17), 4339–4352. https://doi.org/10.1158/1078-0432.CCR-10-1092.

Bornhäuser, M., Mohr, B., Oelschlaegel, U., Bornhäuser, P., Jacki, S., Ehninger, G., Thiede, C., 2007. Concurrent JAK2(V617F) mutation and BCR-ABL translocation within committed myeloid progenitors in myelofibrosis. Leukemia 21 (8), 1824–1826. https://doi.org/10.1038/sj.leu.2404730.

Bousquet, M., Brousset, P., 2006. Myeloproliferative disorders carrying the t(8;9) (PCM1-JAK2) translocation. Hum. Pathol. 37 (4), 500. https://doi.org/10.1016/j.humpath.2005.12.002.

Braun, L.M., Zeiser, R., 2020. Immunotherapy in myeloproliferative diseases. Cell 9 (6), 1559. https://doi.org/10.3390/cells9061559.

Bueno, R., Stawiski, E.W., Goldstein, L.D., Durinck, S., De Rienzo, A., Modrusan, Z., Gnad, F., et al., 2016. Comprehensive genomic analysis of malignant pleural mesothelioma identifies recurrent mutations, gene fusions and splicing alterations. Nat. Genet. 48 (4), 407–416. https://doi.org/10.1038/ng.3520.

Carrara, M., Beccuti, M., Cavallo, F., Donatelli, S., Lazzarato, F., Cordero, F., Calogero, R.A., 2013. State of art fusion-finder algorithms are suitable to detect transcription-induced chimeras in normal tissues? BMC Bioinf. 14 (Suppl. 7), S2. January. https://doi.org/10.1186/1471-2105-14-S7-S2.

Cieślik, M., Chinnaiyan, A.M., 2018. Cancer transcriptome profiling at the juncture of clinical translation. Nat. Rev. Genet. 19 (2), 93–109. Nature Publishing Group. https://doi.org/10.1038/nrg.2017.96.

Darman, R.B., Seiler, M., Agrawal, A.A., Lim, K.H., Peng, S., Aird, D., Bailey, S.L., et al., 2015. Cancer-associated SF3B1 hotspot mutations induce cryptic 3' splice site selection through use of a different branch point. Cell Rep. 13 (5), 1033–1045. https://doi.org/10.1016/j.celrep.2015.09.053.

DeBoever, C., Ghia, E.M., Shepard, P.J., Rassenti, L., Barrett, C.L., Jepsen, K., Jamieson, C.H.M., Carson, D., Kipps, T.J., Frazer, K.A., 2015. Transcriptome sequencing reveals potential mechanism of cryptic 3' splice site selection in SF3B1-mutated cancers. Edited by Edwin Wang, PLoS Comput. Biol. 11 (3), e1004105. https://doi.org/10.1371/journal.pcbi.1004105.

Dufva, O., Pölönen, P., Brück, O., Keränen, M.A.I., Klievink, J., Mehtonen, J., Huuhtanen, J., et al., 2020. Immunogenomic landscape of hematological malignancies. Cancer Cell 38, 1–20. https://doi.org/10.1016/j.ccell.2020.06.002.

Dvinge, H., Kim, E., Abdel-Wahab, O., Bradley, R.K., 2016. RNA splicing factors as oncoproteins and tumour suppressors. Nat. Rev. Cancer 16 (7), 413–430. https://doi.org/10.1038/nrc.2016.51.

Fan, J., Slowikowski, K., Zhang, F., 2020. Single-cell transcriptomics in cancer: computational challenges and opportunities. Exp. Mol. Med., 1–14. https://doi.org/10.1038/s12276-020-0422-0.

Gao, Q., Liang, W.W., Foltz, S.M., Mutharasu, G., Jayasinghe, R.G., Cao, S., Liao, W.W., et al., 2018. Driver fusions and their implications in the development and treatment of human cancers. Cell Rep. 23 (1), 227–238.e3. https://doi.org/10.1016/J.CELREP.2018.03.050.

Grabek, J., Straube, J., Bywater, M., Lane, S.W., 2020. MPN: the molecular drivers of disease initiation, progression and transformation and their effect on treatment. Cell 9 (8), 1901. https://doi.org/10.3390/cells9081901.

Griesinger, F., Hennig, H., Hillmer, F., Podleschny, M., Steffens, R., Pies, A., Wörmann, B., Haase, D., Bohlander, S.K., 2005. ABCR-JAK2 fusion gene as the result of a t(9;22)(P24;Q11.2) translocation in a patient with a clinically typical chronic myeloid leukemia. Genes Chromosomes Cancer 44 (3), 329–333. https://doi.org/10.1002/gcc.20235.

Grinfeld, J., Jyoti, N., Joanna Baxter, E., Wedge, D.C., Angelopoulos, N., Cantrill, R., Anna L. Godfrey, et al., 2018. Classification and personalized prognosis in myeloproliferative neoplasms. N. Engl. J. Med. 379 (15), 1416–1430. https://doi.org/10.1056/NEJMoa1716614.

Haque, A., Engel, J., Teichmann, S.A., Lönnberg, T., 2017. A practical guide to single-cell RNA-sequencing for biomedical research and clinical applications. Genome Med. 9 (1), 1–12. https://doi.org/10.1186/S13073-017-0467-4.

Harbour, J.W., Roberson, E.D.O., Anbunathan, H., Onken, M.D., Worley, L.A., Bowcock, A.M., 2013. Recurrent mutations at codon 625 of the splicing factor SF3B1 in uveal melanoma. Nat. Genet. 45 (2), 133–135. https://doi.org/10.1038/ng.2523.

Hoeller, S., Walz, C., Reiter, A., Dirnhofer, S., Tzankov, A., 2011. PCM1–JAK2-fusion: a potential treatment target in myelodysplastic–myeloproliferative and other hemato-lymphoid neoplasms. Expert Opin. Ther. Targets 15 (1), 53–62. https://doi.org/10.1517/14728222.2011.538683.

Holmström, M.O., Riley, C.H., Svane, I.M., Hasselbalch, H.C., Andersen, M.H., 2016. The CALR exon 9 mutations are shared neoantigens in patients with CALR mutant chronic myeloproliferative neoplasms. Leukemia 30 (12), 2413–2416. https://doi.org/10.1038/leu.2016.233.

Holmström, M.O., Hasselbalch, H.C., Andersen, M.H., 2017a. The JAK2 V617F and CALR exon 9 mutations are shared immunogenic neoantigens in hematological malignancy. OncoImmunology 6 (11), e1358334. https://doi.org/10.1080/2162402X.2017.1358334.

Holmström, M.O., Hjortsø, M.D., Ahmad, S.M., Met, Ö., Martinenaite, E., Riley, C., Straten, P., Svane, I.M., Hasselbalch, H.C., Andersen, M.H., 2017b. The JAK2V617F mutation is a target for specific T cells in the JAK2V617F-positive myeloproliferative neoplasms. Leukemia 31 (2), 495–498. https://doi.org/10.1038/leu.2016.290.

Hussein, K., Van Dyke, D.L., Tefferi, A., 2009. Conventional cytogenetics in myelofibrosis: literature review and discussion. Eur. J. Haematol. 82 (5), 329–338. https://doi.org/10.1111/j.1600-0609.2009.01224.x.

Hwang, B., Lee, J.H., Bang, D., 2018. Single-cell RNA sequencing technologies and bioinformatics pipelines. Exp. Mol. Med. 50 (8). Nature Publishing Group. https://doi.org/10.1038/s12276-018-0071-8.

Jin, Y., Mertens, F., Kullendorff, C.-M., Panagopoulos, L., 2006. Fusion of the tumor-suppressor gene CHEK2 and the gene for the regulatory subunit B of protein phosphatase 2 PPP2R2A in childhood teratoma. Neoplasia 8 (5), 413–418. https://doi.org/10.1593/neo.06139.

Kitano, S., Ito, A., Kitano, S., Kim, Y., Inoue, M., Fuse, M., Tada, K., Yoshimura, K., 2015. Cancer neoantigens: a promising source of immunogens for cancer immunotherapy. J. Clin. Cell. Immunol. 06 (02), 1–7. https://doi.org/10.4172/2155-9899.1000322.

Kjaer, L., Holmström, M.O., Cordua, S., Andersen, M.H., Svane, I.M., Thomassen, M., Kruse, T.A., Pallisgaard, N., Skov, V., Hasselbalch, H.C., 2018. Sorted peripheral blood cells identify CALR mutations in B- and T-lymphocytes. Leuk. Lymphoma 59 (4), 973–977. https://doi.org/10.1080/10428194.2017.1359743.

Kleppe, M., Kwak, M., Koppikar, P., Riester, M., Keller, M., Bastian, L., Hricik, T., et al., 2015. JAK-STAT pathway activation in malignant and nonmalignant cells contributes to MPN pathogenesis and therapeutic response. Cancer Discov. 5 (3), 316–331. https://doi.org/10.1158/2159-8290.CD-14-0736.

Kumar, S., Vo, A.D., Qin, F., Li, H., 2016. Comparative assessment of methods for the fusion transcripts detection from RNA-seq data. Sci. Rep. 6 (January), 21597. https://doi.org/10.1038/srep21597.

Lacronique, V., Boureux, A., Valle, V.D., Poirel, H., Quang, C.T., Mauchauffé, M., Berthou, C., et al., 1997. A TEL-JAK2 fusion protein with constitutive kinase activity in human leukemia. Science (New York, N.Y.) 278 (5341), 1309–1312. http://www.ncbi.nlm.nih.gov/pubmed/9360930.

Lasho, T.L., Finke, C.M., Hanson, C.A., Jimma, T., Knudson, R.A., Ketterling, R.P., Pardanani, A., Tefferi, A., 2012. SF3B1 mutations in primary myelofibrosis: clinical, histopathology and genetic correlates among 155 patients. Leukemia 26 (5), 1135–1137. https://doi.org/10.1038/leu.2011.320.

Lau, W.W.Y., Hannah, R., Green, A.R., Göttgens, B., 2015. The JAK-STAT signaling pathway is differentially activated in CALR-positive compared with JAK2V617F-positive ET patients. Blood 125 (10), 1679–1681. American Society of Hematology. https://doi.org/10.1182/blood-2014-12-618074.

Meng, X., Huang, Z., Teng, F., Xing, L., Jinming, Y., 2015. Predictive biomarkers in PD-1/PD-L1 checkpoint blockade immunotherapy. In: Cancer Treatment Reviews. W.B. Saunders Ltd. https://doi.org/10.1016/j.ctrv.2015.11.001.

Mertens, F., Johansson, B., Fioretos, T., Mitelman, F., 2015. The emerging complexity of gene fusions in cancer. Nat. Rev. Cancer 15 (6), 371–381. https://doi.org/10.1038/nrc3947.

Milosevic, J.D., Puda, A., Malcovati, L., Berg, T., Hofbauer, M., Stukalov, A., Klampfl, T., et al., 2012. Clinical significance of genetic aberrations in secondary acute myeloid leukemia. Am. J. Hematol. 87 (11), 1010–1016. https://doi.org/10.1002/ajh.23309.

Mitelman, F., Johansson, B., Mertens, F., 2007. The impact of translocations and gene fusions on cancer causation. Nat. Rev. Cancer 7 (4), 233–245. https://doi.org/10.1038/nrc2091.

Mody, R.J., Wu, Y.M., Lonigro, R.J., Cao, X., Roychowdhury, S., Vats, P., Frank, K.M., et al., 2015. Integrative clinical sequencing in the management of refractory or relapsed cancer in youth. JAMA J. Am. Med. Assoc. 314 (9), 913–925. https://doi.org/10.1001/jama.2015.10080.

Murati, A., Adélaïde, J., Gelsi-Boyer, V., Etienne, A., Rémy, V., Fezoui, H., Sainty, D., et al., 2006. T(5;12)(Q23–31;P13) with ETV6-ACSL6 gene fusion in polycythemia vera. Leukemia 20 (6), 1175–1178. https://doi.org/10.1038/sj.leu.2404194.

Nam, A.S., Kim, K.T., Chaligne, R., Izzo, F., Ang, C., Taylor, J., Robert M. Myers, et al., 2019. Somatic mutations and cell identity linked by genotyping of transcriptomes. Nature 571 (7765), 355–360. https://doi.org/10.1038/s41586-019-1367-0.

Nangalia, J., Massie, C.E., Baxter, E.J., Nice, F.L., Gundem, G., Wedge, D.C., Avezov, E., et al., 2013. Somatic *CALR* mutations in myeloproliferative neoplasms with nonmutated *JAK2*. N. Engl. J. Med. 369 (25), 2391–2405. https://doi.org/10.1056/NEJMoa1312542.

Obeng, E.A., Chappell, R.J., Seiler, M., Chen, M.C., Campagna, D.R., Schmidt, P.J., Rebekka K. Schneider, et al., 2016. Physiologic expression of Sf3b1K700E causes impaired erythropoiesis, aberrant splicing, and sensitivity to therapeutic spliceosome modulation. Cancer Cell 30 (3), 404–417. https://doi.org/10.1016/j.ccell.2016.08.006.

Olweus, J., Lund-Johansen, F., 2019. Finding neo (antigens, that is). Blood 134 (2), 108–109. American Society of Hematology. https://doi.org/10.1182/blood.2019001334.

Papaemmanuil, E., Cazzola, M., Boultwood, J., Malcovati, L., Vyas, P., Bowen, D., Pellagatti, A., et al., 2011. Somatic *SF3B1* mutation in myelodysplasia with ring sideroblasts. N. Engl. J. Med. 365 (15), 1384–1395. https://doi.org/10.1056/NEJMoa1103283.

Peeters, P., Raynaud, S.D., Cools, J., Wlodarska, I., Grosgeorge, J., Philip, P., Monpoux, F., et al., 1997. Fusion of TEL, the ETS-variant gene 6 (ETV6), to the receptor-associated kinase JAK2 as a result of t(9;12) in a lymphoid and t(9;15;12) in a myeloid leukemia. Blood 90 (7), 2535–2540. http://www.ncbi.nlm.nih.gov/pubmed/9326218.

Prestipino, A., Emhardt, A.J., Aumann, K., O'Sullivan, D., Gorantla, S.P., Duquesne, S., Melchinger, W., et al., 2018. Oncogenic JAK2 V617F causes PD-L1 expression, mediating immune escape in myeloproliferative neoplasms. Sci. Transl. Med. 10 (429), eaam7729. https://doi.org/10.1126/scitranslmed.aam7729.

Pronier, E., Cifani, P., Merlinsky, T.R., Berman, K.B., Somasundara, A.V.H., Rampal, R.K., LaCava, J., et al., 2018. Targeting the CALR interactome in myeloproliferative neoplasms. JCI Insight 3 (22). https://doi.org/10.1172/jci.insight.122703.

Psaila, B., Wang, G., Rodriguez-Meira, A., Li, R., Heuston, E.F., Murphy, L., Yee, D., et al., 2020. Single-cell analyses reveal megakaryocyte-biased hematopoiesis in myelofibrosis and identify mutant clone-specific targets. Mol. Cell 78 (3), 477–492.e8. https://doi.org/10.1016/j.molcel.2020.04.008.

Rampal, R., Al-Shahrour, F., Abdel-Wahab, O., Patel, J.P., Brunel, J.-P., Mermel, C.H., Bass, A.J., et al., 2014. Integrated genomic analysis illustrates the central role of JAK-STAT pathway activation in myeloproliferative neoplasm pathogenesis. Blood 123 (22), e123–e133. https://doi.org/10.1182/blood-2014-02-554634.

Rice, K.L., Lin, X., Wolniak, K., Ebert, B.L., Berkofsky-Fessler, W., Buzzai, M., Sun, Y., et al., 2011. Analysis of genomic aberrations and gene expression profiling identifies novel lesions and pathways in myeloproliferative neoplasms. Blood Cancer J. 1 (10), e40. https://doi.org/10.1038/bcj.2011.39.

Robinson, D.R., Wu, Y.M., Lonigro, R.J., Vats, P., Cobain, E., Everett, J., Cao, X., et al., 2017. Integrative clinical genomics of metastatic cancer. Nature 548 (7667), 297–303. https://doi.org/10.1038/nature23306.

Rodriguez-Meira, A., Buck, G., Clark, S.A., Povinelli, B.J., Alcolea, V., Louka, E., McGowan, S., et al., 2019. Unravelling intratumoral heterogeneity through high-sensitivity single-cell mutational analysis and parallel RNA sequencing. Mol. Cell 73 (6), 1292–1305.e8. https://doi.org/10.1016/j.molcel.2019.01.009.

Rossi, D., Bruscaggin, A., Spina, V., Rasi, S., Khiabanian, H., Messina, M., Fangazio, M., et al., 2011. Mutations of the SF3B1 splicing factor in chronic lymphocytic leukemia: association with progression and fludarabine-refractoriness. Blood 118 (26), 6904–6908. https://doi.org/10.1182/blood-2011-08-373159.

Schischlik, F., Kralovics, R., 2017. Mutations in myeloproliferative neoplasms—their significance and clinical use. Expert Rev. Hematol. 10 (11), 961–973. https://doi.org/10.1080/17474086.2017.1380515.

Schischlik, F., Jäger, R., Rosebrock, F., Hug, E., Schuster, M., Holly, R., Fuchs, E., et al., 2019. Mutational landscape of the transcriptome offers putative targets for immunotherapy of myeloproliferative neoplasms. Blood 134 (2), 199–210. https://doi.org/10.1182/blood.2019000519.

Schubert, C., Allhoff, M., Tillmann, S., Maié, T., Costa, I.G., Lipka, D.B., Schemionek, M., et al., 2019. Differential roles of STAT1 and STAT2 in the sensitivity of JAK2V617F- vs. BCR-ABL-positive cells to interferon alpha. J. Hematol. Oncol. 12 (1), 12–36. https://doi.org/10.1186/s13045-019-0722-9.

Schumacher, T.N., Schreiber, R.D., 2015. Neoantigens in cancer immunotherapy. Science (New York, N.Y.) 348 (6230), 69–74. https://doi.org/10.1126/science.aaa4971.

Seghatoleslami, M., Ketabchi, N., Ordo, A., Asl, J.M., Golchin, N., Saki, N., 2016. Coexistence of P190 BCR/ABL transcript and CALR 52-Bp deletion in chronic myeloid leukemia blast crisis: A case report. Mediterr. J. Hematol. Infect. Dis. 8 (1), e2016002. https://doi.org/10.4084/MJHID.2016.002.

Skov, V., Thomassen, M., Riley, C.H., Jensen, M.K., Bjerrum, O.W., Kruse, T.A., Hasselbalch, H.C., Larsen, T.S., 2012. Gene expression profiling with principal component analysis depicts the biological continuum from essential thrombocythemia over polycythemia vera to myelofibrosis. Exp. Hematol. 40 (9), 771–780.e19. https://doi.org/10.1016/j.exphem.2012.05.011.

Storlazzi, T., Clelia, L.P., Paoli, C., Daniele, G., Lasho, T., Tefferi, A., Vannucchi, A.M., 2014. Complex karyotype in a polycythemia vera patient with a novel SETD1B/GTF2H3 fusion gene. Am. J. Hematol. 89 (4), 438–442. https://doi.org/10.1002/ajh.23659.

Tanaka, H., Takeuchi, M., Takeda, Y., Sakai, S., Abe, D., Ohwada, C., Sakaida, E., et al., 2010. Identification of a novel TEL–Lyn fusion gene in primary myelofibrosis. Leukemia 24 (1), 197–200. https://doi.org/10.1038/leu.2009.167.

Vainchenker, W., Kralovics, R., 2017. Genetic basis and molecular pathophysiology of classical myeloproliferative neoplasms. Blood 129 (6), 667–679. https://doi.org/10.1182/blood-2016-10-695940.

Vannucchi, A.M., Lasho, T.L., Guglielmelli, P., Biamonte, F., Pardanani, A., Pereira, A., Finke, C., et al., 2013. Mutations and prognosis in primary myelofibrosis. Leukemia 27 (9), 1861–1869. https://doi.org/10.1038/leu.2013.119.

Wang, Q., Xia, J., Jia, P., Pao, W., Zhao, Z., 2013. Application of next generation sequencing to human gene fusion detection: computational tools, features and perspectives. Brief. Bioinform. 14 (4), 506–519. https://doi.org/10.1093/bib/bbs044.

Wong, W.J., Baltay, M., Getz, A., Fuhrman, K., Aster, J.C., Hasserjian, R.P., Pozdnyakova, O., 2019. Gene expression profiling distinguishes prefibrotic from overtly fibrotic myeloproliferative neoplasms and identifies disease subsets with distinct inflammatory signatures. PLoS One 14 (5). https://doi.org/10.1371/journal.pone.0216810.
Yamada, O., Mahfoudhi, E., Plo, I., Ozaki, K., Nakatake, M., Akiyama, M., Yamada, H., Kawauchi, K., Vainchenker, W., 2014. Emergence of a BCR-ABL translocation in a patient with the JAK2V617F mutation: evidence for secondary acquisition of BCR-ABL in the JAK2V617F clone. J. Clin. Oncol. 32 (21), e76–e79. https://doi.org/10.1200/JCO.2012.47.8669.
Yoshida, K., Sanada, M., Shiraishi, Y., Nowak, D., Nagata, Y., Yamamoto, R., Sato, Y., et al., 2011. Frequent pathway mutations of splicing machinery in myelodysplasia. Nature 478 (7367), 64–69. https://doi.org/10.1038/nature10496.
Yoshihara, K., Wang, Q., Torres-Garcia, W., Zheng, S., Vegesna, R., Kim, H., Verhaak, R.G.W., 2014. The landscape and therapeutic relevance of cancer-associated transcript fusions. Oncogene 34, 4845–4854. December. https://doi.org/10.1038/onc.2014.406.
Zhang, S.-J., Rampal, R., Manshouri, T., Patel, J., Mensah, N., Kayserian, A., Hricik, T., et al., 2012. Genetic analysis of patients with leukemic transformation of myeloproliferative neoplasms shows recurrent SRSF2 mutations that are associated with adverse outcome. Blood 119 (19), 4480–4485. https://doi.org/10.1182/blood-2011-11-390252.
Zhou, A., Knoche, E.M., Engle, E.K., Fisher, D.A.C., Oh, S.T., 2015. Concomitant JAK2 V617F-positive polycythemia vera and BCR-ABL-positive chronic myelogenous leukemia treated with Ruxolitinib and Dasatinib. Blood Cancer J. 5 (10), e351. https://doi.org/10.1038/bcj.2015.77.
Zhou, J.-G., Liang, B., Jin, S.-H., Liao, H.-L., Guo-Bo, D., Cheng, L., Hu, M., Gaipl, U.S., 2019. Development and validation of an RNA-seq-based prognostic signature in neuroblastoma. Front. Oncol. 9, 1361. https://doi.org/10.3389/fonc.2019.01361.
Zini, R., Guglielmelli, P., Pietra, D., Rumi, E., Rossi, C., Rontauroli, S., Genovese, E., et al., 2017. CALR mutational status identifies different disease subtypes of essential thrombocythemia showing distinct expression profiles. Blood Cancer J. 7 (12), 638. https://doi.org/10.1038/s41408-017-0010-2.

CHAPTER THREE

Targets in MPNs and potential therapeutics

Gabriel Levy[a,b,c], Cristina Mambet[d,e], Christian Pecquet[a,b,g], Sarah Bailly[a,b,f], Violaine Havelange[b,f], Carmen C. Diaconu[d], and Stefan N. Constantinescu[a,b,g,h,*]

[a]Ludwig Institute for Cancer Research, Brussels, Belgium
[b]SIGN Unit, de Duve Institute, Université Catholique de Louvain, Brussels, Belgium
[c]Department of Pediatric Hematology and Oncology, Cliniques Universitaires Saint-Luc, Université Catholique de Louvain, Brussels, Belgium
[d]Department of Cellular and Molecular Pathology, Stefan S. Nicolau Institute of Virology, Bucharest, Romania
[e]Department of Hematology, Carol Davila University of Medicine and Pharmacy, Bucharest, Romania
[f]Department of Hematology, Cliniques Universitaires Saint Luc, Université Catholique de Louvain, Brussels, Belgium
[g]WELBIO (Walloon Excellence in Life Sciences and Biotechnology), Brussels, Belgium
[h]Ludwig Institute for Cancer Research, Nuffield Department of Medicine, Oxford University, Oxford, United Kingdom
*Corresponding author: e-mail address: stefan.constantinescu@bru.licr.org

Contents

International Review of Cell and Molecular Biology, Volume 366
ISSN 1937-6448
https://doi.org/10.1016/bs.ircmb.2021.06.004

Abstract

Philadelphia-negative classical Myeloproliferative Neoplasms (MPNs), including Polycythemia Vera (PV), Essential Thrombocythemia (ET) and Primary Myelofibrosis (PMF), are clonal hemopathies that emerge in the hematopoietic stem cell (HSC) compartment. MPN driver mutations are restricted to specific exons (14 and 12) of Janus kinase 2 (*JAK2*), thrombopoietin receptor (*MPL/TPOR*) and calreticulin (*CALR*) genes, are involved directly in clonal myeloproliferation and generate the MPN phenotype. As a result, an increased number of fully functional erythrocytes, platelets and leukocytes is observed in the peripheral blood. Nevertheless, the complexity and heterogeneity of MPN clinical phenotypes cannot be solely explained by the type of driver mutation. Other factors, such as additional somatic mutations affecting epigenetic regulators or spliceosomes components, mutant allele burdens and modifiers of signaling by driver mutants, clonal architecture and the order of mutation acquisition, signaling events that occur downstream of a driver mutation, the presence of specific germ-line variants, the interaction of the neoplastic clone with bone marrow microenvironment and chronic inflammation, all can modulate the disease phenotype, influence the MPN clinical course and therefore, might be useful therapeutic targets.

1. Introduction

The constitutively activated Janus kinase (JAK)/Signal Transducers and Activators of Transcription (STAT) signaling pathway is fundamental for Myeloproliferative Neoplasms (MPN) pathogenesis (Fig. 1), irrespective of driver mutation, being encountered even in "triple-negative" MPN cases (without JAK2, thrombopoietin receptor (MPL/TPOR) and calreticulin (CALR) mutations) (Rampal et al., 2014). Importantly, several benign conditions characterized by polyclonal hematopoiesis, namely Hereditary Erythrocytosis and Hereditary Thrombocytosis, can mimic "triple-negative" MPNs, leading to diagnostic and therapeutic challenges (Rumi and Cazzola, 2017).

Allogeneic hematopoietic stem cell (HSC) transplantation is the only curative treatment option for MPNs (Zhang et al., 2020). However, due to its considerable risk of mortality and morbidity, this procedure is usually applied to younger Primary Myelofibrosis (PMF) patients exhibiting unfavorable prognostic scores (Kröger et al., 2015). The other therapeutic strategies are mainly palliative, focusing on cytoreduction, prevention of thrombotic events, alleviation of constitutional symptoms, anemia and splenomegaly (Barbui et al., 2018). The increasing knowledge of the complex molecular mechanisms that contribute to MPN development and progression (Grinfeld et al., 2017; Vainchenker and Kralovics, 2017) creates the

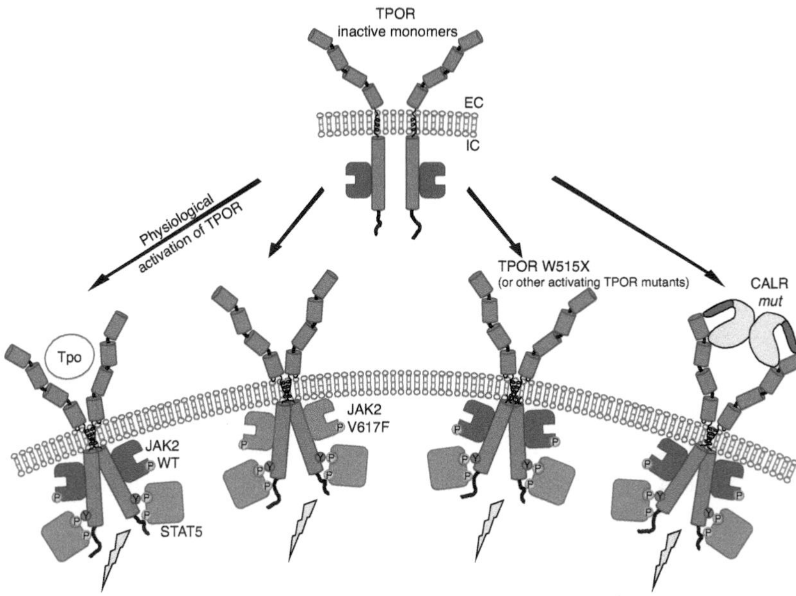

Fig. 1 The "MPN driver mutations" restricted to specific exons of JAK2, MPL/TPOR and CALR genes are involved directly in clonal myeloproliferation and generate the MPN phenotype. *JAK2,* Janus kinase 2; *TPOR,* thrombopoietin receptor (MPL/TPOR); *CALR mut,* calreticulin mutants; *Tpo,* thrombopoietin; *STAT5,* signal transducers and activators of transcription 5; *Y,* tyrosine; *P,* phosphorylation; *EC,* extracellular; *IC,* intracellular.

premises for more personalized therapies. In this chapter, we will address the therapeutic targets and related inhibitors that showed promising results in recent years, some of them being already approved in clinical practice, while most are tested in clinical trials or at least on cellular systems and preclinical animal models.

2. JAK2 mutations

2.1 Behind JAK2V617F: The molecular background

Among all driver mutations of MPNs, *JAK2*V617F is the most prevalent. It accounts for up to 96% of Polycythemia Vera (PVs), 50% of Essential Thrombocythemia (ETs) and 50–60% of Primary Myelofibrosis (PMFs) cases (Baxter et al., 2005; James et al., 2005; Kralovics et al., 2005; Levine et al., 2005; Passamonti et al., 2006). Although predominantly homozygous or carrying homozygous subclones early in the disease in PVs, *JAK2*V617F is found to be mainly heterozygous in ETs (Dupont et al., 2007; Li et al., 2014; Scott et al., 2007). It has been shown in transgenic mice that *JAK2*V617F gene dosage induces a graded continuous phenotype, starting with ET at low gene dosage, erythrocytosis (PV phenotype) at higher gene dosage and myelofibrosis at very high gene dosage (Tiedt et al., 2008). These results have still to be translated to the human disease setting where either heterozygous or homozygous status can be detected; thus possibly gene dosage data could be interpreted as different levels of JAK2V617F biochemical activity, with low levels inducing ET, medium levels PV and very high levels myelofibrosis. Interestingly, since JAK2V617F expression eventually down-modulated cell-surface TPOR (Moliterno et al., 2006; Pecquet et al., 2012), the PV phenotype can be isolated to medium doses of JAK2V617F biochemical activity due to down-modulation of active TPOR complexes, while the erythropoietin receptor (EPOR) is not down-modulated (Pecquet et al., 2012).

The activation of JAK2 by the V617F mutation remains incompletely understood but seems to rely on a conformational change in the Janus kinase homology 2 (JH2) domain, leading to activation of the adjacent kinase domain (JH1) (Bandaranayake et al., 2012; Dusa et al., 2010; Leroy et al., 2016a, 2019). Furthermore, recently it has been demonstrated that JAK2V617F induces inside-out dimerization of dimeric cytokine receptors (Leroy et al., 2019; Wilmes et al., 2020).

2.2 JAK2V617F inhibition strategies

Since the discovery of the *JAK2*V617F mutation in 2004 (reported in 2005) (Baxter et al., 2005; James et al., 2005; Kralovics et al., 2005; Levine et al., 2005), several drugs have been developed to specifically target the JAK2V617F mutant (Fig. 2). However, no molecule with the ability to only target the mutant has been found to date. Among several JAK2 kinase domain inhibitors tested, only ruxolitinib and fedratinib have been approved by the FDA and EMA (Harrison et al., 2012; Pardanani et al., 2015a; Vannucchi et al., 2015; Verstovsek et al., 2012) because they combine adequate JAK inhibition and are not hampered by (too many) deleterious side effects.

Ruxolitinib is a non-specific JAK1 and JAK2 inhibitor, which inactivates both the wild type (WT) and JAK2V617F. It targets the ATP-binding site of JAKs in their active conformation (Type I inhibitor). Quintás-Cardama et al. (2010) first showed that ruxolitinib inhibited interleukin-6 signaling and proliferation of JAK2V617F-carrying Ba/F3 cells and preferentially suppressed erythroid progenitor colony formation from *JAK2*V617F PV patients compared with healthy donors in primary cultures. In a mouse model of *JAK2*V617F-driven MPN, oral ruxolitinib markedly reduced

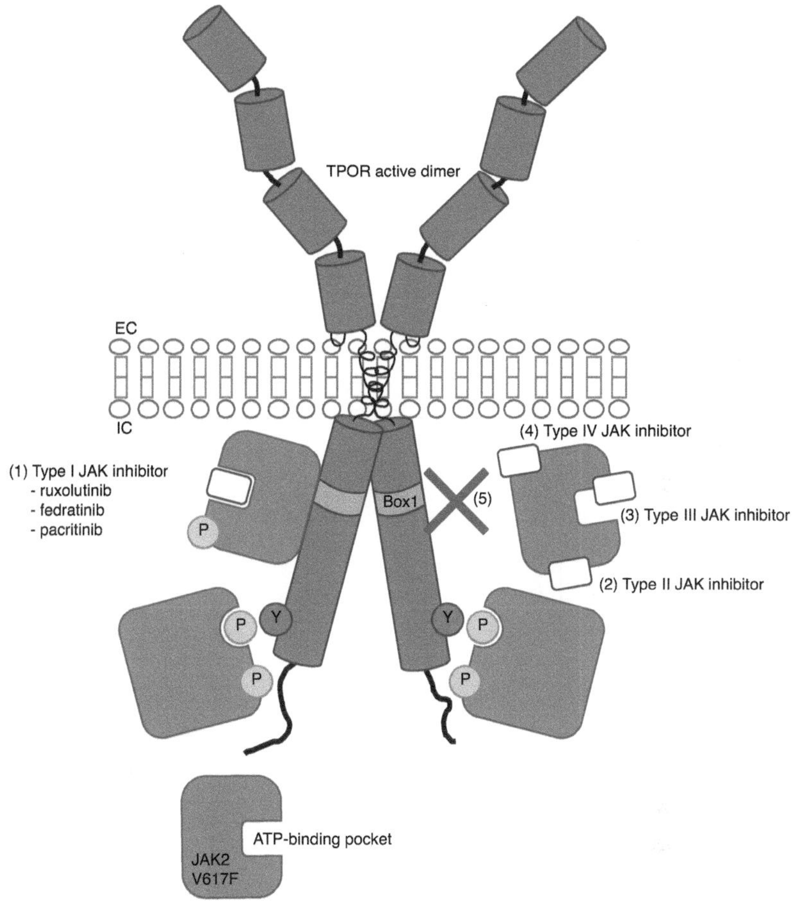

Fig. 2 JAK2 inhibition strategies. Type I inhibitors target the ATP-binding site of JAKs in their active conformation (1). Type II inhibitors interact with JAK2 kinase domain in the inactive conformation (2). Type III and IV inhibitors are allosteric inhibitors targeting sites close (3) or distant (4) of the ATP-binding pocket. Other strategies could target interaction of JAK2V617F and TPOR or of JAK2 and TPOR mutants (5). *JAK*, Janus kinase; *TPOR*, thrombopoietin receptor; *Y*, tyrosine; *P*, phosphorylation; *EC*, extracellular; *IC*, intracellular.

splenomegaly and circulating levels of inflammatory cytokines, and preferentially eliminated neoplastic cells, resulting in significantly prolonged survival without myelosuppressive or immunosuppressive effects.

In the same year 2010, Verstovsek et al. (2010) published the results of a phase 1–2 dose-escalation study (Study 251) for ruxolitinib in MF patients (Table 1). The results of the Study 251 were unprecedented at the time

Table 1 Ongoing studies for targeted therapies in MPNs.

Study ID number	Brief title of the study	Status (first posted)	Condition	Number of patients	Study type	Study phase	Intervention/drugs	Activity	Primary outcome measures
NCT00509899	Open label ruxolitinib (INCB018424) in patients with myelofibrosis and post polycythemia vera/essential thrombocythemia myelofibrosis (expansion of Study 251)	Completed (August 1, 2007)	- MF - PV - Thrombocytosis	154	Interventional	Phase 1 Phase 2	Ruxolitinib (INCB018424)	JAK1 and JAK2 inhibitor	Safety and tolerability of INCB018424
NCT00952289	Controlled myelofibrosis study with oral JAK inhibitor treatment: the COMFORT-I trial	Completed (August 6, 2009)	MF (primary, post-ET or post-PV)	309	Interventional	Phase 3	Ruxolitinib vs placebo	JAK1 and JAK2 inhibitor	Spleen reduction ≥35% (at week 24)
NCT00934544	Controlled myelofibrosis study with oral Janus-associated kinase (JAK) inhibitor treatment-II: the COMFORT-II trial	Completed (July 8, 2009)	MF (primary, post-ET or post-PV)	219	Interventional	Phase 3	Ruxolitinib vs best available therapy	JAK1 and JAK2 inhibitor	Spleen reduction ≥35% (at week 48)
NCT01437787	Phase III study of SAR302503 in intermediate-2 and high risk patients with myelofibrosis (JAKARTA)	Completed (September 21, 2011)	MF (primary, post-ET or post-PV)	289	Interventional	Phase 3	Fedratinib (SAR302503) vs placebo	JAK2 (and FLT3) inhibitor	Spleen reduction ≥35% (at cycle 6)
NCT01523171	Phase II, open label, single arm study of SAR302503 in myelofibrosis patients previously treated with ruxolitinib (JAKARTA2)	Completed (February 1, 2012)	MF (primary, post-ET or post-PV)	97	Interventional	Phase 2	Fedratinib (SAR302503) after ruxolitinib	(JAK1 and) JAK2 inhibitors	Spleen reduction ≥35% (at cycle 6)

NCT01773187	Acritinib versus best available therapy to treat myelofibrosis (PERSIST-1)	Terminated (January 23, 2013)	MF (primary, post-ET or post-PV)	327	Interventional	Phase 3	Pacritinib vs best available therapy	JAK2 (and IRAK1) inhibitor	Spleen reduction $\geq$35% (at week 24)
NCT02055781	Oral pacritinib versus best available therapy to treat myelofibrosis with thrombocytopenia (PERSIST-2, PAC326)	Terminated (February 5, 2014)	MF (primary, post-ET or post-PV)	311	Interventional	Phase 3	Pacritinib vs best available therapy	JAK2 (and IRAK1) inhibitor	Spleen reduction $\geq$35% (at week 24)
NCT03165734	A phase 3 study of pacritinib in patients with primary myelofibrosis, post polycythemia vera myelofibrosis, or post-essential thrombocythemia myelofibrosis (PACIFICA)	Recruiting	MF (primary, post-ET or post-PV)	348 (estimated)	Interventional	Phase 3	Pacritinib vs physician's choice	JAK2 (and IRAK1) inhibitor	Spleen reduction $\geq$35%
NCT04173494	A study of momelotinib versus Danazol in symptomatic and anemic myelofibrosis patients (MOMENTUM)	Recruiting	MF (primary, post-ET or post-PV)	180 (estimated)	Interventional	Phase 3	Momelotinib vs Danazol	JAK1 and JAK2 (and ACVR1) inhibitor	Total symptom score (TSS) response rate at week 24 measured using the myelofibrosis symptom assessment form v4.0
NCT03566446	CALR exon 9 mutant peptide vaccine to patients with CALR-mutant myeloproliferative neoplasms	Active, not recruiting	- MPN, unclassifiable - ET - MF	10 (estimated)	Interventional	Phase 1	CALR long 36 peptide (vaccines)	Anti CALR mutants vaccine	Adverse events evaluated by CTCAE 4.03 (time frame: 1 year)

Continued

Table 1 Ongoing studies for targeted therapies in MPNs.—cont'd

Study ID number	Brief title of the study	Status (first posted)	Condition	Number of patients	Study type	Study phase	Intervention/drugs	Activity	Primary outcome measures
NCT02158858	A phase 2 study of CPI-0610 with and without ruxolitinib in patients with myelofibrosis (MANIFEST)	Recruiting	MF	271 (estimated, phase 2) 10 (phase 1)	Interventional	Phase 1 Phase 2	CPI-0610 +/− ruxolitinib	BET inhibitors	- Phase 1: frequency of dose-limiting toxicities (DLTs) associated with CPI-0610 administration during the first cycle (first 21 days) of treatment - Phase 2: spleen response and red blood cell transfusion independency
NCT01712308	Sotatercept in treating patients with myeloproliferative neoplasm-associated myelofibrosis or anemia	Recruiting	- MF - Myelodysplastic/ myeloproliferative neoplasm, and - Anemia	60 (estimated)	Interventional	Phase 2	Sotatercept	TGF-β superfamily inhibitor	- Incidence of foxicities - Anemia response
NCT03194542	A safety and efficacy study to evaluate luspatercept in subjects with myeloproliferative neoplasm-associated myelofibrosis who have anemia with and without red blood cell-transfusion dependence	Recruiting	- Primary MF and - Anemia	103 (estimated)	Interventional	Phase 2	Luspatercept	TGF-β superfamily inhibitor	Anemia response

NCT04103645	Intra-patient dose escalation study to investigate safety and feasibility of vactosertib in treating anemic MPN patients	Recruiting	- MPN and - Anemia	37 (estimated)	Interventional	Phase 2	Vactosertib (TEW-7197)	TGF-β superfamily inhibitor	- Number of adverse events - Change in symptoms of the disease - Change in spleen size - Change in transfusion dependency - Change in hemoglobin values - Change in EPO levels
NCT04218071	Actuate 1901: 9-ING-41 in myelofibrosis	Recruiting	MF	58 (estimated)	Interventional	Phase 2	9-ING-41 +/− ruxolitinib	GSK-3β inhibitor	Response rate (time frame: 3–24 months)
NCT01981850	A phase 2 study of PRM-151 in participants with myelofibrosis	Completed (November 13, 2013)	MF (primary, post-ET or post-PV)	125	Interventional	Phase 2	PRM-151 +/− ruxolitinib	Recombinant serum amyloid P (SAP)	Overall response rate according to International Working Group consensus criteria for treatment response in myelofibrosis with myeloid metaplasia (Time frame: 6 months)

Continued

Table 1 Ongoing studies for targeted therapies in MPNs.—cont'd

Study ID number	Brief title of the study	Status (first posted)	Condition	Number of patients	Study type	Study phase	Intervention/drugs	Activity	Primary outcome measures
NCT03065400	PD-1 inhibition in advanced myeloproliferative neoplasms	Completed (February 7, 2017)	- MF (primary, post-ET or post-PV) - MPN in accelerated/blast phase	10	Interventional	Phase 2	Pembrolizumab	PD-1	At least one clinical improvement (CI, PR, CR) by combined European Leukemia Net -International Working Group (ELN-IWG) criteria after six cycles of pembrolizumab therapy
NCT04097821	Platform study of novel ruxolitinib combinations in myelofibrosis patients (ADORE)	Recruiting	MF (primary, post-ET or post-PV)	243 (estimated)	Interventional	Phase 1 Phase 2	- Siremadlin +/− ruxolitinib - Crizanlizumab +/− ruxolitinib - MBG453 (sabatolimab) +/− ruxolitinib - LTT462 +/− ruxolitinib - NIS793 +/− ruxolitinib	- p53-MDM2 interaction inhibitor (siremadlin) - P-selectin blocker (crizanlizumab) - TIM-3 inhibitor (MBG453) - ERK1/2 inhibitor (LTT462) - TGF-β inhibitor (NIS793)	- Incidence and severity of dose limiting toxicities within the first 2 cycles - Response rate at the end of cycle 6 or cycle 8
NCT03662126	KRT-232 versus best available therapy for the treatment of subjects with myelofibrosis who are relapsed or refractory to JAK inhibitor treatment	Recruiting	MF (primary, post-ET or post-PV)	385 (estimated)	Interventional	Phase 2 Phase 3	KRT-232 vs best available therapy	MDM2 inhibitor	Spleen reduction $\geq$35% (at week 24)

NCT03669965	KRT-232 compared to ruxolitinib in patients with phlebotomy-dependent polycythemia vera	Active, not recruiting	PV	295 (estimated)	Interventional	Phase 2	KRT-232	MDM2 inhibitor	Response at week 32 defined as having achieved both: absence of phlebotomy eligibility beginning at the week 8 visit and continuing through week 32, reduction in spleen volume $\geq$35% at week 32
NCT04113616	An open-label, multicenter, phase 1b/2 study of the safety and efficacy of KRT-232 combined with low-dose cytarabine or decitabine in patients with acute myeloid leukemia	Recruiting	- AML - AML secondary to MPNs	160 (estimated)	Interventional	Phase 1 Phase 2	KRT-232 with cytarabine or decitabine	MDM2 inhibitor	- Number of dose-limiting toxicities of KRT-232 in combination with cytarabine or decitabine (Time frame: 28 days), - Proportion of patients achieving complete remission (CR) or complete remission with partial hematological improvement (CRh)
NCT01693601	Panobinostat and ruxolitinib in myelofibrosis (PRIME trial)	Completed (September 26, 2012)	MF (primary, post-ET or post-PV)	20	Interventional	Phase 1 Phase 2	Panobinostat +ruxolitinib	HDAC inhibitor	- Number of patients - With adverse events - That achieve at least stable disease

Continued

Table 1 Ongoing studies for targeted therapies in MPNs.—cont'd

Study ID number	Brief title of the study	Status (first posted)	Condition	Number of patients	Study type	Study phase	Intervention/drugs	Activity	Primary outcome measures
NCT01761968	Long-term study evaluating the effect of givinostat in patients with chronic myeloproliferative neoplasms	Active, not recruiting	MPNs	90 (estimated)	Interventional	Phase 2	Givinostat	HDAC inhibitor	Long-term safety and efficacy (Time frame: 3 months)
NCT03935555	Assess the safety, tolerability oral PU-H71 in subjects taking ruxolitinib	Recruiting	MF (primary, post-ET or post-PV)	24 (estimated)	Interventional	Phase 1	PU-H71	HSP90 inhibitor	- Safety and tolerability - Treatment response (Time frame: week 24)
NCT03373877	Evaluation of ruxolitinib in combination with PU-H71 for treatment of myelofibrosis	Terminated (December 14, 2017)	MF (primary, post-ET or post-PV)	4	Interventional	Phase 1	PU-H71 + ruxolitinib	HSP90 inhibitor	- Incidence of adverse events (Time frame: 12 months) - Maximum tolerated dose of PU-H71 (Time frame: 7 months) - Recommended phase 2 dose of PU-H71 (Time frame: 12 months) - Pharmacokinetic profile of PU-H71 (Time frame: 12 months)

NCT01393509	The first-in-human phase I trial of PU-H71 in patients with advanced malignancies	Active, not recruiting	MPNs	47	Interventional	Phase 1	PU-H71	HSP90 inhibitor	- Safety (Time frame: 2 years) - Tolerability (Time frame: 2 years) - Pharmacokinetic - Maximum tolerated dose (Time frame: 2 years)
NCT03136185	IMG-7289 in patients with myelofibrosis	Recruiting	MF (primary, post-ET or post-PV)	75 (estimated)	Interventional	Phase 2	IMG-7289	LSD1 inhibitor	- Safety (Time frame: 28 days) - Tolerability (Time frame: 28 days)
NCT04262141	IMG-7289 in patients with essential thrombocythemia or polycythemia vera	Recruiting	- ET - PV	24 (estimated)	Interventional	Phase 2	IMG-7289	LSD1 inhibitor	Hematologic response rates (Time frame: 24 weeks)
NCT03222609	A study evaluating tolerability and efficacy of navitoclax alone or in combination with ruxolitinib in participants with myelofibrosis	Recruiting	MF (primary, post-ET or post-PV)	164 (estimated)	Interventional	Phase 2	Navitoclax (ABT-263) +/− ruxolitinib	Bcl-2 inhibitor	Spleen reduction ≥35% (at week 24)
NCT04354727	A study of APG-1252 in patients with myelofibrosis who progressed after initial therapy	Recruiting	MF	60 (estimated)	Interventional	Phase 1 Phase 2	Palcitoclax (APG-1252) +/− ruxolitinib	Bcl-2/xL inhibitor	- Dose-limiting toxicity (Time frame: 28 days) - Spleen volume measurement reduction

Continued

Table 1 Ongoing studies for targeted therapies in MPNs.—cont'd

Study ID number	Brief title of the study	Status (first posted)	Condition	Number of patients	Study type	Study phase	Intervention/drugs	Activity	Primary outcome measures
NCT02098161	LCL161 in treating patients with primary myelofibrosis, post-polycythemia vera myelofibrosis, or post-essential thrombocytosis myelofibrosis	Active, not recruiting	MF (primary, post-ET or post-PV)	50 (estimated)	Interventional	Phase 2	LCL161	SMAC mimetic	Objective response rate (after 84 days (three courses) of treatment)
NCT02593760	A study to evaluate the efficacy and safety of vismodegib in combination with ruxolitinib for the treatment of intermediate- or high-risk myelofibrosis	Completed (November 1, 2015)	MF (primary, post-ET or post-PV)	10	Interventional	Phase 1	Vismodegib or placebo + ruxolitinib	Hedgehog pathway inhibitor	- Spleen reduction $\geq$35% (at week 24) - Complete and partial remisions rate (at week 24)
NCT01243073	Open label study to evaluate the activity of imetelstat in patients with essential thrombocythemia or polycythemia vera	Completed (November 18, 2010)	- ET - PV	20	Interventional	Phase 2	Imetelstat (GRN163L)	Telomerase inhibitor	Hematologic response
NCT02426086	Study to evaluate activity of two dose levels of imetelstat in participants with intermediate-2 or high-risk myelofibrosis previously treated with Janus kinase inhibitor	Completed (April 24, 2015)	Primary MF	107	Interventional	Phase 2	Imetelstat (GRN163L)	Telomerase inhibitor	- Spleen reduction $\geq$35% (at week 24) - Reduction in total symptom score (TSS) $\geq$50% (at week 24)

PV, polycythemia vera; *ET*, essential thrombocythemia; *MF*, myelofibrosis; *MPN*, myeloproliferative neoplasm; *AML*, acute myeloid leukemia; *BET*, bromodomain and extra-terminal motif; *HDAC*, histone deacetylase; *SMAC*, second mitochondria-derived activators of caspases.

(Vannucchi, 2010) with both prolonged reduction in blood parameters and spleen volume and massive improvements in symptoms and quality of life. These latter results were biologically accompanied by a reduction in signal transduction and proinflammatory cytokine levels. In the Study 251, ruxolitinib achieved improvements irrespectively of JAK2 mutational status or subtype of MF (i.e., primary, post-PV or post-ET MF). As seen in rodent models, only a modest decrease in mutant *JAK2*V617F allele burden was observed. After the COMFORT-I and -II phase 3 controlled trials (Harrison et al., 2012; Verstovsek et al., 2012) ruxolitinib was approved in 2012 by the FDA and the EMA for the treatment of MF. While bone marrow transplantation remains the only curative treatment for MF (Kröger et al., 2015; Zhang et al., 2020), other studies have since confirmed the efficacy of ruxolitinib and it is considered as a first-line therapy for patients with intermediate or high-risk MF, including primary, post-PV or post-ET MF.

For PV, ruxolitinib was approved by the FDA in 2014 (Vannucchi et al., 2015) (and by the EMA in 2015) and was the first FDA-approved treatment for patients with PV, who have an inadequate response or are intolerant to hydroxyurea.

With respect to ET treatment, ruxolitinib has not been shown to be more efficient or better tolerated than best available therapies (Finazzi, 2017; Harrison et al., 2017) and is therefore not currently recommended. Indeed, while ruxolitinib has shown similar efficacy and, for a subset of patients with particularly troublesome symptoms, has even been shown to be appeasing and could therefore be considered when possible, its relatively high costs combined with potential toxicities, particularly infectious, relating to the long-term exposure required in patients with ET, are probably the main barriers to its routine implementation.

2.3 Other JAK inhibitors in MF

Since the approval of ruxolitinib, many other JAK inhibitors have been assessed for the treatment of MPNs (Table 1), mainly with the aim of being more targeted to JAK2 and lessening the side effects of ruxolitinib, especially cytopenias.

After it was granted priority review and orphan drug designation, fedratinib has been the second JAK inhibitor to be approved by the FDA for the treatment of MF, both in first and second line after ruxolitinib. It was granted marketing authorization with the same indications by the EMA in December 2020. Fedratinib is an oral, selective inhibitor with activity against WT and mutant JAK2, rationally designed while developing a series of pyrimidine-based inhibitors to target JAK2. Fedratinib was assessed as first-line MF treatment in the placebo-controlled, phase 3 JAKARTA study and as post-ruxolitinib treatment in the single-arm, phase 3 JAKARTA2 study (Harrison et al., 2013, 2019a). Especially, fedratinib was well tolerated in patients with platelets counts between 50 and 100.000/mm (Zhang et al., 2020) (compared with ruxolitinib). However, FDA approved fedratinib in August 2019 with a warning about the risk of severe and fatal encephalopathy, including Wernicke encephalopathy. Therefore, patients should have normal thiamine levels before starting and periodic monitoring of thiamine levels during treatment is also recommended (Gangat et al., 2019; Pardanani et al., 2015b; Talpaz and Kiladjian, 2021).

Recently, among other molecules, pacritinib, a JAK2/FLT3/IRAK1/CSF1R inhibitor, has revealed a very potent JAK2 inhibitory activity without myelosuppressive effects. It has also been shown to reduce spleen size and improve clinical outcome in patients with MF in two phase 3 studies (PERSIST-1, PERSIST-2) (Mascarenhas et al., 2018; Mesa et al., 2017) as well as in a Phase 2 dose-finding study (PAC203) (O'Sullivan et al., 2019) including patients with severe thrombocytopenia. Pacritinib is currently being evaluated in the PACIFICA trial, which was designed to evaluate its efficacy vs physician's choice in patients with MF and severe thrombocytopenia (NCT03165734) (Harrison et al., 2019b).

Globally, many clinical trials in MPNs, and especially MF, are being conducted lately with a focus on the potential benefits of alternative JAK inhibitors (momelotinib in the MOMENTUM clinical trial NCT04173494) or agents for patients with cytopenias with or without transfusion dependence. The challenge with JAK2 inhibitors in the context of ET and PV is that the disease burden is quite low compared to the potential side effects of aggressive therapies.

Although difficult to assess due to the lack of uniform criteria, failure of ruxolitinib is reported in many reports and is estimated to lie around 50% between 6 months to 1.5 years after initiation of the treatment (Fonseca et al., 2013; Palandri et al., 2020; Pemmaraju et al., 2020). Ruxolitinib failure is linked to a variety of causes, including suboptimal response, treatment-related toxicities and progression to secondary AML. Moreover, most type I inhibitors can induce therapy resistance due to several mechanisms:

- enhancement of JAK2 activation loop phosphorylation despite kinase inhibition (Andraos et al., 2012), most likely by exposing the activation loop tyrosines to other kinases,

- heterodimerization between tumor overexpressed JAK2 and JAK1 or TYK2 that results in reactivation of JAK2 signaling (Koppikar et al., 2012),
- theoretically certain JAK2 non-canonical mutations may exhibit longer half-life and hence less sensitivity to JAK2 inhibitors; this is the case for JAK2 R867Q or JAK2, S755R/R938Q described in hereditary thrombocytosis (Marty et al., 2014),

Type II JAK inhibitors, that interact with a JAK2-inactive conformation, might be efficient in Type I JAK inhibitor resistance and might exhibit increased specificity and efficacy due to their interaction with a less conserved kinase site (the DFG-pocket) (Andraos et al., 2012). NVP-BBT594 and NVP-CHZ868 have been developed and showed to be effective in preclinical models but were not further developed (Andraos et al., 2012; Meyer et al., 2015). Due to their binding efficiency, type II JAK2 inhibitors might induce profound cytopenia, limiting their potential use in PV or ET (Vainchenker et al., 2018).

Allosteric inhibitors would theoretically be more specific than Type I and Type II inhibitors targeting sites closer (type III) or distant (type IV) to the ATP-binding pocket, such as the pseudokinase domain of JAK2V617F. Selective JAK2V617F inhibitors would be desirable, since JAK2 WT is essential for normal hematopoiesis and its efficient inhibition will induce cytopenia (Leroy et al., 2016a, 2019; Leroy and Constantinescu, 2017).

However, two recent studies have shown that secondary mutations in JAK2V617F that restore wild-type JAK2 function for most cytokine receptors are inhibiting signaling of IFN gamma through its receptor complex as JAK2V617F conformation is close to the active conformation of wild-type JAK2 when associated with the ligand-stimulated IFN gamma receptor (Hammarén et al., 2018; Leroy et al., 2019). The IFN gamma receptor complex is presumed to adopt a tetramer structure (Marsters et al., 1995) and JAK2 activation within this tetramer would seemingly require the same pseudokinase residues as JAK2V617F (Hammarén et al., 2018; Leroy et al., 2019). Thus, the conformation that stabilizes JAK2V617F might be the same as the conformation that allows activation of the IFN gamma receptor complex where the IFNγR1 is coupled with JAK1 and IFNγR2 is coupled with JAK2 (Marsters et al., 1995), suggesting that specific JAK2V617F inhibitors would potentially inhibit IFN gamma signaling. Hence, other approaches, such as antifibrinolytic agents, telomerase inhibitors, bromodomain and extra-terminal motif (BET) inhibitors, and numerous combinatorial strategies are also being explored. We will focus on those strategies which might be called "targeted" in the following subchapter dedicated to "other targeted therapies."

One other approach in relation to JAK2 in MPNs could target the differences between interaction of JAK2V617F or WT with TPOR, EPOR, G-CSFR or other receptors. While Box 1 is conserved, recent X-ray crystal structure data suggested unique interactions for each receptor (Ferrao et al., 2018; Moraga et al., 2015). Detachment of JAK2V617F from TPOR or of JAK2 from TPOR mutants would interrupt the activating events leading to MPNs.

2.4 Concerns emerging from JAK2 inhibition in myelofibrosis

Extensive experience with ruxolitinib showed that in rare patients latent viral infections can be re-activated and as a consequence it has become routine practice to test for viral infections before the start of treatment (Harrison et al., 2012; Verstovsek et al., 2012).

Following sporadic observations of co-occurrence of B-cell non-Hodgkin's lymphoma during treatment of MF patients with ruxolitinib, a recent study reported a significantly higher incidence of lymphoma (5.8% of 69 patients with MF and ruxolitinib) vs only 0.36% of 557 MF patients treated with conventional therapy (Porpaczy et al., 2018). This 16-fold increased risk was confirmed in another cohort of 929 patients with MPNs (15-fold increase). Lymphomas were of aggressive B-cell type, extranodal, or leukemic with high MYC expression and did not carry the MPN driver mutations; the median time from initiation of ruxolitinib therapy to diagnosis was 25 months. Preclinical studies using mice devoid of STAT1 revealed that JAK2 inhibition *in vivo* led to appearance of aberrant B-cells in such a context, suggesting that inhibition of STAT1 in the presence of a pre-existent B-cell clone may increase risk of lymphoma development (Porpaczy et al., 2018). It remains to be confirmed whether long term treatment with ruxolitinib requires extensive and sensitive searches for preexisting aberrant B-cell clones.

Another concern is represented by reports that during ruxolitinib treatment, subclones carrying the driver mutation (most frequently JAK2V617F) and loss of function mutations in ASXL1 appear to be selected (Newberry et al., 2017). In an unpublished case from VH and SNC we were able to detect after 1 year of ruxolitinib treatment an allele frequency of 35% of a pathogenic ASXL1 mutation, whereas before treatment this mutation was <3%. The reasons behind such a selection are unknown but suggest that loss of ASXL1 may help the fitness of the *JAK2*V617F clone under ruxolitinib.

3. TPOR/MPL inhibition strategies

The thrombopoietin receptor is less frequently mutated in MPNs (2% of ET and 5% of PMF) (Vainchenker and Kralovics, 2017) and TPOR mutations do not occur in PV. The most prevalent mutations are W515K/L/R/S/A (Pardanani et al., 2006; Pikman et al., 2006) and S505N (Ding et al., 2004, 2009), which lead to dimerization of the receptor in an active conformation in absence of ligand and activation of downstream signaling. Both mutations occur in *TPOR* exon 10, which roughly corresponds to the transmembrane and amphipathic juxtamembrane region of the receptor. S505N leads to dimerization of TM domains in an active interface due to the strong amino-acid polarity between the two Asn and to unraveling of the helicity of the TM domain above H499 (Ding et al., 2009; Leroy et al., 2016b). On the other hand, mutations of W515 of the RWQFP motif lead to suppression of an inactivating anchor in the lipid bilayer, enabling dimerization of the upstream TM helices in an active configuration (Defour et al., 2013; Leroy et al., 2016b).

The TPOR TM dimer can adopt several activating interfaces (Staerk et al., 2011) which induce various levels of activation of the receptor and might lead to different phenotypes of the disease. While in the human, only one interface is activated by Ala-substitutions, novel mutations have been described lately in the TM domain, that also activate the receptor (Levy et al., 2019).

Other much rarer mutations outside the TM or JM domains activate TPOR, such as S204P/F in the extracellular domain and Y591F/N in the cytosolic domain (Cabagnols et al., 2016; Feenstra et al., 2016). Tyrosine 591 is involved in the negative regulation of TPOR signaling (Drachman and Kaushansky, 1997; Pecquet et al., 2010). When phosphorylated it might serve as a docking site for SH2 domain-containing proteins that negatively regulate TPOR. When mutated to Phe or Asn, phosphorylation at this site is no longer possible and this negative regulation is lost (Sangkhae et al., 2014).

To our knowledge, there is no treatment targeting TPOR in MPNs to date (Fig. 3). Tryptophan 491 located at the outer part of the lipid bilayer has recently been shown to have a role in activation of the receptor by either S505N or W515K. When mutated to Ala (W491A), it suppressed ligand-independent activation of TPOR S505N or W515K. Since this region is on the extracellular side of the juxtamembrane-transmembrane domains, one would imagine that it would be an easy target for TPOR

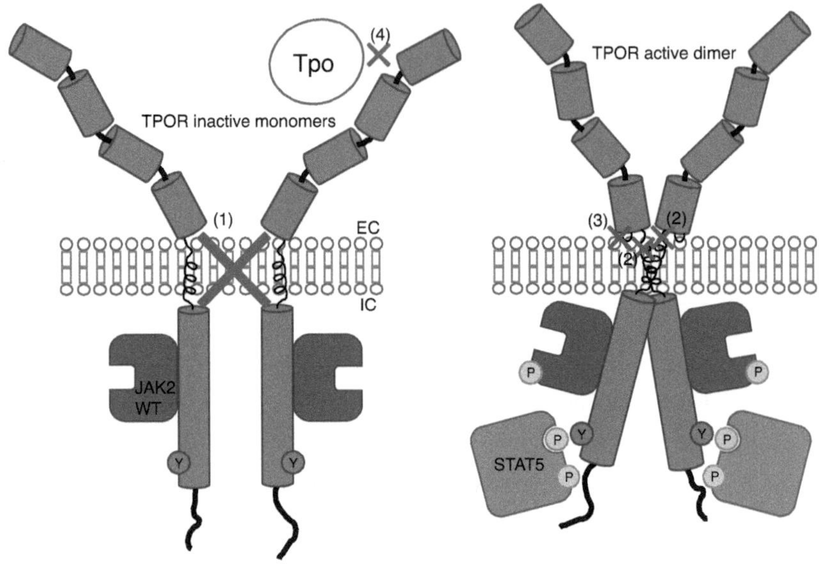

Fig. 3 TPOR inactivating strategies could target: (1) dimerization of the receptor in an active conformation by ligand or by oncogenic mutations, (2) the extracellular juxtamembrane domain at the vicinity or at W491 and H499, where stabilization of active TPOR dimers occurs. Other molecules could aim at preventing the change in the tilt-angle of the extracellular domain necessary for activation of the receptor (3) or prevent Tpo binding (4). *TPOR*, thrombopoietin receptor; *Tpo*, thrombopoietin; *Y*, tyrosine; *P*, phosphorylation; *EC*, extracellular; *IC*, intracellular.

antagonist (antibodies, small molecules, nanobodies, diabodies and others), which might even have specificity for the mutants encountered in PV or MF and not disrupt TPOR wild-type signaling. Other compounds targeting TPOR, but which might lack specificity, could also target the region around H499, as it changes conformation from a random coil in the inactive state to an alpha-helix in the active state. Because TPOR activation also occurs through a change in the tilt angle of the TM domain with the cell surface (Defour et al., 2013), small molecules binding to the outer part of the JM or TM regions could also prevent activation by maintaining the high natural tilt of the ligand-free TPOR. Similarly, increasing the distance between TPOR monomers with diabodies might prevent activation, as suggested for EPOR by Moraga et al. (2015). Targeting TPOR with more or less specificity for the mutants might finally trigger internalization and degradation or antibody-dependent cytotoxicity.

Although potentially beneficial for a small number of patients, it may also be of interest to identify small molecules that bind to the intracellular region adjacent to the membrane close to and without overlapping with W515, that may adopt a common structural change upon W515L/K/R/S/A mutation.

Finally, antagonist antibodies that prevent Tpo binding have been described (Wang et al., 2016). They prevent Tpo effects and could be useful in the rare cases of Tpo-overexpression. Nevertheless, such antibodies do not inhibit TPOR activation by MPN driver mutations but may decrease the myeloproliferative phenotype as was shown in a Tpo-knockout mouse model of retrovirally-induced MPN by CALR mutants (Marty et al., 2016).

4. CALR mutations

4.1 CALR inhibition strategies

Mutations in the last exon of calreticulin (*CALR*), exon 9, are responsible for approximatively 25% of ET and MF, and occur at a higher frequency than TPOR mutations (Klampfl et al., 2013; Nangalia and Green, 2017). While more than 50 mutations are described, all lead to the same +1 frame-shift in exon 9 that is responsible for the synthesis of a CALR molecule with a novel C-terminus devoid of the KDEL motif, but rich in positively-charged amino acids. Two mutations are predominant, a 52 bp deletion (p.L367fs*46), also called type 1, and a 5 bp insertion (p.K385fs*47), also called type 2. The remaining mutations are classified as type 1-like and type 2-like following the size of the remaining WT sequence. Type 1 *CALR* mutations are predominant in MF, whereas the frequency of both types is more similar in ET. While *CALR* mutations are usually heterozygous, homozygous mutations have been observed, more particularly for type 2 mutations (Klampfl et al., 2013; Nangalia and Green, 2017).

CALR is not a signaling molecule, but a chaperone of the endoplasmic reticulum (ER) involved in quality control of N-glycosylated protein and calcium storage (Michalak et al., 2009). The WT exon 9-coded sequence and the calcium-binding sites are nearly entirely lost in the del52 mutant while ins5 retains around 50% of negative charges (Klampfl et al., 2013). The new C-terminus changes the function of the molecule and allows CALR mutants to tightly bind to TPOR through an interaction between their lectin domain and N-glycosylated residues of the TPOR extracellular domain. They stabilize a dimeric state and transport TPOR to the cell surface where such dimers are able to signal through JAK2 (Araki et al., 2016; Chachoua et al., 2016; Elf et al., 2016; Marty et al., 2010; Pecquet et al., 2019). As a rogue chaperone CALR is nevertheless not sufficient for oncogenic transformation and TPOR remains absolutely required for the induction of cytokine-independent growth of factor dependent cell lines by CALR mutants (Chachoua et al., 2016). At ASH 2018, Pecquet et al. furthermore reported that CALR mutants are also secreted and found in the serum of MPNs patients (Pecquet et al., 2018). Secreted CALR mutants remain capable of inducing TPOR-induced cell proliferation, a rogue cytokine propriety possibly playing an auto- and paracrine role for mutated HSCs as well as influencing the bone marrow and spleen microenvironment in MPN CALR-mutated patients.

4.2 Potential therapies targeting CALR mutants

The necessary interaction of CALR mutant and TPOR to induce oncogenic transformation has opened a new field for the research of targeted therapies (Figs. 1 and 4). Both the soluble extracellular domain of TPOR (Chachoua et al., 2016) and a peptide electrostatically complementary to

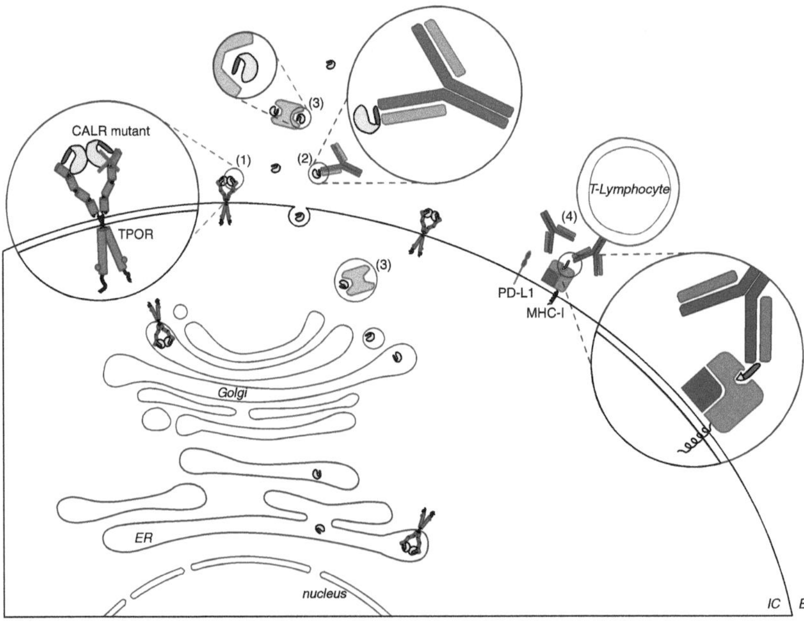

Fig. 4 CALR mutants inactivating strategies could target (1) interaction of CALR with TPOR. Antibodies (2) or small molecules (3) directed at CALR mutants could be used to target and remove CALR mutated proteins. Used as a novel antigen, the C-terminal of the mutated CALR could as well be used to develop vaccines or T-cell-mediated immunotherapy (4). *TPOR*, thrombopoietin receptor; *CALR*, calreticulin; *MHC-I*, major histocompatibility complex I; *PD-L1*, programmed death-ligand 1; *ER*, endoplasmic reticulum; *EC*, extracellular; *IC*, intracellular.

the positively charged sequence of CALR mutants prevent TPOR activation in CALR-mutated target cells (Nivarthi et al., 2016; Pronier et al., 2018), thus small molecules destabilizing TPOR-CALR interaction could be synthetized. These molecules would nevertheless need to be able to inhibit both cell surface and intracellular complexes, as part of the signaling through TPOR-CALR complexes might occur at the intracellular level (Pecquet et al., 2019). This would indeed be a hinge for the development of antibodies preventing such interactions, while antibodies targeting mutated CALR could be used to target and remove circulating and paracrine CALR mutated proteins as well as to eliminate CALR expressing cells, lowering the burden of pathogenic clones. Used as a novel antigen, the C-terminal of the mutated CALR could also be used to develop vaccines or T-cell-mediated immunotherapy. If such therapies would however depend on HLA presentation of the novel peptide, which might be hampered in MPN patients (Arshad and Cresswell, 2018), a trial for a CALR Exon 9 Mutant Peptide Vaccine to Patients With CALR-mutant Myeloproliferative Neoplasms has already been announced by a single center in Denmark, although it is not yet open (NCT03566446) (Table 1). Other development of such targeted antibodies is also ongoing but remains in preclinical stages. On the other hand, immunization with peptides derived from the novel C-tail of mutant CALR did not lead to antibody response in mice transplanted with a mixture of bone marrow from wild-type vs del52 KI mice, while mice transplanted with wild type bone marrow responded (Balligand et al., 2019). These data suggest that the presence of the antigenic sequence in the HSCs may lead to tolerance.

5. Other targeted therapies

The development of targeted therapies aims to actually inhibit causative alterations specifically in the tumor. While the three driver mutations are found as single mutations in MPNs, they can be associated with numerous additional acquired mutations (Grinfeld et al., 2018; Lundberg et al., 2014), which can modulate the phenotype or the evolution of the disease. Furthermore, understanding the biology of MPN has revealed how certain physiological pathways are disrupted in the diseases (Fig. 5). Alongside the rapid development of targeted therapies, this has led to an increasing number of trials in MPNs, in combination or not with JAK inhibitors. While there is indeed no standard of care for patients with intermediate or high-risk MF who

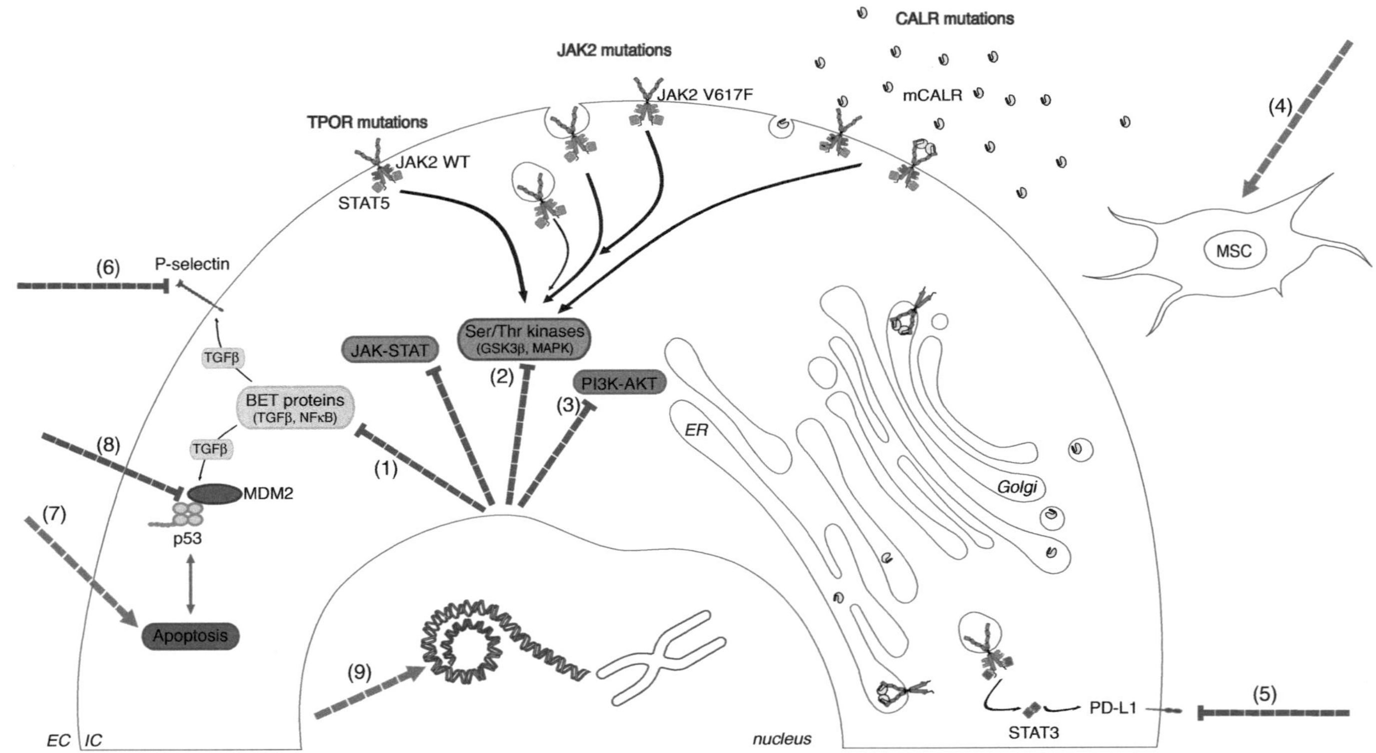

Fig. 5 See figure legend on opposite page.

have failed, become intolerant or resistant to JAK inhibitors, combination of JAK inhibitors with various drugs may lead to clinical benefits in patients.

We will describe here (Table 1) some of the ongoing trials for targeted therapies and also mention potential therapeutic targets that were addressed in preclinical studies.

5.1 Bromodomain and extra-terminal family of proteins

Bromodomain-containing proteins are chromatin-reading proteins with a wide variety of functions, from chromatin remodeling to transcriptional mediation and co-activation (Belkina and Denis, 2012). They contain two tandem bromodomains and an extra-terminal domains. Their main activity is to bind to acetylated histones via the bromodomains. Over the past 15 years, BET proteins have been shown to regulate key oncogenic pathways, especially in myeloid malignancies (Zuber et al., 2011). Such targeted pathways include those of NFκB and TGFβ signaling, which are important drivers of pro-inflammatory cytokine expression and bone marrow fibrosis, respectively, and are involved in myelofibrosis pathogenesis (Le Bousse-Kerdilès, 2012). BET proteins may also play a role in post-MPN secondary AML cells acting as enhancers for the transcription of pro-oncogenes such as c-MYC, BCL2 and CDK4/6 (Fiskus et al., 2014a).

BET inhibitors have been shown to inhibit growth *in vitro* and *in vivo* and induce apoptosis in cultured and primary AML cells (Dawson et al., 2014; Fiskus et al., 2014b; Saenz et al., 2017). Preclinical studies in MPNs suggest

Fig. 5 Other targeted therapies in MPNs could rely on inhibition of downstream pathologically-activated signaling pathways in MPNs (JAK/STAT, PI3K-Akt and Serine/Threonine kinases such as GSK3β) by specific pharmacological inhibitors alone or in combination. BET proteins regulate several pathogenic pathways in MPNs, including NFκB and TGFβ. BET inhibitors and other inhibitors of the TGFβ pathway (1) could be used in MF patients. Other trials aim at targeting Ser/Thr kinases as GSK-3β (2) or the PI3K/AKT/mTOR (3) pathway, as their inhibitors have anti-neoplastic and/or anti-fibrotic properties. The niche plays a role in MPN progression and treatment acting on MSCs or other of its components could be used to decrease MF (4). New strategies have also emerged with the development of immune checkpoint inhibitors (5) or P-selectin inhibitors (6). Targeting apoptosis (7) or increasing p53 levels in MPN cells is another strategy that might be induced by type I interferon treatment, but which is also addressed today by disrupting the interaction between MDM2 and p53 (8). Finally, certain trials have considered targeting histones through histone deacetylases (HDAC) or histone demethylases inhibitors (9). *BET*, bromodomain and extra-terminal motif; *Ser/Thr*, Serine/threonine; *MF*, myelofibrosis; *MSC*, mesenchymal stromal cell; *ER*, endoplasmic reticulum; *EC*, extracellular; *IC*, intracellular.

that a combination of BET inhibitors and JAK inhibitors may result in synergistic reduction of splenomegaly, bone marrow (BM) fibrosis and mutant cell burden (Kleppe et al., 2018).

CPI-0610, a selective and potent small molecule BET inhibitor, has shown clinical activity and a wide therapeutic window in a Phase 1 lymphoma study (Abramson et al., 2015; Blum et al., 2018). Preliminary data of the MANIFEST trial (Mascarenhas et al., 2019a) indicated that CPI-0610 alone or "add-on" to ruxolitinib is generally well-tolerated and provides clinical benefits in MF patients with inadequate responses or who were refractory to ruxolitinib. Alongside, improvement in bone marrow fibrosis and anemia revealed the potential for meaningful disease modification and the trial is currently running for gaining more significant power.

5.2 Inactivation of the TGF-β pathway

Members of the TGF-β superfamily, which include activin and the growth differentiation factor 11 (GDF11), are secreted by bone marrow stromal cells and play a role in proliferation and differentiation of the erythroid precursors. Sotatercept, a TGF-β superfamily inhibitor, which acts as a ligand trap to inhibit activin and GDF11 (soluble activin receptor 2A-IgG Fc fusion), improved anemia in preclinical models of β-thalassemia and Diamond-Blackfan anemia and is being evaluated for treatment of patients with anemia in MPN (NCT01712308). Luspatercept, another ligand trap inhibitor of TGF-β (fusion between activin receptor type 2B extracellular domain and an IgG fragment), leading to inhibition of the SMAD signaling pathway, also demonstrated clinical efficacy for the treatment of anemia (Suragani et al., 2014) in β-thalassemia (Cappellini et al., 2020) and myelodysplastic syndromes (Platzbecker et al., 2017). It has been approved in 2019 by the FDA and in 2020 by the EMA for the treatment of anemia in blood transfusion-dependent patients with β-thalassemia (Markham, 2020) and is currently being evaluated in association with ruxolitinib in a phase 2 multicentric trial for the treatment of anemia in patients with PMF (NCT03194542). The differences between luspatercept and sotatercept have been summarized recently and appear to involve the fact that luspatercept does not alter GDF11 signaling like sotatercept (Guerra et al., 2019).

Based on inhibition of the same signaling pathway, a phase 2, intra-patient dose escalation study is currently open to assess the potential of using the TGFβ receptor inhibitor vactosertib (TEW-7197) in patients with MPNs and anemia (NCT04103645).

5.3 Targeting fibrosis

Following the hypothesis that reduction in bone marrow fibrosis might also help to restore hematopoiesis and improve cytopenias, studies have aimed at identifying fibrosis-driving cells and inhibit their activity.

9-ING-41 is a potent selective GSK-3β inhibitor with clinical anticancer activity, as well as significant pre-clinical ability to reverse pathologic fibrosis in multiple models of pulmonary and pleural fibrosis. The Actuate 1901 Study (NCT04218071) is expected to start soon in patients with advanced and poor prognosis MF. This phase 2 study tackles fibrosis and will investigate 9-ING-41 as a single agent or in combination with ruxolitinib. As 9-ING-41 could potentially act as an anti-neoplastic agent and anti-fibrotic agent in patients with MF, it may reduce the dose of ruxolitinib needed for optimal therapeutic response and/or reverse myelosuppression.

As chronic inflammation driven by an abnormal production of cytokines and other soluble proteins by both malignant and non-malignant cells was suggested to play an important role in MPN progression, leading to a progressive impairment of bone marrow niche (Kleppe et al., 2015; Koschmieder et al., 2016), alternative therapies that address bone marrow components and their products were proposed and investigated mainly in preclinical studies.

Serum amyloid P (SAP, pentraxin-2), also known as fibrocyte inhibitor, is a constitutive, anti-inflammatory, innate immune plasma protein, that has been shown to successively reduce fibrotic disease of the kidney. SAP levels were found to be lower in patients with MF. In xenograft mice models, increasing SAP levels prolonged survival and delayed bone marrow fibrosis. The recently completed NCT01981850 study investigated PRM-151, a recombinant human SAP, either as single agent or in combination with ruxolitinib, in patients with PMF and post ET/PV MF. Decreased bone marrow fibrosis was one of the outcomes measured in this phase 2 study and results are pending.

It has been demonstrated in mouse models of bone marrow fibrosis that $Gli1^{+}$ mesenchymal stromal cells (MSCs) represent the progenitors of myofibroblasts, being involved in myelofibrotic transformation. Treatment with Gli antagonist 61 (GANT61) of MSCs isolated from mouse models as well as from patients with myelofibrosis inhibited myofibroblast differentiation and, consequently, alleviated bone marrow fibrosis, indicating that Gli proteins may serve as promising therapeutic targets (Schneider et al., 2017). Also, the MPN local neuropathy induced by alteration of sympathetic nerve fibers that innervate bone marrow nestin positive MSCs having a role in normal HSC regulation can be targeted therapeutically by β3-adrenergic agonists. In this respect, treatment with the human β_3-adrenergic agonist mirabegron increased the number of nestin positive MSCs by restoring their sympathetic regulation and leading to a reduction of bone marrow fibrosis (Arranz et al., 2014). Recently, it was suggested that the Wnt inhibitor Dickkopf-related protein 1 (Dkk-1), measured in plasma of MPN patients, could differentiate between ET and prefibrotic PMF (Mambet et al., 2018). As Dkk-1 might be associated with bone marrow niche alterations and especially neoangiogenesis, it might represent a future therapeutic target (Mambet et al., 2018).

Other avenues for the treatment of MPNs have also emerged with the development of the immune checkpoint inhibitors. Of interest, JAK2V617F activates persistent TPOR-dependent STAT3 signaling that induces transcriptional activation at the promoter of Programmed Death Ligand 1 (PD-L1) leading to enhanced PD-L1 expression at the surface of monocytes, megacaryocytes and platelets (Prestipino et al., 2018) and thus reducing T-cell responses against the mutated clone. Data in a xenotransplantation model suggest that inhibition of PD1 may be a promising strategy in MPN treatment (Prestipino et al., 2018; Veletic et al., 2018; Wang et al., 2019) and Mascarenhas et al. completed in May 2020 the NCT03065400 trial (May 28, 2020 being the final data collection date for primary outcome measure, clinicaltrials.gov/ct2/show/NCT03065400) studying the anti-PD-1 pembrolizumab in patients with accelerated or blast phase disease (MPN-AP/BP) refractory to or intolerant to conventional therapies such as decitabine, and for whom hematopoietic stem cell transplant was no therapeutic option (exploratory cohort).

In another international trial comparing different approaches in different arms (NCT04097821) and now open to enrolment of MF patients, ruxolitinib is being tested in one arm in combination with another immune checkpoint drug, MBG453, an inhibitor of the inhibitory T-cell receptor T-cell immunoglobulin and mucin domain-containing protein 3 (TIM-3; hepatitis A virus cellular receptor 2; HAVCR2), with potential immune checkpoint inhibitory and antineoplastic activities. The second arm is investigating the adjunction of crizanlizumab, a humanized monoclonal antibody that binds P-selectin and blocks interaction with its ligands (including leukocyte PSGL-1). Indeed, it is now accepted that reduced GATA1 content in megakaryocytes as a consequence of ribosomal deficiency is a hallmark of primary MF and a driving event in the disease. Among others, GATA1 deficiency leads to abnormal activation of the TGF-β/P-selectin axis in megakaryocytes of primary MF patients, which in turn contributes to the development of MF (Zetterberg et al., 2013). Consequently, deletion of the P-selectin gene prevented development of myelofibrosis in a GATA1-low mouse model of MF (Spangrude et al., 2015) and inhibition of P-selectin is thought to constitute a new target in MF treatment (Ling et al., 2018; Zingariello et al., 2019).

5.4 The p53 pathway

Finally, the third drug tested in the trial in combination with ruxolitinib is HDM201 (siremadlin), a potent and selective orally bioavailable inhibitor of p53-MDM2 (HDM2 in human) interaction, that activates p53 and induces robust p53-dependent cell cycle arrest and apoptosis in human p53 wild-type tumor cells (Jeay et al., 2018). Of interest, in patients with chronic-phase MPN, *p53* is usually wild-type, and malignant hematopoietic stem cells (HSCs) express low levels of the p53 protein (Nakatake et al., 2011). It was hypothesized that since interferon is an effective therapy in patients with MPNs (Gowin et al., 2012), this may be due, at least in part, to interferon-linked increase in p53 expression in malignant HSCs. Furthermore, TGF-β1 induces expression of HDM2 which might drive high levels of MDM2. p53 is indeed reduced in MPN cells secondarily to the overexpression of its major regulators MDM2 and MDM4 (HDM4/HDMX), which are expressed at levels higher than normal in MF $CD34^+$ cells (Lu et al., 2012; Nakatake et al., 2011). Hence, targeting the interaction of MDM2 and p53 might be of interest in MF, as it has already been suggested with idasanutlin (Lu et al., 2012). Idasanutlin is an MDM2-p53 binding cell cycle inhibitor, the first long-acting MDM2 antagonist investigated in clinical trials. *In vitro*, it binds selectively and with high affinity to the p53 site on the surface of the MDM2 molecule and can effectively displace p53 from MDM2, leading to stabilization, accumulation of p53 protein and activation of the p53 pathway. The phase 1 study of idasanutlin in PV was presented by Mascarenhas et al. in 2017 at ASH (Mascarenhas et al., 2017) and published in 2019 (Mascarenhas et al., 2019b). It showed promising hematologic, symptomatic, pathologic and molecular responses, indicating idasanutlin is a promising novel agent for PV. Moreover, *in vitro* combination of low doses of pegylated interferon alfa-2a with nutlin-3 resulted in selective and significant inhibition of JAK2V617F-positive CD34+ PV colony formation compared to normal colony formation. The global phase 2 trial that was evaluating idasanutlin for hydroxyurea-resistant or intolerant PV patients was however discontinued by its sponsor in April 2020 as the sponsor decided to stop the development of idasanutlin in the PV indication. If idasanutlin presents indeed a much increased activity compared to its precursor nutlins, some authors fear that it might display similar weaknesses, which are limited elimination of cancer cells and the generation of p53-mutated drug-resistant subpopulations (Skalniak et al., 2018). It is important to emphasize that secondary AML on MPNs frequently carry TP53 mutations or overexpression of MDM4, and that leukemic transformation is associated with acquired homozygosity or compound heterozygosity for mainly missense TP53 mutations (Goyal et al., 2020; Lundberg et al., 2014).

Nevertheless, other trials with nutlins keep ongoing. First results of the Phase 2 trial on the use of KRT-232, a novel oral small molecule inhibitor of MDM2 (NCT03662126), were presented at EHA 2020 annual congress and demonstrated a dose-dependent relationship for safety and efficacy as well as promising clinical activity and tolerability in patients with poor prognosis MF. In the meantime, KRT-232 is also studied in a global, open-label Phase 2a/2b study to determine its efficacy and safety in the treatment of patients with phlebotomy-dependent PV (NCT03669965). As nutlins have also showed durable response in the treatment of selected patients with AML (Yee et al., 2014), another multicenter phase 1b/2 trial (NCT04113616) is recruiting for the treatment with KRT-232 of patients with AML secondary to MPN.

Finally, other drugs have been proposed in combination with ruxolitinib and are under evaluation. Because JAK2V617F-expressing cells that co-express TPOR are synergically inhibited by the combination of JAK2 and PI-3 kinase or mTOR inhibitors (Bartalucci et al., 2013; Choong et al., 2013), one trial investigated the combination of ruxolitinib and BKM120 (Buparlisib, a PI3K inhibitor) in MF. However, it was terminated prematurely by its sponsor in 2019. Yacoub et al. (Abstract Book, 2020) presented interesting results at the EHA 2020 annual congress for the addition to ruxolitinib in MF patients with a suboptimal response of parsaclisib, a potent, highly selective, next-generation PI3Kδ inhibitor. The addition of parsaclisib appeared to be well tolerated and resulted in further reduction in spleen volume and improvement in symptom burden.

Other trials are considering or have considered different histone deacetylase (HDAC) inhibitors such as panobinostat (NCT01693601), vorinostat (Andersen et al., 2013) or givinostat (NCT01761968) (Rambaldi et al., 2010). HDAC inhibitors have multiple potential mechanisms of action in MPN cells, especially through down-regulation of JAK2 via inhibition of the chaperone protein function of heat shock protein 90 (HSP90) (Akada et al., 2012; Fiskus et al., 2011; Guerini et al., 2007; Wang et al., 2009).

Givinostat and vorinostat are active in patients with PV and ET, alleviating efficiently symptoms as pruritus and inducing spleen and hematologic responses in a substantial proportion of patients, apparently regardless of *JAK2* mutational status (Andersen et al., 2013; Rambaldi et al., 2010) and although the reduction of the allele burden of mutated *JAK* was modest. Secondary effects as asthenia and gastrointestinal complaints were however present. Nevertheless, givinostat continues to be developed for PV (NCT01761968).

HSP90 can also be directly targeted and inhibitors have been developed as antineoplastic drugs. PU-H71 is one of them and started to be studied for various solid tumors and non-Hodgkin lymphoma (Speranza et al., 2018). Although evidence of its efficiency *in vivo* still seems scarce, some trials are studying its efficiency in MPNs in combination with ruxolitinib (NCT03935555, NCT03373877, NCT01393509).

5.5 LSD1: Inhibition of lysine specific demethylase 1A

Another proposed epigenetic therapy consists of targeting histone demethylases. Lysine-specific demethylase 1A (LSD1) is a flavin-dependent monoamine oxidase, which demethylates mono- and di-methylated lysines, specifically histone 3 lysines 4 and 9. LSD1 has been shown to play an important role in cell proliferation and in various tumors. Elevated levels of LSD1 have been found in diverse cancers and were related with cellular effects such as epithelial-mesenchymal transition (EMT), cell proliferation and differentiation, stem cell biology and malignant transformation (Fang et al., 2019). LSD1 overexpression was also found in hematologic malignancies, such as AML or MPNs, with a critical role for self-renewal of malignant myeloid cells and differentiation of myeloid progenitors (Magliulo et al., 2018; Niebel et al., 2014). IMG-7289 (bomedemstat) is an orally available LSD1 inhibitor. After IMG-7289 reduced peripheral cell counts, spleen size, inflammatory cytokines, mutant allele frequencies and marrow fibrosis in mouse models of MPN (Jutzi et al., 2018), trials were started in the human. Preliminary results from the open-label Phase 2b study NCT03136185 of IMG-7289 in myelofibrosis showed a trend toward improvement in symptom burden in most patients and modest reduction in spleen volumes in a subset of patients, despite under-dosing and slow dose escalation (Pettit et al., 2019). Hence this study keeps on recruiting with more aggressive dosing and IMG-7289 is now also being evaluated in patients with ET and PV (NCT04262141).

In addition, another target being considered for MF treatment is the apoptosis pathways through the B-cell lymphoma 2 (Bcl-2) protein, as preclinical studies have demonstrated cytotoxic activity of navitoclax, a novel small-molecule that targets and binds with high affinity to multiple anti-apoptotic Bcl-2 family proteins (including Bcl-xL, Bcl-2, and Bcl-W) against myeloproliferative neoplasm-derived cell lines (Lu et al., 2010). In a phase 2 single-arm, multicenter, open-label study (NCT03222609) assessing the efficacy and safety of navitoclax combined with ruxolitinib in patients with MF, navitoclax was well tolerated with clinically meaningful spleen responses, allelic burden reductions, improvement in total symptom score and encouraging improvements in bone marrow fibrosis (in comparison to treatment with ruxolitinib alone in the patients who had been on therapy before) (Harrison et al., 2019c). Because the development of navitoclax was initially hampered by dose-limiting thrombocytopenia (Mason et al., 2007), an on-target effect resulting from inhibition of Bc-xL, other more specific Bcl-2 inhibitors have been developed. Among them, palcitoclax (APG-1252) has shown reduced toxicity on platelets (Lakhani et al., 2020) and is currently being tested in the treatment of MF (NCT04354727).

Evasion from programmed cell death is a hallmark of cancer and can be achieved in cancer cells by overexpression of inhibitor of apoptosis proteins (IAPs). Second mitochondria-derived activators of caspases (SMAC) directly bind to IAPs and promote apoptosis. SMAC mimetics have been investigated in various types of cancer and are hypothesized to be particularly effective in diseases with high inflammation and NFκB activation. Given that elevated TNFα levels with strong NFκB activation are a feature of patients with MF (Fleischman et al., 2011; Tefferi and Vainchenker, 2011), the SMAC mimetic LCL-161 was investigated in MF patients in the phase 2 NCT02098161 trial and showed interesting activity, especially in patients, who had failed on JAK inhibitors (Pemmaraju et al., 2019). Moreover, further arguments indicate that JAK2 signaling might even be required for the differential sensitivity of mutant cells to SMAC mimetics, suggesting that the addition of JAK2 inhibitors may reduce the ability of SMAC mimetics to selectively target JAK2V617F mutant cells (Craver et al., 2020).

Finally, other targets have also been considered, as the sonic hedgehog pathway, that could be targeted by vismodegib (NCT02593760), therapy directed at CD123 (tagraxofusp) or telomerase inhibitor such as imetelstat.

If telomerase inhibition by imetelstat was able to induce hematologic response in all 18 ET patients in the NCT01243073 trial (Baerlocher et al., 2015), caution was advised because the treatment was accompanied by significant myelosuppression (Tefferi et al., 2015). In patients with relapsed or refractory MF, imetelstat did not appear to significantly increase overall survival in all patients of the NCT02426086 trial (preliminary results) but might have been more efficient in the subgroup of patients with shorter telomeres at diagnosis (Mascarenhas et al., Oral and Poster Abstracts n°347 at 62nd ASH Annual Meeting 2020).

6. Conclusion

In conclusion, while the number of treatments directed at MPNs approved by government agencies remains very low, efforts are under way to identify new treatment strategies for MPNs, with a particular focus on MPN-specific altered pathways. It should however be noted, that none of above-mentioned treatments have so far proven effective in curing MPNs. In this race to treat rare diseases, the number of patients included in each trial is also relatively small, requiring international efforts. Finally, it should be noted that indications for enrolment in clinical trials may particularly not be an easy choice in MPNs, neither for the patients nor for the clinician, as evolution of the diseases remains slow for many patients and should be compared with the possible burden of treatment-associated toxicities.

Acknowledgments

Funding to SNC is acknowledged from Ludwig Institute for Cancer Research, from Fondation contre le cancer, from Salus Sanguinis and Fondation "Les avions de Sébastien," and from projects Action de recherche concertée (ARC) 16/21-073 and WelBio (Walloon Excellence in Life Sciences and Biotechnology) F 44/8/5—MCF/UIG—10955. Funding to CCD, CM and SNC from Program COP, A.1.1.4. ID:9_37_798 SMIS 106774, MyeloAL-EDiaProT 149/26.10.2016 is acknowledged. VH was supported by the FNRS and Fondation contre le cancer, Belgium, Belgium. SB is the recipient of an Aspirant PhD Fellowship for clinicians from the FNRS. GL is the recipient of the PhD Fellowship of the Foundation "Les avions de Sébastien."

References

Abramson, J.S., Blum, K.A., Flinn, I.W., et al., 2015. BET inhibitor CPI-0610 is well tolerated and induces responses in diffuse large B-cell lymphoma and follicular lymphoma: preliminary analysis of an ongoing phase 1 study. Blood 126, 1491.

Akada, H., Akada, S., Gajra, A., et al., 2012. Efficacy of vorinostat in a murine model of polycythemia vera. Blood 119, 3779–3789.

Andersen, C.L., McMullin, M.F., Ejerblad, E., et al., 2013. A phase II study of vorinostat (MK-0683) in patients with polycythaemia vera and essential thrombocythaemia. Br. J. Haematol. 162, 498–508.

Andraos, R., Qian, Z., Bonenfant, D., et al., 2012. Modulation of activation-loop phosphorylation by JAK inhibitors is binding mode dependent. Cancer Discov. 2, 512–523.

Araki, M., Yang, Y., Masubuchi, N., et al., 2016. Activation of the thrombopoietin receptor by mutant calreticulin in CALR-mutant myeloproliferative neoplasms. Blood 127, 1307–1316.

Arranz, L., Sánchez-Aguilera, A., Martín-Pérez, D., et al., 2014. Neuropathy of haematopoietic stem cell niche is essential for myeloproliferative neoplasms. Nature 512, 78–81.

Arshad, N., Cresswell, P., 2018. Tumor-associated calreticulin variants functionally compromise the peptide loading complex and impair its recruitment of MHC-I. J. Biol. Chem. 293, 9555–9569.

Baerlocher, G.M., Leibundgut, E., Ottmann, O.G., et al., 2015. Telomerase inhibitor imetelstat in patients with essential thrombocythemia. N. Engl. J. Med. 373, 920–928.

Balligand, T., Achouri, Y., Pecquet, C., et al., 2019. Knock-in of murine Calr del52 induces essential thrombocythemia with slow-rising dominance in mice and reveals key role of Calr exon 9 in cardiac development. Leukemia 34, 510–521.

Bandaranayake, R.M., Ungureanu, D., Shan, Y., Shaw, D.E., Silvennoinen, O., Hubbard, S.R., 2012. Crystal structures of the JAK2 pseudokinase domain and the pathogenic mutant V617F. Nat. Struct. Mol. Biol. 19, 754–759.

Barbui, T., Tefferi, A., Vannucchi, A.M., et al., 2018. Philadelphia chromosome-negative classical myeloproliferative neoplasms: revised management recommendations from European LeukemiaNet. Leukemia 32, 1057–1069.

Bartalucci, N., Tozzi, L., Bogani, C., et al., 2013. Co-targeting the PI3K/mTOR and JAK2 signalling pathways produces synergistic activity against myeloproliferative neoplasms. J. Cell. Mol. Med. 17, 1385–1396.

Baxter, E.J., Scott, L.M., Campbell, P.J., et al., 2005. Acquired mutation of the tyrosine kinase JAK2 in human myeloproliferative disorders. Lancet 365, 1054–1061.

Belkina, A.C., Denis, G.V., 2012. BET domain co-regulators in obesity, inflammation and cancer. Nat. Rev. Cancer 12, 465–477.

Blum, K.A., Abramson, J., Maris, M., et al., 2018. A phase I study of CPI-0610, a bromodomain and extra terminal protein (BET) inhibitor in patients with relapsed or refractory lymphoma. Ann. Oncol. 29.

Abstract Book, 2020. 25th congress of the European Hematology Association virtual edition, 2020. HemaSphere 4, 1–1168.

Cabagnols, X., Favale, F., Pasquier, F., et al., 2016. Presence of atypical thrombopoietin receptor (MPL) mutations in triple-negative essential thrombocythemia patients. Blood 127, 333–342.

Cappellini, D.M., Viprakasit, V., Taher, A.T., et al., 2020. A phase 3 trial of Luspatercept in patients with transfusion-dependent β-thalassemia. N. Engl. J. Med. 382, 1219–1231.

Chachoua, I., Pecquet, C., El-Khoury, M., et al., 2016. Thrombopoietin receptor activation by myeloproliferative neoplasm associated calreticulin mutants. Blood 127, 1325–1335.

Choong, M.L., Pecquet, C., Pendharkar, V., et al., 2013. Combination treatment for myeloproliferative neoplasms using JAK and pan-class I PI3K inhibitors. J. Cell. Mol. Med. 17, 1397–1409.

Craver, B.M., Nguyen, T., Nguyen, J., et al., 2020. The SMAC mimetic LCL-161 selectively targets JAK2V617F mutant cells. Exp. Hematol. Oncol. 9, 1.

Dawson, M.A., Gudgin, E.J., Horton, S.J., et al., 2014. Recurrent mutations, including NPM1c, activate a BRD4-dependent core transcriptional program in acute myeloid leukemia. Leukemia 28, 311–320.

Defour, J.P., Itaya, M., Gryshkova, V., et al., 2013. Tryptophan at the transmembrane-cytosolic junction modulates thrombopoietin receptor dimerization and activation. Proc. Natl. Acad. Sci. U. S. A. 110, 2540–2545.

Ding, J., Komatsu, H., Wakita, A., et al., 2004. Familial essential thrombocythemia associated with a dominant-positive activating mutation of the c-MPL gene, which encodes for the receptor for thrombopoietin. Blood 103, 4198–4200.

Ding, J., Komatsu, H., Iida, S., et al., 2009. The Asn505 mutation of the c-MPL gene, which causes familial essential thrombocythemia, induces autonomous homodimerization of the c-Mpl protein due to strong amino acid polarity. Blood 114, 3325–3328.

Drachman, J.G., Kaushansky, K., 1997. Dissecting the thrombopoietin receptor: functional elements of the Mpl cytoplasmic domain. Proc. Natl. Acad. Sci. U. S. A. 94, 2350–2355.

Dupont, S., Masse, A., James, C., et al., 2007. The JAK2 617V>F mutation triggers erythropoietin hypersensitivity and terminal erythroid amplification in primary cells from patients with polycythemia vera. Blood 110, 1013–1021.

Dusa, A., Mouton, C., Pecquet, C., Herman, M., Constantinescu, S.N., 2010. JAK2 V617F constitutive activation requires JH2 residue F595: a pseudokinase domain target for specific inhibitors. PLoS One 5, e11157.

Elf, S., Abdelfattah, N.S., Chen, E., et al., 2016. Mutant calreticulin requires both its mutant C-terminus and the thrombopoietin receptor for oncogenic transformation. Cancer Discov. 6, 368–381.

Fang, Y., Liao, G., Yu, B., 2019. LSD1/KDM1A inhibitors in clinical trials: advances and prospects. J. Hematol. Oncol. 12, 129.

Feenstra, J.D., Nivarthi, H., Gisslinger, H., et al., 2016. Whole-exome sequencing identifies novel MPL and JAK2 mutations in triple-negative myeloproliferative neoplasms. Blood 127, 325–332.

Ferrao, R.D., Wallweber, H.J., Lupardus, P.J., 2018. Receptor-mediated dimerization of JAK2 FERM domains is required for JAK2 activation. Elife 7, e38089. https://doi.org/10.7554/eLife.38089. PMID: 30044226. PMCID: PMC6078494.

Finazzi, G., 2017. Ruxolitinib in ET: not all MPN are equal. Blood 130, 1873–1874.

Fiskus, W., Verstovsek, S., Manshouri, T., et al., 2011. Heat shock protein 90 inhibitor is synergistic with JAK2 inhibitor and overcomes resistance to JAK2-TKI in human myeloproliferative neoplasm cells. Clin. Cancer Res. 17, 7347–7358.

Fiskus, W., Sharma, S., Qi, J., et al., 2014a. BET protein antagonist JQ1 is synergistically lethal with FLT3 tyrosine kinase inhibitor (TKI) and overcomes resistance to FLT3-TKI in AML cells expressing FLT-ITD. Mol. Cancer Ther. 13, 2315–2327.

Fiskus, W., Sharma, S., Qi, J., et al., 2014b. Highly active combination of BRD4 antagonist and histone deacetylase inhibitor against human acute myelogenous leukemia cells. Mol. Cancer Ther. 13, 1142–1154.

Fleischman, A.G., Aichberger, K.J., Luty, S.B., et al., 2011. TNFα facilitates clonal expansion of JAK2V617F positive cells in myeloproliferative neoplasms. Blood 118, 6392–6398.

Fonseca, E., Silver, R.T., Kazis, L.E., Iqbal, S., Rose, M., Khan, N., 2013. Ruxolitinib discontinuation in patients with myelofibrosis: an analysis from clinical practice. Blood 122, 2833.

Gangat, N., Phelps, A., Lasho, T.L., et al., 2019. A prospective evaluation of vitamin B1 (thiamine) level in myeloproliferative neoplasms: clinical correlations and impact of JAK2 inhibitor therapy. Blood Cancer J. 9, 11.

Gowin, K., Thapaliya, P., Samuelson, J., et al., 2012. Experience with PEGylated interferon -2a in advanced myeloproliferative neoplasms in an international cohort of 118 patients. Haematologica 97, 1570–1573.

Goyal, H., Chachoua, I., Pecquet, C., Vainchenker, W., Constantinescu, S.N., 2020. A p53-JAK-STAT connection involved in myeloproliferative neoplasm pathogenesis and progression to secondary acute myeloid leukemia. Blood Rev. 42, 100712.

Grinfeld, J., Nangalia, J., Green, A.R., 2017. Molecular determinants of pathogenesis and clinical phenotype in myeloproliferative neoplasms. Haematologica 102, 7–17.

Grinfeld, J., Nangalia, J., Baxter, E.J., et al., 2018. Classification and personalized prognosis in myeloproliferative neoplasms. N. Engl. J. Med. 379, 1416–1430.

Guerini, V., Barbui, V., Spinelli, O., et al., 2007. The histone deacetylase inhibitor ITF2357 selectively targets cells bearing mutated JAK2V617F. Leukemia 22, 740–747.

Guerra, A., Oikonomidou, P.R., Sinha, S., et al., 2019. Lack of Gdf11 does not improve anemia or prevent the activity of RAP-536 in a mouse model of beta-thalassemia. Blood 134, 568–572.

Hammarén, H.M., Virtanen, A.T., Abraham, B., Peussa, H., Hubbard, S.R., Silvennoinen, O., 2018. Janus kinase 2 activation mechanisms revealed by analysis of suppressing mutations. J. Allergy Clin. Immunol. 143, 1549–1559.e6.

Harrison, C., Kiladjian, J.-J., Al-Ali, H., et al., 2012. JAK inhibition with ruxolitinib versus best available therapy for myelofibrosis. N. Engl. J. Med. 366, 787–798.

Harrison, C.N., Schaap, N.P.M., Zweegman, S., et al., 2013. Efficacy and safety of fedratinib (SAR302503/TG101348) in patients with intermediate- or high-risk myelofibrosis (MF), post-polycythemia Vera (PV) MF, or post-essential thrombocythemia (ET) MF previously treated with ruxolitinib: interim results from a phase II study (JAKARTA-2). Blood 122, 661.

Harrison, C.N., Mead, A.J., Panchal, A., et al., 2017. Ruxolitinib vs best available therapy for ET intolerant or resistant to hydroxycarbamide. Blood 130, 1889–1897.

Harrison, C.N., Schaap, N., Vannucchi, A.M., et al., 2019a. Fedratinib induces spleen responses and reduces symptom burden in patients with myeloproliferative neoplasm (MPN)-associated myelofibrosis (MF) and low platelet counts, who were either ruxolitinib-Naïve or were previously treated with ruxolitinib. Blood 134, 668.

Harrison, C.N., Gerds, A.T., Kiladjian, J.-J., et al., 2019b. Pacifica: a randomized, controlled phase 3 study of pacritinib vs. physician's choice in patients with primary myelofibrosis, post polycythemia vera myelofibrosis, or post essential thrombocytopenia myelofibrosis with severe thrombocytopenia (platelet count <50,000/mL). Blood 134, 4175.

Harrison, C.N., Garcia, J.S., Mesa, R.A., et al., 2019c. Results from a phase 2 study of Navitoclax in combination with ruxolitinib in patients with primary or secondary myelofibrosis. Blood 134, 671.

James, C., Ugo, V., Le Couedic, J.P., et al., 2005. A unique clonal JAK2 mutation leading to constitutive signalling causes polycythaemia vera. Nature 434, 1144–1148.

Jeay, S., Ferretti, S., Holzer, P., et al., 2018. Dose and schedule determine distinct molecular mechanisms underlying the efficacy of the p53–MDM2 inhibitor HDM201. Cancer Res. 78, 6257–6267.

Jutzi, J.S., Kleppe, M., Dias, J., Staehle, H.F., Shank, K., Teruya-Feldstein, J., Gambheer, S.M.M., Dierks, C., Rienhoff Jr., H.Y., Levine, R.L., Pahl, H.L., 2018. LSD1 inhibition prolongs survival in mouse models of MPN by selectively targeting the disease clone. Hemasphere 2 (3), e54. https://doi.org/10.1097/HS9.0000000000000054. PMID: 31723778. PMCID: PMC6745991.

Klampfl, T., Gisslinger, H., Harutyunyan, A.S., et al., 2013. Somatic mutations of calreticulin in myeloproliferative neoplasms. N. Engl. J. Med. 369, 2379–2390.

Kleppe, M., Kwak, M., Koppikar, P., et al., 2015. JAK-STAT pathway activation in malignant and nonmalignant cells contributes to MPN pathogenesis and therapeutic response. Cancer Discov. 5, 316–331.

Kleppe, M., Koche, R., Zou, L., et al., 2018. Dual targeting of oncogenic activation and inflammatory signaling increases therapeutic efficacy in myeloproliferative neoplasms. Cancer Cell 33, 785–787.

Koppikar, P., Bhagwat, N., Kilpivaara, O., et al., 2012. Heterodimeric JAK-STAT activation as a mechanism of persistence to JAK2 inhibitor therapy. Nature 489, 155–159.

Koschmieder, S., Mughal, T.I., Hasselbalch, H.C., et al., 2016. Myeloproliferative neoplasms and inflammation: whether to target the malignant clone or the inflammatory process or both. Leukemia 30, 1018–1024.

Kralovics, R., Passamonti, F., Buser, A.S., et al., 2005. A gain-of-function mutation of JAK2 in myeloproliferative disorders. N. Engl. J. Med. 352, 1779–1790.

Kröger, N., Giorgino, T., Scott, B.L., et al., 2015. Impact of allogeneic stem cell transplantation on survival of patients less than 65 years of age with primary myelofibrosis. Blood 125, 3347–3350.

Lakhani, N.J., Rasco, D.W., Zeng, Q., et al., 2020. First-in-human study of palcitoclax (APG-1252), a novel dual Bcl-2/Bcl-xL inhibitor, demonstrated advantages in platelet safety while maintaining anticancer effect in U.S. patients with metastatic solid tumors. J. Clin. Oncol. 38, 3509.

Leroy, E., Constantinescu, S.N., 2017. Rethinking JAK2 inhibition: towards novel strategies of more specific and versatile Janus kinase inhibition. Leukemia 31, 1023–1038.

Leroy, E., Dusa, A., Colau, D., et al., 2016a. Uncoupling JAK2 V617F activation from cytokine-induced signalling by modulation of JH2 alphaC helix. Biochem. J. 473, 1579–1591.

Leroy, E., Defour, J.P., Sato, T., et al., 2016b. His499 regulates dimerization and prevents oncogenic activation by asparagine mutations of the human thrombopoietin receptor. J. Biol. Chem. 291, 2974–2987.

Le Bousse-Kerdilès, M.C., 2012. Primary myelofibrosis and the "bad seeds in bad soil" concept. Fibrogenesis Tissue Repair 5 (Suppl. 1), S20. https://doi.org/10.1186/1755-1536-5-S1-S20. PMID: 23259918. PMCID: PMC3368798.

Leroy, E., Balligand, T., Pecquet, C., Mouton, C., Colau, D., Shiau, A.K., Dusa, A., Constantinescu, S.N., 2019. Differential effect of inhibitory strategies of the V617 mutant of JAK2 on cytokine receptor signaling. J. Allergy Clin. Immunol. 144 (1), 224–235. https://doi.org/10.1016/j.jaci.2018.12.1023. Epub 2019 Jan 29. PMID: 30707971.

Levine, R.L., Wadleigh, M., Cools, J., et al., 2005. Activating mutation in the tyrosine kinase JAK2 in polycythemia vera, essential thrombocythemia, and myeloid metaplasia with myelofibrosis. Cancer Cell 7, 387–397.

Levy, G., Carillo, S., Papoular, B., et al., 2019. A novel activating TPOR mutation in essential thrombocytopenia points to an activating effect of tryptophan and aromatics at the outset of the transmembrane domain. HemaSphere 3, 664.

Li, J., Kent, D.G., Godfrey, A.L., et al., 2014. JAK2V617F homozygosity drives a phenotypic switch in myeloproliferative neoplasms, but is insufficient to sustain disease. Blood 123, 3139–3151.

Ling, T., Crispino, J.D., Zingariello, M., Martelli, F., Migliaccio, A., 2018. GATA1 insufficiencies in primary myelofibrosis and other hematopoietic disorders: consequences for therapy. Expert Rev. Hematol. 11, 169–184.

Lu, M., Wang, J., Li, Y., et al., 2010. Treatment with the Bcl-xL inhibitor ABT-737 in combination with interferon α specifically targets JAK2V617F-positive polycythemia vera hematopoietic progenitor cells. Blood 116, 4284–4287.

Lu, M., Wang, X., Li, Y., et al., 2012. Combination treatment in vitro with Nutlin, a small-molecule antagonist of MDM2, and PEGylated interferon-α 2a specifically targets JAK2V617F-positive polycythemia vera cells. Blood 120, 3098–3105.

Lundberg, P., Karow, A., Nienhold, R., et al., 2014. Clonal evolution and clinical correlates of somatic mutations in myeloproliferative neoplasms. Blood 123, 2220–2228.

Magliulo, D., Bernardi, R., Messina, S., 2018. Lysine-specific demethylase 1A as a promising target in acute myeloid leukemia. Front. Oncol. 8, 255.

Mambet, C., Necula, L., Mihai, S., et al., 2018. Increased Dkk-1 plasma levels may discriminate disease subtypes in myeloproliferative neoplasms. J. Cell. Mol. Med. 22, 4005–4011.

Markham, A., 2020. Luspatercept: first approval. Drugs 80, 85–90.
Marsters, S.A., Pennica, D., Bach, E., Schreiber, R.D., Ashkenazi, A., 1995. Interferon gamma signals via a high-affinity multisubunit receptor complex that contains two types of polypeptide chain. Proc. Natl. Acad. Sci. U. S. A. 92, 5401–5405.
Marty, C., Lacout, C., Martin, A., et al., 2010. Myeloproliferative neoplasm induced by constitutive expression of JAK2V617F in knock-in mice. Blood 116, 783–787.
Marty, C., Saint-Martin, C., Pecquet, C., et al., 2014. Germ-line JAK2 mutations in the kinase domain are responsible for hereditary thrombocytosis and are resistant to JAK2 and HSP90 inhibitors. Blood 123, 1372–1383.
Marty, C., Pecquet, C., Nivarthi, H., et al., 2016. Calreticulin mutants in mice induce an MPL-dependent thrombocytosis with frequent progression to myelofibrosis. Blood 127, 1317–1324.
Mascarenhas, J., Lu, M., Virtgaym, E., et al., 2017. Open label phase I study of single agent Oral RG7388 (idasanutlin) in patients with polycythemia vera and essential thrombocythemia. Blood 130, 254.
Mascarenhas, J., Hoffman, R., Talpaz, M., et al., 2018. Pacritinib vs best available therapy, including ruxolitinib, in patients with myelofibrosis: a randomized clinical trial. JAMA Oncol. 4, 652.
Mascarenhas, J., Kremyanskaya, M., Hoffman, R., et al., 2019a. MANIFEST, a phase 2 study of CPI-0610, a bromodomain and extraterminal domain inhibitor (BETi), as monotherapy or "add-on" to ruxolitinib, in patients with refractory or intolerant advanced myelofibrosis. Blood 134, 670.
Mascarenhas, J., Lu, M., Kosiorek, H., et al., 2019b. Oral idasanutlin in patients with polycythemia vera. Blood 134, 525–533.
Mason, K.D., Carpinelli, M.R., Fletcher, J.I., et al., 2007. Programmed anuclear cell death delimits platelet life span. Cell 128, 1173–1186.
Mesa, R.A., Vannucchi, A.M., Mead, A., Egyed, M., Szoke, A., Suvorov, A., Jakucs, J., Perkins, A., Prasad, R., Mayer, J., Demeter, J., Ganly, P., Singer, J.W., Zhou, H., Dean, J.P., Te Boekhorst, P.A., Nangalia, J., Kiladjian, J.J., Harrison, C.N., 2017. Pacritinib versus best available therapy for the treatment of myelofibrosis irrespective of baseline cytopenias (PERSIST-1): an international, randomised, phase 3 trial. Lancet Haematol. 4 (5), e225–e236. https://doi.org/10.1016/S2352-3026(17)30027-3. Epub 2017 Mar 20. PMID: 28336242. PMCID: PMC8209752.
Meyer, S.C., Keller, M.D., Chiu, S., et al., 2015. CHZ868, a type II JAK2 inhibitor, reverses type I JAK inhibitor persistence and demonstrates efficacy in myeloproliferative neoplasms. Cancer Cell 28, 15–28.
Michalak, M., Groenendyk, J., Szabo, E., Gold, L.I., Opas, M., 2009. Calreticulin, a multi-process calcium-buffering chaperone of the endoplasmic reticulum. Biochem. J. 417, 651–666.
Moliterno, A.R., Williams, D.M., Rogers, O., Spivak, J.L., 2006. Molecular mimicry in the chronic myeloproliferative disorders: reciprocity between quantitative JAK2 V617F and Mpl expression. Blood 108, 3913–3915.
Moraga, I., Wernig, G., Wilmes, S., et al., 2015. Tuning cytokine receptor signaling by re-orienting dimer geometry with surrogate ligands. Cell 160, 1196–1208.
Nakatake, M., Monte-Mor, B., Debili, N., et al., 2011. JAK2(V617F) negatively regulates p53 stabilization by enhancing MDM2 via La expression in myeloproliferative neoplasms. Oncogene 31, 1323–1333.
Nangalia, J., Green, A.R., 2017. Myeloproliferative neoplasms: from origins to outcomes. Blood 130, 2475–2483.
Newberry, K.J., Patel, K., Masarova, L., et al., 2017. Clonal evolution and outcomes in myelofibrosis after ruxolitinib discontinuation. Blood 130, 1125–1131.

Niebel, D., Kirfel, J., Janzen, V., Höller, T., Majores, M., Gütgemann, I., 2014. Lysine-specific demethylase 1 (LSD1) in hematopoietic and lymphoid neoplasms. Blood 124, 151–152.

Nivarthi, H., Chen, D., Cleary, C., et al., 2016. Thrombopoietin receptor is required for the oncogenic function of CALR mutants. Leukemia 30, 1759–1763.

O'Sullivan, J., Taylor, J.A., Gerds, A.T., et al., 2019. Molecular analysis in the pacritinib dose-finding PAC203 study in patients with myelofibrosis refractory or intolerant to ruxolitinib. Blood 134, 4214.

Palandri, F., Breccia, M., Bonifacio, M., et al., 2020. Life after ruxolitinib: reasons for discontinuation, impact of disease phase, and outcomes in 218 patients with myelofibrosis. Cancer 126, 1243–1252.

Pardanani, A.D., Levine, R.L., Lasho, T., et al., 2006. MPL515 mutations in myeloproliferative and other myeloid disorders: a study of 1182 patients. Blood 108, 3472–3476.

Pardanani, A., Tefferi, A., Jamieson, C., Gabrail, N.Y., Lebedinsky, C., Gao, G., Liu, F., Xu, C., Cao, H., Talpaz, M., 2015a. A phase 2 randomized dose-ranging study of the JAK2-selective inhibitor fedratinib (SAR302503) in patients with myelofibrosis. Blood Cancer J. 5 (8), e335. https://doi.org/10.1038/bcj.2015.63. PMID: 26252788. PMCID: PMC4558588.

Pardanani, A., Harrison, C., Cortes, J.E., et al., 2015b. Safety and efficacy of fedratinib in patients with primary or secondary myelofibrosis: a randomized clinical trial. JAMA Oncol. 1, 643–651.

Passamonti, F., Rumi, E., Pietra, D., et al., 2006. Relation between JAK2 (V617F) mutation status, granulocyte activation, and constitutive mobilization of CD34+ cells into peripheral blood in myeloproliferative disorders. Blood 107, 3676–3682.

Pecquet, C., Staerk, J., Chaligne, R., et al., 2010. Induction of myeloproliferative disorder and myelofibrosis by thrombopoietin receptor W515 mutants is mediated by cytosolic tyrosine 112 of the receptor. Blood 115, 1037–1048.

Pecquet, C., Diaconu, C.C., Staerk, J., et al., 2012. Thrombopoietin receptor down-modulation by JAK2 V617F: restoration of receptor levels by inhibitors of pathologic JAK2 signaling and of proteasomes. Blood 119, 4625–4635.

Pecquet, C.B.T., Chachoua, I., et al., 2018. Secreted mutant calreticulins as rogue cytokines trigger thrombopoietin receptor activation specifically in CALR mutated cells: perspectives for MPN therapy. In: 60th ASH Annual Meeting & Exposition. San Diego, CA.

Pecquet, C., Chachoua, I., Roy, A., et al., 2019. Calreticulin mutants as oncogenic rogue chaperones for TpoR and traffic-defective pathogenic TpoR mutants. Blood 133, 2669–2681.

Pemmaraju, N., Carter, B.Z., Kantarjian, H.M., et al., 2019. Final results of phase 2 clinical trial of LCL161, a novel oral SMAC mimetic/IAP antagonist, for patients with intermediate to high risk myelofibrosis. Blood 134, 555.

Pemmaraju, N., Yu, J., Parasuraman, S., Paranagama, D., Kish, J., Visaria, J., Singhal, M., Verstovsek, S., Gerds, A.T., 2020. Ruxolitinib (RUX) retreatment in patients (pts) with myelofibrosis (MF): real-world evidence on pt characteristics and outcomes. J. Clin. Oncol. 38 (15_Suppl), e19535.

Pettit, K., Gerds, A.T., Yacoub, A., et al., 2019. A phase 2a study of the LSD1 inhibitor Img-7289 (bomedemstat) for the treatment of myelofibrosis. Blood 134, 556.

Pikman, Y., Lee, B.H., Mercher, T., et al., 2006. MPLW515L is a novel somatic activating mutation in myelofibrosis with myeloid metaplasia. PLoS Med. 3, e270.

Platzbecker, U., Germing, U., Götze, K.S., et al., 2017. Luspatercept for the treatment of anaemia in patients with lower-risk myelodysplastic syndromes (PACE-MDS): a multicentre, open-label phase 2 dose-finding study with long-term extension study. Lancet Oncol. 18, 1338–1347.

Porpaczy, E., Tripolt, S., Hoelbl-Kovacic, A., et al., 2018. Aggressive B-cell lymphomas in patients with myelofibrosis receiving JAK1/2 inhibitor therapy. Blood 132, 694–706.

Prestipino, A., Emhardt, A.J., Aumann, K., et al., 2018. Oncogenic JAK2(V617F) causes PD-L1 expression, mediating immune escape in myeloproliferative neoplasms. Sci. Transl. Med. 10, eaam7729. https://doi.org/10.1126/scitranslmed.aam7729.

Pronier, E., Cifani, P., Merlinsky, T.R., Berman, K.B., Somasundara, A.V.H., Rampal, R.K., LaCava, J., Wei, K.E., Pastore, F., Maag, J.L., Park, J., Koche, R., Kentsis, A., Levine, R.L., 2018. Targeting the CALR interactome in myeloproliferative neoplasms. JCI Insight 3 (22), e122703. https://doi.org/10.1172/jci.insight.122703. PMID: 30429377. PMCID: PMC6302938.

Quintás-Cardama, A., Vaddi, K., Liu, P., et al., 2010. Preclinical characterization of the selective JAK1/2 inhibitor INCB018424: therapeutic implications for the treatment of myeloproliferative neoplasms. Blood 115, 3109–3117.

Rambaldi, A., Dellacasa, C., Finazzi, G., et al., 2010. A pilot study of the histone-deacetylase inhibitor givinostat in patients with JAK2V617F positive chronic myeloproliferative neoplasms: givinostat treatment for CMNs. Br. J. Haematol. 150, 446–455.

Rampal, R., Al-Shahrour, F., Abdel-Wahab, O., Patel, J.P., Brunel, J.P., Mermel, C.H., Bass, A.J., Pretz, J., Ahn, J., Hricik, T., Kilpivaara, O., Wadleigh, M., Busque, L., Gilliland, D.G., Golub, T.R., Ebert, B.L., Levine, R.L., 2014. Integrated genomic analysis illustrates the central role of JAK-STAT pathway activation in myeloproliferative neoplasm pathogenesis. Blood 123 (22), e123-e33. https://doi.org/10.1182/blood-2014-02-554634. Epub 2014 Apr 16. PMID: 24740812. PMCID: PMC4041169.

Rumi, E., Cazzola, M., 2017. Advances in understanding the pathogenesis of familial myeloproliferative neoplasms. Br. J. Haematol. 178, 689–698.

Saenz, D.T., Fiskus, W., Manshouri, T., et al., 2017. BET protein bromodomain inhibitor-based combinations are highly active against post-myeloproliferative neoplasm secondary AML cells. Leukemia 31, 678–687.

Sangkhae, V., Etheridge, S.L., Kaushansky, K., Hitchcock, I.S., 2014. The thrombopoietin receptor, MPL, is critical for development of a JAK2V617F-induced myeloproliferative neoplasm. Blood 124, 3956–3963.

Schneider, R.K., Mullally, A., Dugourd, A., et al., 2017. Gli1 + mesenchymal stromal cells are a key driver of bone marrow fibrosis and an important cellular therapeutic target. Cell Stem Cell 20, 785.

Scott, L.M., Tong, W., Levine, R.L., et al., 2007. JAK2 exon 12 mutations in polycythemia vera and idiopathic erythrocytosis. N. Engl. J. Med. 356, 459–468.

Skalniak, L., Kocik, J., Polak, J., et al., 2018. Prolonged idasanutlin (RG7388) treatment leads to the generation of p53-mutated cells. Cancer 10, 396.

Spangrude, G.J., Lewandowski, D., Martelli, F., et al., 2015. P-selectin sustains extramedullary hematopoiesis in the gata1 low model of myelofibrosis. Stem Cells (Dayton, Ohio) 34, 67–82.

Speranza, G., Anderson, L., Chen, A.P., et al., 2018. First-in-human study of the epichaperome inhibitor PU-H71: clinical results and metabolic profile. Invest. New Drugs 36, 230–239.

Staerk, J., Defour, J.P., Pecquet, C., et al., 2011. Orientation-specific signalling by thrombopoietin receptor dimers. EMBO J. 30, 4398–4413.

Suragani, R.N.V.S., Cadena, S.M., Cawley, S.M., et al., 2014. Transforming growth factor-β superfamily ligand trap ACE-536 corrects anemia by promoting late-stage erythropoiesis. Nat. Med. 20, 408–414.

Tefferi, A., Vainchenker, W., 2011. Myeloproliferative neoplasms: molecular pathophysiology, essential clinical understanding, and treatment strategies. J. Clin. Oncol. 29, 573–582.

Talpaz, M., Kiladjian, J.J., 2021. Fedratinib, a newly approved treatment for patients with myeloproliferative neoplasm-associated myelofibrosis. Leukemia 35, 1–17. https://doi.org/10.1038/s41375-020-0954-2.

Tefferi, A., Lasho, T.L., Begna, K.H., et al., 2015. A pilot study of the telomerase inhibitor Imetelstat for myelofibrosis. N. Engl. J. Med. 373, 908–919.

Tiedt, R., Hao-Shen, H., Sobas, M.A., et al., 2008. Ratio of mutant JAK2-V617F to wild-type Jak2 determines the MPD phenotypes in transgenic mice. Blood 111, 3931–3940.

Vainchenker, W., Kralovics, R., 2017. Genetic basis and molecular pathophysiology of classical myeloproliferative neoplasms. Blood 129, 667–679.

Vainchenker, W., Leroy, E., Gilles, L., Marty, C., Plo, I., Constantinescu, S.N., 2018. JAK inhibitors for the treatment of myeloproliferative neoplasms and other disorders. F1000Res 7, 82.

Vannucchi, A.M., 2010. From palliation to targeted therapy in myelofibrosis. N. Engl. J. Med. 363, 1180–1182.

Vannucchi, A.M., Kiladjian, J., Griesshammer, M., et al., 2015. Ruxolitinib versus standard therapy for the treatment of polycythemia vera. N. Engl. J. Med. 372, 426–435.

Veletic, I., Manshouri, T., González, G.M., et al., 2018. PD-1-mediated T cell exhaustion is prevalent among patients with MPN-associated myelofibrosis independent of JAK1/2 inhibition. Blood 132, 5487.

Verstovsek, S., Kantarjian, H., Mesa, R.A., et al., 2010. Safety and efficacy of INCB018424, a JAK1 and JAK2 inhibitor in myelofibrosis. N. Engl. J. Med. 363, 1117–1127.

Verstovsek, S., Mesa, R.A., Gotlib, J., et al., 2012. A double-blind, placebo-controlled trial of ruxolitinib for myelofibrosis. N. Engl. J. Med. 366, 799–807.

Wang, Y., Fiskus, W., Chong, D.G., et al., 2009. Cotreatment with panobinostat and JAK2 inhibitor TG101209 attenuates JAK2V617F levels and signaling and exerts synergistic cytotoxic effects against human myeloproliferative neoplastic cells. Blood 114, 5024–5033.

Wang, X., Haylock, D., Hu, C.S., et al., 2016. A thrombopoietin receptor antagonist is capable of depleting myelofibrosis hematopoietic stem and progenitor cells. Blood 127, 3398–3409.

Wang, J.-C., Chen, C., Kundra, A., et al., 2019. Programmed cell death receptor (PD-1) ligand (PD-L1) expression in Philadelphia chromosome-negative myeloproliferative neoplasms. Leuk. Res. 79, 52–59.

Wilmes, S., Hafer, M., Vuorio, J., et al., 2020. Mechanism of homodimeric cytokine receptor activation and dysregulation by oncogenic mutations. Science 367, 643–652.

Yee, K., Martinelli, G., Vey, N., et al., 2014. Phase 1/1b study of RG7388, a potent MDM2 antagonist, in acute myelogenous leukemia (AML) patients (pts). Blood 124, 116.

Zetterberg, E., Verrucci, M., Martelli, F., et al., 2013. Abnormal P-selectin localization during megakaryocyte development determines thrombosis in the gata1low model of myelofibrosis. Platelets 25, 539–547.

Zhang, L., Yang, F., Feng, S., 2020. Allogeneic hematopoietic stem-cell transplantation for myelofibrosis. Ther. Adv. Hematol. 11, 2040620720906002. https://doi.org/10.1177/2040620720906002. PMID: 32110286. PMCID: PMC7019406.

Zingariello, M., Martelli, F., Verachi, P., et al., 2019. Novel targets to cure primary myelofibrosis from studies on gata1 low mice. IUBMB Life 72, 131–141.

Zuber, J., Shi, J., Wang, E., et al., 2011. RNAi screen identifies Brd4 as a therapeutic target in acute myeloid leukaemia. Nature 478, 524–528.

CHAPTER FOUR

Mutational landscape of blast phase myeloproliferative neoplasms (MPN-BP) and antecedent MPN

Sergiu Pasca[†,‡], Helen T. Chifotides[†], Srdan Verstovsek, and Prithviraj Bose*

Leukemia Department, The University of Texas MD Anderson Cancer Center, Houston, TX, United States
*Corresponding author: e-mail address: pbose@mdanderson.org

Contents

Abstract

Myeloproliferative neoplasms (MPN) have an inherent tendency to evolve to the blast phase (BP), characterized by ≥20% myeloblasts in the blood or bone marrow. MPN-BP portends a dismal prognosis and currently, effective treatment modalities are scarce, except for allogeneic hematopoietic stem cell transplantation (allo-HSCT) in selected

[†] The two authors contributed equally to this review and share lead authorship.
[‡] Present address: Iuliu Haţieganu University of Medicine and Pharmacy, Cluj-Napoca, Romania.

International Review of Cell and Molecular Biology, Volume 366
ISSN 1937-6448
https://doi.org/10.1016/bs.ircmb.2021.02.008

patients, particularly those who achieve complete/partial remission. The mutational landscape of MPN-BP differs from *de novo* acute myeloid leukemia (AML) in several key aspects, such as significantly lower frequencies of *FLT3* and *DNMT3A* mutations, and higher incidence of *IDH1/2* and *TP53* in MPN-BP. Herein, we comprehensively review the impact of the three signaling driver mutations (*JAK2* V617F, *CALR* exon 9 indels, *MPL* W515K/L) that constitutively activate the JAK/STAT pathway, and of the other somatic non-driver mutations (epigenetic, mRNA splicing, transcriptional regulators, and mutations in signal transduction genes) that cooperatively or independently promote MPN progression and leukemic transformation. The MPN subtype, harboring two or more high-molecular risk (HMR) mutations (epigenetic regulators and mRNA splicing factors) and "triple-negative" PMF are among the critical factors that increase risk of leukemic transformation and shorten survival. Primary myelofibrosis (PMF) is the most aggressive MPN; and polycythemia vera (PV) and essential thrombocythemia (ET) are relatively indolent subtypes. In PV and ET, mutations in splicing factor genes are associated with progression to myelofibrosis (MF), and in ET, *TP53* mutations predict risk for leukemic transformation. The advent of targeted next-generation sequencing and improved prognostic scoring systems for PMF inform decisions regarding allo-HSCT. The emergence of treatments targeting mutant enzymes (e.g., IDH1/2 inhibitors) or epigenetic pathways (BET and LSD1 inhibitors) along with new insights into the mechanisms of leukemogenesis will hopefully lead the way to superior management strategies and outcomes of MPN-BP patients.

1. Introduction

Myeloproliferative neoplasms in blast phase (MPN-BP), defined by the presence of blasts ≥20% in the bone marrow or peripheral blood (Mesa et al., 2007; Mudireddy et al., 2018), evolve from chronic phase Philadelphia chromosome-negative (Ph^-) MPN (Bose et al., 2021), usually via an intervening accelerated phase (MPN-AP) with blasts in the range ≥10-19% (Shahin et al., 2021; Tam et al., 2009). The classic Ph^- MPN comprise primary myelofibrosis (PMF), which is the most aggressive, and two relatively indolent subtypes, polycythemia vera (PV) and essential thrombocythemia (ET). The reported frequencies of leukemic transformation range from 10–20% for PMF to 2.3–14.4% for PV and 0.7–3% for ET at 10 years (Cerquozzi and Tefferi, 2015; Vallapureddy et al., 2019; Yogarajah and Tefferi, 2017). Clinically, MPN-BP prognosis is dismal—inferior to *de novo* acute myeloid leukemia (AML)—with a median overall survival (OS) ranging approximately from 2 to 8 months in different studies (Bose et al., 2020; Heaney and Soriano, 2013; Lancman et al., 2018; Mascarenhas et al., 2020c; Palandri et al., 2020a,b; Rampal et al., 2018; Tefferi et al., 2018e; Yogarajah and Tefferi, 2017). Currently, there is no standard of care conferring prolonged survival to patients with MPN-BP except for allogeneic hematopoietic stem cell transplantation (allo-HSCT) in selected cases (Alchalby et al., 2014; Chihara et al., 2016; Lancman et al., 2018; Shah et al., 2021; Takagi et al., 2016; Tam et al., 2008).

The risk of transformation to MPN-BP is significantly impacted by several factors that have been incorporated in the Dynamic International Prognostic Scoring System (DIPSS) for PMF (Passamonti et al., 2010), for example: advanced age, subtype of MPN (PMF has the highest risk of leukemic transformation), circulating blasts $\geq$3% (Masarova et al., 2020; Palandri et al., 2020a; Vallapureddy et al., 2019), unfavorable karyotype, moderate to severe anemia, and leukocytosis (Cerquozzi and Tefferi, 2015; Grinfeld et al., 2018; Milosevic and Kralovics, 2013; Yogarajah and Tefferi, 2017). Importantly, as shown in several studies, the mutational profile during chronic phase MPN (MPN-CP) constitutes a significant factor in leukemic transformation (Grinfeld et al., 2018; Iurlo et al., 2019; Tefferi and Pardanani, 2015). The impact of the mutational profile in disease progression is also evidenced by the incorporation of high-molecular risk (HMR) mutations in the Mutation-enhanced International Prognostic Scoring System for transplant-eligible patients aged $\leq$70 years (MIPSS70), the karyotype-enhanced MIPSS70-plus (MIPSS70+), the MIPSS70+ version 2.0, and the Genetically Inspired Prognostic Scoring System (GIPSS) for PMF (Bose and Verstovsek, 2016, 2020; Vallapureddy et al., 2019). MPN-BP is characterized by a distinct mutational profile as compared to *de novo* AML (Dunbar et al., 2020; Milosevic et al., 2012; Rampal et al., 2014a). Particularly striking is the rarity of *FLT3*, *DNMT3A* and *NPM1* mutations and the higher incidence of *JAK2*, *IDH1/2*, *TP53* and *TET2* mutations in MPN-BP compared to *de novo* AML wherein *FLT3*, *NPM1* and *DNMT3A* are frequently detected (Abdel-Wahab et al., 2010; Beer et al., 2010; Gupta et al., 2020; Iurlo et al., 2019; Kantarjian et al., 2021a; Milosevic et al., 2012; Tefferi, 2011; Zhang et al., 2012). The challenging nature of MPN-BP is reflected by the paucity of effective treatments and the fact that the outcomes have not improved substantially in the last 15 years (Iurlo et al., 2019; Tefferi and Pardanani, 2015; Yogarajah and Tefferi, 2017); in contrast, a number of novel treatments have emerged for other subtypes of AML in the last few years (Daver et al., 2020; Kantarjian et al., 2021a,b; Short et al., 2020).

The aforementioned findings demonstrate that *de novo* AML and MPN-BP are distinct diseases and hence, should be managed differently. Although the mechanisms of leukemic transformation have not been fully elucidated, targeted next generation sequencing has provided insights into the mutational spectrum of MPN and MPN-BP (Skov, 2020), and although some novel regimens were explored in recent years, satisfactory treatment of MPN-BP remains a major unmet medical need (Mannelli, 2021; Yogarajah and Tefferi, 2017). The present review aims to shed light on the possible molecular pathogenetic mechanisms and provides an overview of pathogenetically relevant mutations, their associations and impact on clonal dynamics, and prognosis in MPN-BP.

2. Importance of activated JAK/STAT signaling and driver mutations

The somatic mutation V617F in the Janus kinase 2 (*JAK2*) gene and its prominent role in the classic Ph^- MPN was first discovered in 2005 (Baxter et al., 2005; James et al., 2005; Kralovics et al., 2005; Levine et al., 2005). *JAK2* V617F leads to an alteration of the JH2 pseudokinase domain of JAK2 that disrupts its auto-inhibitory activity on the JH1 kinase domain and constitutively activates the enzyme (Chen and Mullally, 2014). Acquisition of *JAK2* V617F in exon 14 or other activating *JAK2* mutations, e.g., in exon 12, directly activate JAK2 signaling, which leads to cytokine-independent proliferation of hematopoietic stem and progenitor cells (Quintás-Cardama and Verstovsek, 2013; Quintás-Cardama et al., 2011). The central role of JAK/signal transducer and activator of transcription (STAT) signaling dysregulation in MPN pathogenesis, irrespective of *JAK2* mutational status or disease phenotype, has been established; JAK/STAT pathway genes were upregulated, due to JAK2 activation, in MPN patients who were homozygous, heterozygous, and wild type for *JAK2* V617F (Kleppe et al., 2015; Rampal et al., 2014b). Studies conducted in *JAK2*-wild-type MPN patients demonstrated that mutations in the myeloproliferative leukemia (*MPL*) gene (encoding the thrombopoietin receptor, which regulates megakaryopoiesis), viz., W515K/L (Pikman et al., 2006) and indels in exon 9 of the calreticulin (*CALR*) gene (Klampfl et al., 2013; Nangalia et al., 2013) account for the activated JAK/STAT signaling in most of these cases. CALR is a calcium-binding endoplasmic reticulum chaperone protein; mutant CALR induces JAK/STAT activation by binding to MPL, the thrombopoietin receptor (Araki et al., 2016; Chachoua et al., 2016; Elf et al., 2016; Marty et al., 2016; Nivarthi et al., 2016). Thus, MPNs are highly dependent on the JAK/STAT pathway (Quintás-Cardama and Verstovsek, 2013; Quintás-Cardama et al., 2011; Rampal et al., 2014b; Skoda et al., 2015), and JAK/STAT signaling remains active in MPN-BP (Lee et al., 2012), but the mutational landscape is more complex compared to that of MPN-CP.

In contrast to *de novo* AML where less than 10% of the patients harbor *JAK2* V617F, the majority of patients who have AML that evolved from an antecedent MPN harbor *JAK2* V617F (Dunbar et al., 2020; McNamara et al., 2018). In the cases of MPN-BP retaining *JAK2* V617F, the leukemic clone may have arisen from clonal evolution of a common progenitor that drove the disease in the chronic phase (Dunbar et al., 2020; Rampal and Mascarenhas, 2014). In a recent study, McKerrell et al. (2017) reported the presence of *JAK2* V617F^{+} hematopoietic clones in 9 out of 12 patients, 4.6 to 15.2 years (median 10.2 years) before the patients were diagnosed with MPN; these findings demonstrate a latency period of *JAK2* V617F clonal hematopoiesis in most MPNs, with varying rates of proliferation. Notably, Williams et al. (2020) recently demonstrated that *JAK2* V617F can be acquired during childhood or even earlier in life, and it can take decades before an MPN becomes clinically manifest.

Hallmark features in the bone marrow of MPN patients are megakaryocytes with moderate to high hyperplasia and variable pleiomorphism, features driving fibrotic progression (Malherbe et al., 2016). *JAK2* V617F^{+} megakaryocytes induced abnormal myeloproliferation of stem/progenitor cells and development of myelofibrosis (MF) features in mice through thrombopoietin/MPL signaling (Zhan et al., 2016; Zhang et al., 2018); and activation of JAK/STAT in megakaryocytes sustained myeloproliferation *in vivo* (Woods et al., 2019). A murine model expressing *JAK2* V617F-mutated long-term hematopoietic stem cells developed a PV phenotype, and all the mice underwent myelofibrotic progression with marked megakaryocytic abnormalities (Mullally et al., 2012). Another study, based on immunostaining of bone marrow biopsies from *CALR*-mutated patients with PMF and ET, showed that *CALR* mutations were preferentially expressed in the megakaryocyte lineage compared to erythroid and myeloid lineages, and that *CALR* expression was significantly upregulated during megakaryocytic differentiation (Vannucchi et al., 2014).

Notably, the megakaryocyte morphology and distribution changed to small, condensed, and non-clumped in the majority of the patients who underwent leukemic transformation, as compared to large and clustered during chronic phase PMF in a cohort of patients at a single institution (Mesa et al., 2005). The abnormal megakaryocytic nuclear morphology and formation of clusters/sheets that are characteristic of PMF may be attributed to unique somatic mutations that the megakaryocytes acquire, apart from the driver mutations (*JAK2*, *CALR*, *MPL*), such as *ASXL1*, *EZH2*, *TET2*, *IDH1*, and *IDH2*, frequently encountered in advanced MPN (Guo et al., 2017). Genes primarily regulating chromatin remodeling, chromosome alignment and stability were up-regulated in the megakaryocytes of MPN patients who manifested increased reticulin deposition in the bone marrow (Guo et al., 2017). In a cohort of 81 patients with MPN, the megakaryocytes exhibited increased proliferation, attenuated apoptosis (enhanced by *CALR* mutations), and dysregulated epigenetic regulation, along with higher expression of the anti-apoptotic protein, B-cell lymphoma extra-large (BCL-X_L) (Malherbe et al., 2016). Moreover, megakaryocyte apoptosis was prevented via the JAK pathway through preservation of BCL-X_L and MCL-1 expression levels (Kodama et al., 2012).

2.1 Targeting the anti-apoptotic/apoptotic machinery, epigenetic modifiers, and JAK/STAT

Expression of BCL-X_L is elevated in MPN cells compared to healthy controls (Petiti et al., 2020; Silva et al., 1998) because BCL-X_L is a downstream effector of activated JAK/STAT (Chen and Mullally, 2014). In AML cell lines, sensitivity to venetoclax (BCL-2 inhibitor) correlated negatively with BCL-X_L levels: HL-60 cells that overexpressed BCL-X_L were fully resistant to venetoclax-induced apoptosis (Pan et al., 2014), and resistance of AML cell lines to venetoclax after chronic exposure was driven by upregulation of BCL-X_L compared to the parental AML cells (Lin et al., 2016). Consistent with resistance to venetoclax mediated by BCL-X_L (Bose et al., 2017; Lin et al., 2016; Pan et al., 2014), venetoclax would not be expected to be effective in treatment of MPN-BP. However, venetoclax in combination with hypomethylating agents has demonstrated notable clinical efficacy in newly diagnosed (DiNardo et al., 2020a) and relapsed/refractory (DiNardo et al., 2020b) AML patients; and has become the standard of care in older/unfit AML patients (Kantarjian et al., 2021a,b). A few recent studies in cohorts with MPN-BP included several patients who underwent allo-HSCT after treatment with venetoclax and hypomethylating agents (Gangat et al., 2020, 2021; Garcia-Horton et al., 2021;Tremblay et al., 2020). However, 18 patients with MPN-BP who were treated with a combination of venetoclax and hypomethylating agents were not able to undergo allo-HSCT in a recent retrospective study reported by Masarova et al. (2021). Notably, a few recent preclinical and clinical studies demonstrated that *IDH1/2* mutations conferred sensitivity of human AML cells to venetoclax and prolonged survival of AML patients treated with venetoclax monotherapy or in combination with hypomethylating agents (Chyla et al., 2018; DiNardo et al., 2020a,c; Konopleva et al., 2016; Pollyea et al., 2019a), in accordance with the strong BCL-2 dependence of *IDH1/2*-mutated AML cells for survival (Chan et al., 2015). Accordingly, *IDH1/2*-mutated MPN-BP patients treated with venetoclax in combination with hypomethylating agents were among the responders who underwent allo-HSCT in three recent studies (Garcia-Horton et al., 2021; Gangat et al., 2020, 2021). Currently, a phase 1/2 clinical study, evaluating venetoclax/ivosidenib with or without azacitidine in *IDH1*-mutated patients with advanced malignancies, including MPN-AP/BP, is underway (Lachowiez et al., 2021).

In preclinical studies, Waibel et al. (2013) showed that concurrent inhibition of BCL-2/BCL-X_L (with ABT-737: BCL-2/BCL-X_L antagonist, a preclinical analog of navitoclax) and JAK2 (with ruxolitinib) overcame resistance to JAK2-inhibitor monotherapy in *JAK2* V617F-expressing MPN cells. Another recent preclinical study, conducted by Petiti et al. (2020), demonstrated synergism of ABT-737 in combination with ruxolitinib in leukocytes of patients with PMF (a significantly higher apoptotic rate was noted with the drug combination). Notably, the addition of navitoclax (potent BCL-2/BCL-X_L inhibitor, clinical analog of ABT-737) to ruxolitinib induced significant responses (improvements in splenomegaly, symptoms and bone marrow fibrosis grade) in patients who had MF-CP and suboptimal response to ruxolitinib, in an ongoing phase 2 clinical trial (Harrison et al., 2021). Clinical trials combining hypomethylating agents with ruxolitinib in primary and secondary MF (Masarova et al., 2018), MPN-AP and MPN-BP (Bose et al., 2020; Mascarenhas et al., 2020c; Rampal et al., 2018) have been conducted and showed that the combinations of ruxolitinib with hypomethylating agents provide viable treatment alternatives to intensive chemotherapy for elderly or unfit patients with MPN-AP/BP.

Another mechanism conferring an advantage to the leukemic clone is the increased genomic instability caused by *JAK2* V617F (Plo et al., 2008). Notably, in cells from MPN patients, *JAK2* V617F induced overexpression of human double minute 2 (HDM2), resulting in degradation of p53 (Nakatake et al., 2012); this biology forms the basis of clinical trials assessing HDM2 inhibitors, which activate p53-dependent apoptosis of tumor cells, in MF (Al-Ali et al., 2020; Verstovsek et al., 2021a) and other malignancies (Konopleva et al., 2020). Navtemadlin (formerly KRT-232), a small-molecule HDM2 inhibitor, in combination with low-dose cytarabine or decitabine, is currently being assessed in a phase 1b/2 clinical trial (NCT04113616) in relapsed/

refractory AML patients, including MPN-BP and *JAK2*-positive AML, but excluding *TP53*-mutated patients (*TP53* mutations are more common in MPN-BP compared to MPN-CP; Milosevic and Kralovics, 2013; Rampal et al., 2014a). In addition, navtemadlin is currently being studied in *TP53*-wild type MF patients refractory/resistant to prior JAK inhibitor treatment. The phase 2 portion of this trial is now complete (Al-Ali et al., 2020), and the trial has entered its pivotal phase 3 portion in which navtemadlin is compared to best available therapy excluding JAK inhibitors (Verstovsek et al., 2021a); navtemadlin is additionally being evaluated as an "add on" to ruxolitinib in MF patients who had a suboptimal response to ruxolitinib in a phase 1b/2 study (Mascarenhas et al., 2020d; NCT04485260).

2.2 Targeting telomerase

Other mechanisms, such as deregulation of telomerase activity, provide a proliferative advantage to AML cells (Kishtagari and Watts, 2017; Yamada and Kawauchi, 2013). Telomerase, which is upregulated through the JAK-STAT pathway in hematologic malignancies (Spanoudakis et al., 2011; Yamada and Kawauchi, 2013), is inhibited by imetelstat. Treatment with imetelstat impaired maturation and inhibited the secretion of fibrogenic factors by MPN megakaryocytes *ex vivo* (Mosoyan et al., 2017). Wang et al. (2018) showed that imetelstat selectively depleted MF stem and progenitor cells in immunodeficient mice, thus enabling potential restoration of normal hematopoiesis (Wang et al., 2018). Imetelstat yielded encouraging clinical efficacy in a single-center, pilot study in patients with PMF in the chronic phase (Tefferi et al., 2015) and appeared to improve the median OS (29.9 months for the cohort treated with the higher dose, 9.4 mg/kg every 3 weeks) of patients with intermediate-2 or high-risk MF relapsed/refractory to JAK inhibitors in a larger, multi-center phase 2 clinical trial (IMbark™) (Mascarenhas et al., 2021). A pivotal phase 3 trial (IMpactMF) was recently launched to further evaluate the possible survival advantage of imetelstat in JAK inhibitor-refractory MF patients (Mascarenhas et al., 2020a).

3. Driver mutations and beyond in MPN and leukemic transformation

3.1 Driver mutations

Besides universal activation of the JAK/STAT pathway and the classic JAK/STAT activating mutations (*JAK2*, *MPL*, *CALR*) (Rampal et al., 2014b), there is mounting evidence that other key oncogenic mutations cooperate in MPN-CP and MPN-BP pathogenesis (Bartels et al., 2021; Dunbar et al., 2021; Grinfeld et al., 2017, 2018; Milosevic and Kralovics, 2013; Tefferi and Pardanani, 2015). In a recent study that identified distinct genetic subgroups from a cohort of 2035 patients with MPN (the vast majority had ET), only 45% of the patients harbored mutated *JAK2*, *MPL*, and *CALR* as the sole abnormalities (Grinfeld et al., 2018). Notably, "triple-negative" patients (*JAK2-/CALR-/MPL*-wild type) with PMF (10–15% of the cases), the vast majority of whom harbor mutations in non-driver (usually epigenetic and splicing) genes, have the highest incidence of leukemic transformation (Rumi et al., 2014; Tefferi et al., 2014c, 2016a) and the worst OS (median 2.3 vs 15.9, and 5.9 years for "triple-negative" vs *CALR*-, and *JAK2*-mutated PMF) (Tefferi et al., 2014a). In patients with PMF, *CALR* mutations conferred longer OS and leukemia free survival (LFS) as compared to *JAK2* and *MPL* mutations (Rumi et al., 2014; Tefferi et al., 2014a). The superior outcome associated with type 1/type 1-like *CALR* mutations, even in the presence of the HMR mutations *ASXL1/SRSF2*, was also validated in a cohort of 386 patients with PMF (Tefferi et al., 2018a). The association of type 1/type 1-like *CALR* mutations with superior OS in patients with MF was also demonstrated in another study (median survival 26.4 years vs 7.2 years for type 1/type 1-like *CALR* vs *JAK2* V617F mutations, respectively) (Guglielmelli et al., 2015). The favorable impact of type 1/type 1-like *CALR* mutations is reflected by allocation of points for their absence in the MIPSS70, MIPSS70+, MIPSS70 + version 2.0, GIPSS, and MYSEC-PM prognostic models (Bose and Verstovsek, 2020). Apart from harboring *CALR* mutations, a high *JAK2* V617F allele burden ($\geq$50%) was associated with significantly better OS compared to low allele burden in patients with PMF (Guglielmelli et al., 2009; Rozovski et al., 2017; Tefferi et al., 2008).

3.2 Non-driver mutations

The non-driver oncogenic mutations occur in genes classified as epigenetic regulators (*ASXL1, EZH2, TET2, IDH1/2, DNMT3A*), mRNA splicing (*SRSF2, SF3B1, U2AF1, ZRSR2*) (Hautin et al., 2020), and transcriptional (*TP53, RUNX1, IKZF1*) regulators as well as mutations in signal transduction genes (*NRAS/KRAS, PTPN11*), and negative regulators of JAK/STAT signaling (*CBL, LNK* or *SH2B3*) (Milosevic and Kralovics, 2013; Vainchenker and Kralovics, 2017). After *JAK2* and *CALR*, the most frequently encountered mutated genes in MPN are involved in epigenetic regulation and RNA splicing (Grinfeld et al., 2017, 2018; Lundberg et al., 2014). Several non-driver mutations, such as *ASXL1*, *TP53*, *IDH1/2*, *EZH2* and *TET2*, are more prevalent in MPN-BP than MPN-CP (Tefferi and Pardanani, 2015). Vannucchi et al. (2013) studied two cohorts of 879 PMF patients; the authors found that *ASXL1*, *EZH2*, *IDH1*, *IDH2* and *SRSF2*, harbored by 25–30% of all PMF patients, were associated with increased risk of leukemic transformation and premature death. The role of these five HMR mutations, alongside *TP53* and *TET2* in leukemic transformation was corroborated in subsequent reports (Li et al., 2018; Schieber et al., 2019; Vainchenker and Kralovics, 2017; Vallapureddy et al., 2019). However, only *ASXL1* remained prognostically significant (hazard ratio 2.02) in the context of the International Prognostic Scoring System (IPSS) (Vannucchi et al., 2013). In another study, the detection of *ASXL1*, *TP53*, *IDH1/2*, *RUNX1*, and *SRSF2* mutations in *JAK2* V617F^{+} PV/ET patients at baseline and acquisition of new mutations during follow-up were associated with higher probability of leukemic transformation (83% of the patients progressed to AML); in contrast, the strongest predictor of progression to MF in PV/ET patients was a gradual increase of the *JAK2* V617F allele burden during cytoreductive therapy (hazard ratio 10.8) (Senín et al., 2018).

3.3 Clonal evolution of MPN-BP

Despite the immense progress that has been made in understanding the clonal evolution of MPNs, the specific genetic and epigenetic events as well as the complex clonal dynamics associated with leukemic transformation have not been elucidated (Dunbar et al., 2020). The following models have been proposed for transformation of MPN-CP to MPN-BP: (a) evolution of the malignant MPN-BP clone from the ancestral MPN-CP clone after acquisition of additional mutations, and (b) independent founder clones give rise to MPN-CP and MPN-BP clones after undergoing separate genetic events (Beer et al., 2010; Dunbar et al., 2020; Jain and Rampal, 2020; Meggendorfer et al., 2020; Milosevic and Kralovics, 2013; Mylonas et al., 2020; Rampal and Mascarenhas, 2014). The latter model is supported by studies on paired samples from a substantial number of patients demonstrating that a *JAK2* V617F^{+} MPN-CP transformed to *JAK2* V617F^{-} MPN-BP in a subset of patients (Beer et al., 2009, 2010; Campbell et al., 2006; Jamal et al., 2008; Li et al., 2008; Pan et al., 2020; Tam et al., 2008; Theocharides et al., 2007).

Consistent with the aforementioned models (Dunbar et al., 2020; Rampal and Mascarenhas, 2014), mutations in the epigenetic modifier genes *TET2* and *DNMT3A* were shown to have been primarily acquired prior to *JAK2* V617F (Vainchenker and Kralovics, 2017) or to have co-existed with the driver mutation in separate clones (biclonal) whereas *ASXL1*, *IDH1*, and *EZH2* mutations occurred before, after or concurrently with *JAK2* V617F in separate subclones (Lundberg et al., 2014). Abdel-Wahab et al. (2010) conducted studies on paired MPN-CP and MPN-BP samples and showed that *TET2* mutations were acquired during leukemic transformation in several cases and in others preceded *JAK2*, whereas *ASXL1* mutations were detected in both MPN-CP and MPN-BP clones from the same patients. The same study on paired samples led to the conclusion that *ASXL1*, *TET2* and *IDH1* mutations were common in MPN-BP and that the mutational order of events in MPN-CP and MPN-BP differed in various patients (Abdel-Wahab et al., 2010). The importance of mutational order in MPN was further corroborated in a study by Ortmann et al. (2015), showing that the order of *JAK2* and *TET2* acquisition influenced clinical presentation, hematopoietic stem/progenitor cell biology, and clonal evolution of MPN. These investigators showed that either *TET2* or *JAK2* could be the first mutation in the three MPN subtypes; however, the patients who acquired *JAK2* before *TET2* were significantly more likely to present with PV than ET (Ortmann et al., 2015). In another study, it was concluded that *TET2* preceded *JAK2* V617F in the MPN clone, given that *TET2*-mutant/*JAK2*-wild type and *TET2*-mutant/*JAK2*-mutant clones but not *JAK2*-mutant/*TET2*-wild type alleles were identified in the patients (Delhommeau et al., 2009). Based on these findings, it was inferred that a *JAK2*-wild type, *TET2*-mutant clone could transform to MPN-BP in patients with *JAK2*-mutant/*TET2*-mutant MPN (Delhommeau et al., 2009). Notably, in a preclinical study, *Jarid2* deletion preceded by other mutations (*JAK2* V617F or *IDH2* R140Q) promoted MPN acceleration or drove leukemic transformation in hematopoietic progenitor cells of an immunodeficient mouse model (Celik et al., 2018). Besides the tumor suppressor *JARID2* (Celik et al., 2018), Marinaccio et al. (2021) showed that liver kinase B1 (LKB1)/serine threonine kinase 11 (*STK11*) also is a hematopoietic tumor suppressor in MPN. In this study, deletion of *Stk11* in a *MPL* W515L mouse model of MPN induced intense fibrosis, osteosclerosis, accumulation of immature blasts (>20%) in the bone marrow, and bone marrow failure as well as rapid lethality; in addition, loss of *Stk11* enhanced engraftment of primary human MPN cells in immunocompromised mice;

these findings suggest the possible role of *STK11* in collaboration with activated JAK/STAT signaling in leukemic transformation (Marinaccio et al., 2021).

A recent study demonstrated that *JAK2* V617F was acquired earlier in PMF and PV as opposed to ET where the mutation was a later event (Grinfeld et al., 2018). The same study showed that the other two driver mutations (*CALR* and *MPL*) were more frequently acquired early in the MPN phase whereas *TP53*, *PPM1D* and *NRAS* mutations occurred much later (Grinfeld et al., 2018). The presence of *PPM1D* mutations in "triple negative" ET further underscored the importance of clonality in MPN (Grinfeld et al., 2018). The influence of timing that driver and non-driver mutations were acquired in disease phenotype was also demonstrated by Nangalia et al. (2015) who genotyped specimens from MPN patients over time: patients who acquired *DNMT3A* mutations first tended to present with ET whereas the vast majority of patients who had *JAK2* V617F arise in wild type *DNMT3A* clones presented with PV. Regarding driver mutations and phenotype, although *JAK2* V617F is found in PV (95%), PMF and ET (50-60%), PV is its classic phenotypic consequence, characterized by proliferation of all three main myeloid lineages (predominantly erythroid, but also granulocytic and megakaryocyte/platelet) (James et al., 2005; Mullally et al., 2010). *CALR* is the second most common driver mutation in PMF (25–30%) and ET (20–25%), whereas *MPL* is encountered in PMF and ET at much lower frequencies (2–5%); the megakaryocyte/platelet lineage is the main target of the *CALR* and *MPL* mutations (Grinfeld et al., 2017; Vainchenker and Kralovics, 2017).

Besides the order and type of mutation acquisition (for example, driver mutation first, epigenetic-modifying mutation second or in the reverse order) (Dunbar et al., 2021), the number of mutations play an important role in disease phenotype and progression; for example, in most patients with PV and ET, one driver mutation was found, whereas three or more somatic mutations were found in the majority of PMF cases (Vainchenker and Kralovics, 2017). Newberry et al. (2017) demonstrated that MF patients who exhibited clonal evolution had significantly shorter survival after ruxolitinib discontinuation compared to those who did not acquire new mutations (6 vs 16 months, respectively). In other studies, a higher number of somatic mutations correlated with shorter LFS (Pan et al., 2020) and OS (Lundberg et al., 2014) in MPN patients. In particular, 80% of PMF patients harbored adverse non-driver mutations, e.g., *ASXL1*, *SRSF2*, *CBL*, *RUNX1*, and these were quantitatively and qualitatively associated with considerably inferior LFS (25% vs 4% 7-year risk) and OS (median 3.6 vs 8.5 years) (Tefferi et al., 2016a). The prognostic value of the number of mutated genes was also shown in an international study of 797 patients with PMF; presence of two or more mutated HMR genes (*ASXL1*, *EZH2*, *SRSF2* and *IDH1/2*) was correlated with significantly shorter median LFS as compared to one or no mutations in these genes (2.6 vs 7, and 12.3 years, respectively) (Guglielmelli et al., 2014). Another study showed that patients with MPN-BP had inferior survival, irrespective of treatment modality, when they harbored ≥4 mutations (McNamara et al., 2018). In a targeted, deep sequencing study in patients with PV and ET, approximately half of the patients harbored non-driver mutations; *ASXL1*, *SRSF2*, and *IDH2* in PV, and *SH2B3*, *SF3B1*, *U2AF1*, *TP53*, *IDH2*, and *EZH2* in ET were identified as "adverse," with a combined prevalence of 15% in each MPN subtype (Tefferi et al., 2016b); these adverse variants/mutations impacted myelofibrosis-free survival (MFFS), LFS and OS in PV and ET (Tefferi et al., 2016b). In another recent study, the presence of mutations in splice factor (*SRSF2*, *U2AF1*) or epigenetic (*IDH1/2*) modifier genes in combination with mutations in driver genes at the time of diagnosis was associated with rapid leukemic transformation (Bartels et al., 2021). Patients with primary and secondary MF harbor *ASXL1* and *TET2* mutations more frequently than those with PV or ET (Abdel-Wahab et al., 2012). Recently, Pan et al. (2020) studied the mutational profiles of 49 patients during MPN-BP that evolved from PV or ET and paired specimens of 29 patients from the same cohort during MPN-CP by targeted next-generation sequencing. The cohort with MPN-BP was classified in short-term, intermediate, and long-term genomic subgroups according to the length of time to develop MPN-BP. The short-term subgroup harbored *RUNX1*, *IDH1/2*, *TET2*, and *U2AF1*, but not *TP53*, with significantly higher variant allele frequencies (VAFs) during MPN-CP. In the intermediate and long term subgroups, 53% and 50% of the patients harbored mutated *TP53*, respectively; and *TP53* VAFs increased as compared to MPN-CP in both subgroups (Pan et al., 2020).

Finally, aging has been associated with hematopoietic stem cell deregulation, loss-of-function, and proliferation (Xie et al., 2014). Age-related *JAK2* V617F clonal hematopoiesis was detected in about 0.5% of the general population (Hinds et al., 2016); however, a recent study that employed digital droplet polymerase chain reaction showed a higher prevalence of *JAK2* V617F (3.2%) in the general population (Cordua et al., 2019). Moreover, age-related clonal hematopoiesis with somatic mutations in three epigenetic regulators (*ASXL1*, *DNMT3A* and *TET2*) was detected in healthy individuals and increased appreciably with age (Genovese et al., 2014; Jaiswal et al., 2014; McKerrell et al., 2015; Young et al., 2016). Another study demonstrated the strong association of low-burden mutant *TP53* incidence with age in MPN patients; also, mutant *TP53* was either undetectable or present at low allele burden at MPN diagnosis, but increased considerably during leukemic transformation in a subset of patients (Kubesova et al., 2018).

3.4 Mutations affecting signal transduction and transcription regulators

Mutations in the tumor suppressor gene *TP53* are generally regarded as an adverse prognostic factor in multiple malignancies and have been implicated in MPN progression (Hidalgo López et al., 2018; Marcellino et al., 2018; McNamara et al., 2018; Milosevic and Kralovics, 2013; Milosevic et al., 2012; Pan et al., 2020; Rampal et al., 2014a). A retrospective study of 108 patients with MPN-BP showed that *TP53* was the only mutation that correlated with shorter OS (McNamara et al., 2018). *TP53* deregulation in patients with PV or ET is a risk factor for evolution to MF or MPN-BP (Marcellino et al., 2018). Harutyunyan et al. (2011) analyzed specimens from patients with MPN and showed that *TP53* mutations were strongly associated with transformation to MPN-BP ($P=0.003$). Moreover, in a large study that included 502 patients with ET, it was concluded that *TP53* mutations predict leukemic transformation (Tefferi et al., 2020). Loss of heterozygosity in *TP53* has been strongly correlated with leukemic transformation in MPN (Lundberg et al., 2014; Marcellino et al., 2018; Milosevic and Kralovics, 2013). Recently, Gupta et al. (2020) demonstrated that *TP53*-mutated patients with MPN-BP who underwent allo-HSCT had a significantly inferior 3-year OS and cumulative incidence of relapse compared to *TP53*-wild type patients.

The potential of concurrent *JAK2* V617F and *TP53* mutations to drive progressive disease is underscored by the fact that overexpression of *JAK2* V617F in *TP53*-wild type mice induced PV and led to leukemic transformation thereafter (Tsuruta-Kishino et al., 2017). Importantly, Rampal et al. (2014a) reported that *TP53* mutations were recurrent (44%) in *JAK2* V617$^+$ MPN-BP specimens (but not in MPN-CP), that the double-mutant *TP53*/*JAK2* V617F clone was dominant and the *TP53* VAF was significantly higher in MPN-BP (57%) than in MPN-CP (7%); furthermore, *JAK2* V617F in combination with *Tp53* loss synergized to induce AML in a mouse model. Consistent with the previous results, a preclinical study showed that *JAK2* V617F and homozygous *TP53* mutations existed concurrently in erythroleukemia and megakaryocytoblastic leukemic cell lines (Zhao et al., 2012). Furthermore, in a recent study, co-occurring *TP53* mutations, *JAK2* V617F and *JAK2* "variants" (e.g., R1063H, L393V, N1108S) were detected in a cohort of MPN patients who displayed higher cumulative incidence of leukemic transformation (Benton et al., 2019).

In a recent single-cell analysis of clonal evolution in MF patients, Mylonas et al. (2020) showed that *NRAS/KRAS* mutations were acquired in the patients who progressed to MPN-AP/BP. Santos et al. (2020) conducted a study on the impact of RAS-pathway mutations in MF. The study showed that patients with *NRAS/KRAS* mutations had a higher incidence of leukemic transformation at 3 years compared to patients without the mutations (18% vs 11%, respectively), and the 3-year OS was lower in the *NRAS/KRAS*-mutated cohort compared to wild type (34% vs 58%, respectively). In addition, *PTPN11* mutations (besides *NRAS/KRAS*) were associated with inferior 3-year OS in the mutant cohort compared to wild type (19% vs 55%) (Santos et al., 2020). In another recent study, Coltro et al. (2020) demonstrated that MF patients who harbored *NRAS/KRAS/CBL* (*RAS/CBL*) mutations had inferior OS and a significantly higher 5-year cumulative incidence of leukemic transformation compared to the wild type cohort. In a targeted next-generation sequencing study, Lasho et al. (2018) showed that *PTPN11* mutations were enriched in MPN-BP. In addition, *RUNX1* and *PTPN11* mutations/variants were associated with shorter survival (hazard ratio 2.1 and 3.0, respectively) in patients with MPN-BP. Guglielmelli et al. (2019) demonstrated that MPN patients harboring *RUNX1* mutations had a median LFS of 2.8 years, whereas the median was not reached (median follow-up 4.9 years) in MPN patients with wild type *RUNX1*. In another preclinical study, activating mutations in *Ptpn11* promoted development of MPN and disease progression in the bone marrow mesenchymal stem/progenitor cells of mice (Dong et al., 2016). Deletions in *IKZF1*, the gene that encodes the transcription factor IKAROS, are significantly more common in MPN-BP compared to MPN-CP (Tefferi and Pardanani, 2015), suggesting a possible driver effect in leukemic transformation via upregulation of JAK/STAT signaling (Jäger et al., 2009; Tefferi et al., 2010; Vainchenker and Kralovics, 2017).

3.5 Mutations in epigenetic regulators

Mutations in epigenetic regulators occur frequently in MPN and have been associated with disease progression and leukemic transformation (Dunbar et al., 2021; Patel et al., 2015; Skov, 2020). DNA methyltransferases

(DNMT) methylate cytosine residues at CpG sites, constituting a transcriptional regulation mechanism. Genes commonly mutated in MPN that are also involved in DNA methylation include *DNMT3A*, *TET2*, *IDH1*, and *IDH2*. *DNMT3A* is a key gene involved in DNA methylation and a driver mutation that is common in AML (Ley et al., 2010). *DNMT3A* mutations are found during both MPN-CP and MPN-BP in about 17% of the patients (Tefferi and Pardanani, 2015), but they are less frequent compared to *de novo* AML (Lasho et al., 2018; Stegelmann et al., 2011; Yogarajah and Tefferi, 2017). The pathogenetic contribution of *DNMT3A* mutations in MPN-BP is not clear as they were also detected in early hematopoietic clones from healthy individuals along with *TET2* and *ASXL1* (Young et al., 2016). A recent study showed that CpG sites in 14 common genes were hypermethylated in patients who had any MPN-CP subtype compared to healthy controls (Pérez et al., 2013). Additional analyses of serial MPN specimens in CP and upon leukemic transformation revealed more hypermethylation in MPN-BP compared to MPN-CP (44% vs 4%, respectively) (Pérez et al., 2013).

Loss-of-function *TET2* mutations result in reduced levels of 5-hydroxymethylcytosine in DNA and ultimately lead to DNA hypermethylation. TET enzymes are critical for myeloid and erythroid differentiation, and homozygous as well as heterozygous *TET2* mutations frequently occur in hematologic malignancies (Delhommeau et al., 2009). *TET2* mutations are found across all MPN subtypes (Grinfeld et al., 2017; Tefferi and Pardanani, 2015) and occur at increased frequency in leukemic transformation (Abdel-Wahab et al., 2010; Couronné et al., 2010). Furthermore, mutant *TET2* was detected in the ancestral founder clones that gave rise to both *JAK2* V617F$^+$ MPN and *JAK2* V617F$^-$ AML clones (Beer et al., 2010; Delhommeau et al., 2009). In a murine model, concomitant *JAK2* V617F and *TET2* loss accelerated the disease compared to MPN induced by *JAK2* V617F only (Kameda et al., 2015). Loss-of-function *TET2* mutations are mutually exclusive with *IDH* mutations because 2-hydroxyglutarate, the abnormal oncometabolite produced by mutant IDH enzymes, inhibits TET enzymes (Figueroa et al., 2010). Interestingly, a recent study demonstrated that JAK2 phosphorylates TET2 within its catalytic core, and that the *JAK2* V617F mutation is associated with increased TET2 activity in erythroid progenitors (Jeong et al., 2019).

IDH1 and *IDH2* mutations lead to accumulation of the oncometabolite 2-hydroxyglutarate, elicit histone hypermethylation, and promote leukemogenesis (Medeiros et al., 2017). Oncogenic *IDH1/2* mutations occur more frequently in MPN-BP as compared to MPN-CP (20–25% vs ~4%, respectively) (Dunbar et al., 2020; Pardanani et al., 2010; Tefferi, 2010). In addition, concomitant presence of *JAK2* V617F and *IDH1/2* mutations induced progressive MPN in a mouse model (McKenney et al., 2018); and *IDH1/2* and *JAK2* V617F mutations were noted in patients with MPN-CP that transformed to MPN-BP (Bartels et al., 2020b; Green and Beer, 2010; Tefferi et al., 2012), revealing a potential underlying mechanism of leukemogenesis. *IDH1/2* mutations were among the original HMR mutations for PMF (Vannucchi et al., 2013) and were included in the newer prognostic scoring systems for PMF (reviewed by Bose and Verstovsek, 2020); and mutant *IDH2* has been associated with inferior survival in both PV and ET (Tefferi et al., 2016b). A hazard ratio of 4.3 was determined for mutant *IDH1* with respect to leukemic transformation in a recent study; and a three-tiered predictive model was developed based on studying 1306 patients with PMF (Vallapureddy et al., 2019). Currently, *IDH1/2* are therapeutically targetable mutations given that the recently approved oral IDH1 and IDH2 inhibitors, ivosidenib (DiNardo et al., 2018; Roboz et al., 2020) and enasidenib (Pollyea et al., 2019b; Stein et al., 2017, 2019), respectively, confer significant clinical benefits and have expanded the therapeutic landscape for patients with *IDH1/2*-mutated AML (Issa and DiNardo, 2021). A preclinical study showed synergism between ruxolitinib and enasidenib in double-mutant *IDH2/JAK2* V617F MPN and MPN-BP patient-derived cells (McKenney et al., 2018). IDH1/2 inhibitors as single agents or in combination with other drugs (such as ruxolitinib, hypomethylating agents, intensive chemotherapy or venetoclax) elicited significant clinical responses, in a retrospective study of 12 patients with newly diagnosed or relapsed/refractory *IDH1/2*-mutated MPN-BP at our institution (Chifotides et al., 2020). In this study, three of the seven patients treated in the frontline setting with IDH1/2-inhibitor combinations achieved complete remission and a median duration of response 17.5 months and median OS 19 months. In the frontline cohort, an elderly patient with *JAK2*-mutated MF received 50 cycles of ruxolitinib/azacitidine. Upon disease progression to MPN-BP and detection of mutated *IDH2*, enasidenib was added to the regimen; treatment with the triple combination resulted in sustained complete remission (molecular and hematological), negative measurable residual disease, and reversal to MF (Chifotides et al., 2020). In addition, stable disease with ≥ 50% blast reduction was noted in 2 and 3 patients in the frontline and relapsed/refractory settings, respectively (Chifotides et al., 2020). Another recent retrospective study showed that treatment with enasidenib monotherapy (and with azacitidine in one case) conferred durable survivals in a small cohort of patients with *IDH2*-mutated MPN-AP/BP (Patel et al., 2020). Furthermore, a small subgroup of patients with *IDH1*-mutated MPN-BP and *JAK2* V617F at baseline showed high response rates after treatment with ivosidenib monotherapy in a phase 1 study (Choe et al., 2020). A phase 2 clinical trial evaluating the combination of enasidenib with ruxolitinib in *IDH2*-mutated patients with MPN-AP/BP and MF-CP (circulating blasts 4-9%) was recently launched (NCT04281498).

ASXL1 is involved in regulation of histone methylation; mutations in the *ASXL1* gene are the second most frequent non-driver mutations in MPN after *TET2* (Grinfeld et al., 2018; Lundberg et al., 2014). *ASXL1* mutations increase the risk of leukemic transformation and death in PMF (Tefferi et al., 2016a; Vannucchi et al., 2013), as well as in PV and ET (Tefferi et al., 2016b). A hazard ratio of 2.0 was calculated for mutated *ASXL1* in a predictive model for leukemic transformation in patients with PMF (Vallapureddy et al., 2019). *ASXL1* was one of the five original HMR mutations for MF (Vannucchi et al., 2013) with an approximate frequency of 20% in patients with PMF and MPN-BP (Abdel-Wahab et al., 2010; Tefferi and Pardanani, 2015). However, Lasho et al. (2018) reported 47% frequency of *ASXL1* mutations in a cohort of patients with MPN-BP treated at the Mayo Clinic, and Vannucchi et al. (2013) detected *ASXL1* mutations in 42–45% of the patients who had high risk PMF and were assessed for risk of leukemic transformation in two large cohorts. *ASXL1* mutations were determined to be adverse in both PMF (Tefferi et al., 2016a) and PV (Tefferi et al., 2016b); and they have been associated with both initiation and progression of disease, clonal evolution, and poor outcomes (Newberry et al., 2017; Vainchenker and Kralovics, 2017). In a recent study that included 479 MF patients, Paz and colleagues (2021) demonstrated that sole *ASXL1* mutations had a moderate impact regarding leukemic transformation and did not have prognostic impact on the risk of death; however, *ASXL1* mutations in association with mutated-*TP53* or high-risk genes resulted in an increased risk of death. In a study employing risk-stratification by *ASXL1* and *CALR* mutational status only, patients with *ASXL1*$^{+}$ *CALR*$^{-}$ PMF had the worst outcomes (median OS 2.3 years) (Tefferi et al., 2014b,c). Concurrent *ASXL1* and *EZH2* mutations in hematopoietic stem and progenitor cells of patients with PMF preceded acquisition of *JAK2* V617F and *CALR* mutations and promoted expansion of pre-leukemic clones and impaired hematopoiesis regardless of MPN drivers (Triviai et al., 2019). Another noteworthy study showed that PV patients, harboring both *JAK2* V617F^{+} and mutated *ASXL1*, experienced cooperative acceleration of disease and shorter LFS compared to *JAK2*

V617F$^+$ *ASXL1*-wild type PV patients (Guo et al., 2019). Recent experiments showed direct interaction of the mutant ASXL1 truncated protein with bromodomain-containing protein 4 (BRD4), thus leading to upregulation of genes involved in myeloid differentiation; these findings indicate that targeting *ASXL1* truncation mutations could be a viable therapeutic approach in myeloid malignancies (Yang et al., 2018).

EZH2 is a histone methyltransferase (H3K27) and one of the catalytic subunits of the polycomb repressive complex 2 (PRC2). PRC2 is an important protein complex that functions as an epigenetic regulator of gene silencing by catalyzing trimethylation of histone H3 lysine 27 (Milosevic and Kralovics, 2013; Shi et al., 2017). JARID2 encodes a cofactor of the PRC2 complex and functions as a hematopoietic tumor suppressor via repression of progenitor cells (Celik et al., 2018); *JARID2* deletions have been associated with leukemic transformation in preclinical (Celik et al., 2018) and clinical (Puda et al., 2012) studies. Loss-of-function mutations in *EZH2* were detected more frequently in MPN-BP compared to MPN-CP (Lasho et al., 2018), and they have been associated with inferior outcomes in MF (Guglielmelli et al., 2011) and ET (Tefferi et al., 2016b). *EZH2* mutations were among the original HMR mutations in PMF (Vannucchi et al., 2013) and were included in the newer prognostic models that incorporate mutational information (reviewed by Bose and Verstovsek, 2020). *EZH2* mutations play a role in MPN disease initiation and progression (Vainchenker and Kralovics, 2017). Cooperation of *Ezh2* loss with *JAK2* V617F was noted in development of MF in *JAK2* V617F-expressing mice (Sashida et al., 2016; Shimizu et al., 2016); however, bromodomain and extra-terminal (BET) protein inhibitors abrogated the MF phenotype in this model, suggesting that these agents may be particularly effective in improving outcomes of MF patients harboring *EZH2* mutations (Sashida et al., 2016). In another mouse model, *EZH2* and *NRAS* G12D mutations cooperatively drove progression of MPN to lethal myeloid leukemias (Gu et al., 2019). Notably, a recent study showed a strong selective co-occurrence advantage of mutations in the epigenetic modifiers *EZH2* and *ASXL1* in progenitors from MF patients (70%) and demonstrated the strong impact of double mutants in promoting pre-leukemic clones (Triviai et al., 2019).

3.5.1 Preclinical and clinical studies on combinations targeting epigenetic modifiers

In preclinical studies, Saenz et al. (2017a,b) demonstrated that the BET protein inhibitor JQ1 or the BET proteolysis targeting chimera (BET-PROTAC) ARV-825 in combination with ruxolitinib synergistically induced apoptosis of patient-derived MPN-BP cells; and the combination of JQ1 with ruxolitinib significantly increased the survival of immunodeficient mice engrafted with patient-derived MPN-BP cells. Kleppe et al. (2018) showed that treatment with the combination of ruxolitinib and a BET protein inhibitor (JQ1) synergistically reduced MPN-driven inflammation and disease burden; and eliminated bone marrow fibrosis in murine models of MF. Fiskus et al. (2021) demonstrated that concomitant treatment of patient-derived MPN-BP cells with a lysine-specific (histone) demethylase 1 (LSD1) inhibitor and a BET protein inhibitor or ruxolitinib was synergistically lethal; and that the combination of LSD1 and BET inhibitors improved the survival of mice engrafted with BET inhibitor-persister/resistant MPN-BP cells. In addition, treatment of a tertiary transplant MPN-BP retroviral transduction murine model with the LSD1 inhibitor CPI-482 significantly improved survival compared to ruxolitinib (Rampal et al., 2020). In other preclinical studies, adaptive or innate resistance of MPN-BP cells to BET protein inhibition was overcome by inhibiting the β-catenin-TCF7L2-JMJD6-c-Myc axis (Saenz et al., 2020); and cotreatment with a BET-PROTAC (ARV-771) and the small molecule tegavivint (disruptor of β-catenin binding) induced synergistic lethality and improved the survival of mice engrafted with BET inhibitor-persister/resistant MPN-BP cells (Saenz et al., 2020) or JAK inhibitor-persister/resistant MPN-BP cells (Saenz et al., 2019). Moreover, dual inhibition of BET proteins and the heat shock protein 90 was synergistically lethal against ruxolitinib-persister/resistant MPN-BP cells (Saenz et al., 2017a).

The combination of the selective BET inhibitor pelabresib (formerly CPI-0610) and ruxolitinib has been evaluated in the ongoing global phase 2 study (MANIFEST) in MF patients who were JAK inhibitor-naïve (Kremyanskaya et al., 2021; Mascarenhas et al., 2020b) or had experienced a suboptimal response to ruxolitinib (Verstovsek et al., 2020). Interim analyses of the data demonstrated significant anemia benefits; and improvement in splenomegaly and symptoms (Kremyanskaya et al., 2021; Mascarenhas et al., 2020b; Verstovsek et al., 2020, 2021b). In light of the promising results reported in the MANIFEST study, the global, randomized phase 3 trial assessing pelabresib in combination with ruxolitinib in the frontline setting (MANIFEST-2) is currently underway.

3.6 Mutations in RNA splicing factors

The spliceosome genes *SRSF2*, *SF3B1* and *U2AF1* are more frequently mutated in PMF as compared to PV and ET (Delic et al., 2016; Hautin et al., 2020; Tefferi and Pardanani, 2015). *SRSF2* mutations have been detected more often in MPN-BP as compared to *de novo* AML, although they are also frequently observed in AML arising from myelodysplastic syndromes (Dunbar et al., 2020; Lindsley et al., 2015; Zhang et al., 2012). The association of *SRSF2* mutations with leukemic transformation and death, particularly in PMF (Vannucchi et al., 2013; Zhang et al., 2012), was reported in several studies. In both multivariate and univariate analyses on paired samples from a cohort of MPN patients, *SRSF2* mutations remained an adverse prognostic factor, and *SRSF2*-mutated MPN patients had a shorter time to MPN-BP compared to wild type patients (51.2 vs 133.8 months, respectively) (Venton et al., 2018). Indeed, in a recent study, a hazard ratio of 3 was allocated to *SRSF2* mutations in a prognostic model for leukemic transformation in PMF patients (Vallapureddy et al., 2019). *SRSF2* mutations were among the original HMR mutations for PMF (Vannucchi et al., 2013) and were included in the newer prognostic models for PMF that incorporate mutational data (reviewed by Bose and Verstovsek, 2020). Mutated *SRSF2* and *SF3B1* had significant associations with LFS in univariate analysis of data from PMF patients (Tefferi et al., 2016a). In addition, mutated *SRSF2* and *SF3B1* were correlated with inferior survival in PV and ET, respectively (Tefferi et al., 2016b). In accordance with the above, *SRSF2* and *SRSF2*/*SF3B1* mutations emerged as adverse factors predicting inferior MFFS and OS in PV and ET, respectively; and they were incorporated in the three-tiered MIPSS-PV and MIPSS-ET prognostic models, respectively (Tefferi et al., 2020). Notably, in two recent studies, Bartels et al. (2018, 2020a) detected spliceosome gene mutations in the bone marrow of patients with PV and PMF who developed worsening fibrosis, a hallmark of disease progression in MPN. Specifically, 22% of the patients with PMF ($N=77$), with fibrotic progression during the course of the disease, acquired *SRSF2* (13/77), *SF3B1* (1/77), and *U2AF1* (3/77) mutations, whereas these mutations were not detected in patients without fibrotic progression; and acquisition of *SRSF2* mutations was statistically significant ($P=0.0192$) (Bartels et al., 2020a). In the PV cohort, 21.4% of the patients who developed fibrosis and progressed to MF harbored mutations in the splicing factor genes (*SRSF2*, *U2AF1*) in the prefibrotic specimens, suggesting that spliceosome gene mutations herald progressive disease in PV; moreover, the cohort that progressed to MF exhibited increased dynamics of subclonal composition (Bartels et al., 2018). In two other studies, Bartels et al. (2020b, 2021) showed the significant correlation of *SRSF2* and *U2AF1* with progression of MPN-CP to MPN-BP and the significantly shorter LFS in patients who also harbored driver mutations as compared to patients without mutations in splicing factor genes.

Przychodzen et al. (2013) suggested that mutated *U2AF1* play a significant role in myeloid leukemogenesis via mis-splicing of tumor-associated genes. *U2AF1* mutations were detected in 16% of patients with PMF and were associated with anemia (Tefferi and Pardanani, 2015; Tefferi et al., 2016a). Notably, patients with PMF (aged ≤70 years) had significantly adverse OS when they harbored the point mutation *U2AF1* Q157 as compared to mutations at S34 or wild type *U2AF1* (median 3.8 vs 6 vs 6.7 years, respectively) (Tefferi et al., 2018b). These results led to the recognition of *U2AF* Q157 as an adverse non-driver mutation and its incorporation in the MIPSS70+ version 2.0 and GIPSS prognostic models for PMF (Bose and Verstovsek, 2020; Tefferi et al., 2018c,d). Additionally, *U2AF1* Q157 was identified as adverse in ET patients regarding LFS and OS and was incorporated in the MIPSS-ET prognostic model (Tefferi et al., 2016b, 2020).

4. Conclusions

MPN-BP remains a challenging disease, arising from all three classic MPN subtypes (though most commonly MF) and has a dismal prognosis due to its complex pathogenesis, its distinct heterogeneous molecular landscape compared to that of *de novo* AML, and the dearth of effective treatments. Founded on partial unraveling of the biological underpinnings of myeloid leukemogenesis, mounting evidence underscores the key roles of the signaling driver mutations (*JAK2*, *CALR*, *MPL*) in concert with other HMR mutations, primarily in epigenetic regulators and RNA splicing factors, in disease progression. Elegant studies on paired bone marrow specimens from the chronic and blast phases have unveiled the heterogeneity and dynamics of the mutational profiles during the two phases, evidenced by acquisition and/or loss of mutations during clonal evolution from MPN-CP to MPN-BP (for example, acquisition of *IDH1/2* and loss of *JAK2* V617F in MPN-BP); these studies indicate that the dominant chronic and blast phase clones may arise from the same ancestral founder clone or from unrelated clones that underwent separately clonal evolution. Several studies have associated *ASXL1*, *IDH1*, *IDH2*, *EZH2*, *SRSF2*, and *U2AF1* Q157 mutations as well as "triple-negative" status for the three driver mutations with leukemogenesis. Co-occurrence of two or more HMR mutations and absence of *CALR* type 1/type 1-like mutations are among the adverse factors incorporated in the newer prognostic models for primary and secondary MF. In PV and ET, mutations in splicing factor genes are associated with progression to MF and worse OS, and in ET, *TP53* mutations predict leukemic transformation.

Molecular profiling of patients with MPN-BP is becoming increasingly more important in light of the necessity to attain remission, followed by allo-HSCT in eligible patients, for a realistic chance of cure. Integration of mutational information into risk stratification and design of targeted therapeutic strategies is becoming imperative in the burgeoning era of treatments for myeloid malignancies, for example, IDH1/2, EZH1/2, splicing and BET inhibitors (Marcellino and Mascarenhas, 2019; Odenike, 2018; Pan et al., 2020; Saenz et al., 2020; Scherber and Mesa, 2020; Schieber et al., 2019; Short et al., 2020; Taylor and Lee, 2019). Ongoing and future clinical trials will guide the development of novel, targeted and synergistic drug combinations (for example, ruxolitinib with enasidenib, BET or LSD1 inhibitors), and personalized therapies to improve prognosis of MPN-BP.

Acknowledgments

This work was supported, in part, by the MD Anderson Cancer Center Support Grant P30 CA016672 from the National Cancer Institute (National Institutes of Health).

Disclosures

S.V. received research funding from Sierra Oncology, Incyte, Roche, NS Pharma, Celgene (now BMS), Gilead, Promedior, CTI BioPharma, AbbVie, Blueprint Medicines, Novartis, PharmaEssentia, Italfarmaco, Protagonist Therapeutics, Constellation Pharmaceuticals, and Kartos Therapeutics. P.B. reports honoraria from Incyte, Celgene (now BMS), CTI BioPharma, Kartos, Novartis, Sierra Oncology, and Blueprint Medicines Corporation; and research support from Incyte, Celgene (now BMS), CTI BioPharma, Kartos, Blueprint Medicines, Constellation, Astellas, Pfizer, NS Pharma, and Promedior.

References

Abdel-Wahab, O., Manshouri, T., Patel, J., Harris, K., Yao, J., Hedvat, C., Heguy, A., Bueso-Ramos, C., Kantarjian, H., Levine, R.L., Verstovsek, S., 2010. Genetic analysis of transforming events that convert chronic myeloproliferative neoplasms to leukemias. Cancer Res. 70, 447–452.

Abdel-Wahab, O., Tefferi, A., Levine, R.L., 2012. Role of *TET2* and *ASXL1* mutations in the pathogenesis of myeloproliferative neoplasms. Hematol. Oncol. Clin. North Am. 26, 1053–1064.

Al-Ali, H.K., Delgado, R.G., Lange, A., Pluta, A., McLornan, D.P., Vachhani, P., Laurent Damaj, G., Jost, P.J., Rejto, L., Hus, M., Grosicki, S., Hernández-Boluda, J.C., Ewing, J., McGreivy, J., Rothbaum, W.P., Qamoos, H., Kilajian, J.-J., Verstovsek, S., 2020. KRT-232, a first-in-class, murine double minute 2 inhibitor (MDM2i), for myelofibrosis (MF) relapsed or refractory (R/R) to Janus-associated kinase inhibitor (JAKi) treatment. HemaSphere 4, S215.

Alchalby, H., Zabelina, T., Stübig, T., Van Biezen, A., Bornhäuser, M., Di Bartolomeo, P., Beelen, D., Cahn, J.Y., Dreger, P., Schroyens, W., De Witte, T., Olavarria, E., Kröger, N., 2014. Allogeneic stem cell transplantation for myelofibrosis with leukemic transformation: a study from the Myeloproliferative Neoplasm Subcommittee of the CMWP of the European Group for Blood and Marrow Transplantation. Biol. Blood Marrow Transplant. 20, 279–287.

Araki, M., Yang, Y., Masubuchi, N., Hironaka, Y., Takei, H., Morishita, S., Mizukami, Y., Kan, S., Shirane, S., Edahiro, Y., Sunami, Y., Ohsaka, A., Komatsu, N., 2016.

Activation of the thrombopoietin receptor by mutant calreticulin in *CALR*-mutant myeloproliferative neoplasms. Blood 127, 1307–1316.

Bartels, S., Faisal, M., Büsche, G., Schlue, J., Kreipe, H., Lehmann, U., 2018. Fibrotic progression in polycythemia vera is associated with early concomitant driver-mutations besides *JAK2*. Leukemia 32, 556–558.

Bartels, S., Faisal, M., Büsche, G., Schlue, J., Hasemeier, B., Schipper, E., Vogtmann, J., Westphal, L., Lehmann, U., Kreipe, H., 2020a. Mutations associated with age-related clonal hematopoiesis in PMF patients with rapid progression to myelofibrosis. Leukemia 34, 1364–1372.

Bartels, S., Vogtmann, J., Schipper, E., Büsche, G., Schlue, J., Lehmann, U., Kreipe, H., 2020b. Combination of myeloproliferative neoplasm driver gene activation of splice factor or epigenetic modifier genes increases risk of rapid blastic progression. Eur. J. Haematol. 106, 520–528.

Bartels, S., Vogtmann, J., Schipper, E., Büsche, G., Schlue, J., Lehmann, U., Kreipe, H., 2021. Combination of myeloprolifeartive neoplasm driver gene activation with mutations of splice factor or epigenetic modifier genes increases risk of rapid blastic progression. Eur. J. Haematol. 106, 520–528.

Baxter, E.J., Scott, L.M., Campbell, P.J., East, C., Fourouclas, N., Swanton, S., Vassiliou, G.S., Bench, A.J., Boyd, E.M., Curtin, N., Scott, M.A., Erber, W.N., Green, A.R., 2005. Acquired mutation of the tyrosine kinase JAK2 in human myeloproliferative disorders. Lancet 365, 1054–1061.

Beer, P.A., Jones, A.V., Bench, A.J., Goday-Fernandez, A., Boyd, E.M., Vaghela, K.J., Erber, W.N., Odeh, B., Wright, C., McMullin, M.F., Cullis, J., Huntly, B.J.P., Harrison, C.N., Cross, N.C.P., Green, A.R., 2009. Clonal diversity in the myeloproliferative neoplasms: independent origins of genetically distinct clones. Br. J. Haematol. 144, 904–908.

Beer, P.A., Delhommeau, F., Lecouédic, J.-P., Dawson, M.A., Chen, E., Bareford, D., Kušec, R., McMullin, M.F., Harrison, C.N., Vannucchi, A.M., Vainchenker, W., Green, A.R., 2010. Two routes to leukemic transformation after a *JAK2* mutation–positive myeloproliferative neoplasm. Blood 115, 2891–2900.

Benton, C.B., Boddu, P.C., DiNardo, C.D., Bose, P., Wang, F., Assi, R., Pemmaraju, N., Kc, D., Pierce, S., Patel, K., Konopleva, M., Ravandi, F., Garcia-Manero, G., Kadia, T.M., Cortes, J., Kantarjian, H.M., Andreeff, M., Verstovsek, S., 2019. Janus kinase 2 variants associated with the transformation of myeloproliferative neoplasms into acute myeloid leukemia. Cancer 125, 1855–1866.

Bose, P., Verstovsek, S., 2016. The evolution and clinical relevance of prognostic classification systems in myelofibrosis. Cancer 122, 681–692.

Bose, P., Verstovsek, S., 2020. Mutational profiling in myelofibrosis: implications for management. Inter. J. Hematol. 111, 192–199.

Bose, P., Gandhi, V.V., Konopleva, M.Y., 2017. Pathways and mechanisms of venetoclax resistance. Leuk. Lymphoma 58, 2026–2039.

Bose, P., Verstovsek, S., Cortes, J.E., Tse, S., Gasior, Y., Jain, N., Jabbour, E.J., Estrov, Z., Alvarado, Y., DiNardo, C.D., Pemmaraju, N., Kornblau, S.M., Kadia, T.M., Daver, N.G., Naqvi, K., Short, N.J., Masarova, L., Villareal, J., Pierce, S.A., Nogueras-Gonzalez, G., Huang, X., Garcia-Manero, G., Kantarjian, H.M., Ravandi, F., 2020. A phase 1/2 study of ruxolitinib and decitabine in patients with post-myeloproliferative neoplasm acute myeloid leukemia. Leukemia 34, 2489–2492.

Bose, P., Masarova, L., Amin, H.M., Verstovsek, S., 2021. Philadelphia chromosome-negative myeloproliferative neoplasms (Chapter 6). In: Kantarjian, H.M., Wolff, R.A., Rieber, A.G. (Eds.), The MD Anderson Manual of Medical Oncology, fourth ed.,. China, McGraw-Hill Education. in press.

Campbell, P.J., Baxter, E.J., Beer, P.A., Scott, L.M., Bench, A.J., Huntly, B.J.P., Erber, W.N., Kusec, R., Larsen, T.S., Giraudier, S., Le Bousse-Kerdiles, M.C., Griesshammer, M., Reilly, J.T., Cheung, B.Y., Harrison, C.N., Green, A.R., 2006. Mutation of *JAK2* in the myeloproliferative disorders: timing, clonality studies, cytogenetic associations, and role in leukemic transformation. Blood 108, 3548–3555.

Celik, H., Koh, W.K., Kramer, A.C., Ostrander, E.L., Mallaney, C., Fisher, A.C., Xiang, J., Wilson, W.C., Martens, A., Kothari, A., Fishberger, G., Tycksen, E., Karpova, D., Duncavage, E.J., Lee, Y., Oh, S.T., Challen, G.A., 2018. JARID2 functions as a tumor suppresor in myeloid neoplasms by repressing self-renewal in hematopoietic progenitor cells. Cancer Cell 34, 741–756.

Cerquozzi, S., Tefferi, A., 2015. Blast transformation and fibrotic progression in polycythemia vera and essential thrombocythemia: a literature review of incidence and risk factors. Blood Cancer J. 5, e366.

Chachoua, I., Pecquet, C., El-Khoury, M., Nivarthi, H., Albu, R.-I., Marty, C., Gryshkova, V., Defour, J.-P., Vertenoeil, G., Ngo, A., Koay, A., Raslova, H., Courtoy, P.J., Choong, M.L., Plo, I., Vainchenker, W., Kralovics, R., Constantinescu, S.N., 2016. Thrombopoietin receptor activation by myeloproliferative neoplasm associated calreticulin mutants. Blood 127, 1325–1335.

Chan, S.M., Thomas, D., Corces-Zimmerman, M.R., Xavy, S., Rastogi, S., Hong, W.-J., Zhao, F., Medeiros, B.C., Tyvoll, D.A., Majeti, R., 2015. Isocitrate dehydrogenase 1 and 2 mutations induce BCL-2 dependence in acute myeloid leukemia. Nat. Med. 21, 178–184.

Chen, E., Mullally, A., 2014. How does *JAK2*V617F contribute to the pathogenesis of myeloproliferative neoplasms. Hematol. Am. Soc. Hematol. Educ. Program 2014 (1), 268–276.

Chifotides, H.T., Masarova, L., Alfayez, M., Daver, N., Alvarado, Y., Jabbour, E., Konopleva, M., Kantarjian, H.M., Patel, K.P., DiNardo, C.D., Verstovsek, S., 2020. Outcome of patients with *IDH1/2*-mutated post-myeloproliferative neoplasm AML in the era of IDH inhibitors. Blood Adv. 4, 5336–5342.

Chihara, D., Kantarjian, H.M., Newberry, K.J., Ravandi, F., Daver, N.G., Bose, P., DiNardo, C.D., Pemmaraju, N., Kadia, T.M., Borthakur, G., Konopleva, M., Garcia-Manero, G., Popat, U.R., Cortes, J.E., Verstovsek, S., 2016. Survival outcome of patients with acute myeloid leukemia transformed from myeloproliferative neoplasms. Blood 128, 1940.

Choe, S., Wang, H., DiNardo, C.D., Stein, E.M., De Botton, S., Roboz, G.J., Altman, J.K., Mims, A.S., Watts, J.M., Pollyea, D.A., Fathi, A.T., Tallman, M.S., Kantarjian, H.M., Stone, R.M., Quek, L., Konteatis, Z., Dang, L., Nicolay, B., Nejad, P., Liu, G., Zhang, V., Liu, H., Goldwasser, M., Liu, W., Marks, K., Bowden, C., Biller, S.A., Attar, E.C., Wu, B., 2020. Molecular mechanisms mediating relapse following ivosidenib monotherapy in IDH1-mutant relapsed or refractory AML. Blood Adv. 4, 1894–1905.

Chyla, B., Daver, N., Doyle, K., McKeegan, E., Huang, X., Ruvolo, V., Wang, Z., Chen, K., Souers, A., Leverson, J., Potluri, J., Boghaert, E., Bhathena, A., Konopleva, M., Popovic, R., 2018. Genetic biomarkers of sensitivity and resistance to venetoclax monotherapy in patients with relapsed acute myeloid leukemia. Am. J. Hematol. 93, E202–E205.

Coltro, G., Rotunno, G., Mannelli, L., Mannarelli, C., Fiaccabrino, S., Romagnoli, S., Bartalucci, N., Ravenda, E., Gelli, E., Sant'Antonio, E., Patnaik, M.M., Tefferi, A., Vannucchi, A.M., Guglielmelli, P., 2020. *RAS/CBL* mutations predict resistance to JAK inhibitors in myelofibrosis and are associated with poor prognostic features. Blood Adv. 4, 3677–3687.

Cordua, S., Kjaer, L., Skov, V., Pallisgaard, N., Hasselbalch, H.C., Ellervik, C., 2019. Prevalence and phenotypes of *JAK2* V617F and *calreticulin* mutations in a Danish general population. Blood 134, 469–479.

Couronné, L., Lippert, E., Andrieux, J., Kosmider, O., Radford-Weiss, I., Penther, D., Dastugue, N., Mugneret, F., Lafage, M., Gachard, N., Nadal, N., Bernard, O.A., Nguyen-Khac, F., 2010. Analyses of *TET2* mutations in post-myeloproliferative neoplasm acute myeloid leukemias. Leukemia 24, 201–203.

Daver, N., Wei, A.H., Pollyea, D.A., Fathi, A.T., Vyas, P., DiNardo, C.D., 2020. New directions for emerging therapies in acute myeloid leukemia: the next chapter. Blood Cancer J. 10, 107.

Delhommeau, F., Dupont, S., Valle, V.D., James, C., Trannoy, S., Massé, A., Kosmider, O., Le Couedic, J.-P., Robert, F., Alberdi, A., Lécluse, Y., Plo, I., Dreyfus, F.J., Marzac, C., Casadevall, N., Lacombe, C., Romana, S.P., Dessen, P., Soulier, J., Viguié, F., Fontenay, M., Vainchenker, W., Bernard, O.A., 2009. Mutation in *TET2* in myeloid cancers. N. Engl. J. Med. 360, 2289–2301.

Delic, S., Rose, D., Kern, W., Nadarajah, N., Haferlach, C., Haferlach, T., Meggendorfer, M., 2016. Application of an NGS-based 28-gene panel in myeloproliferative neoplasms reveals distinct mutation patterns in essential thrombocythaemia, primary myelofibrosis and polycythaemia vera. Br. J. Haematol. 175, 419–426.

DiNardo, C.D., Stein, E.M., de Botton, S., Roboz, G.J., Altman, J.K., Mims, A.S., Swords, R., Collins, R.H., Mannis, G.N., Pollyea, D.A., Donnellan, W., Fathi, A.T., Pigneux, A., Erba, H.P., Prince, G.T., Stein, A.S., Uy, G.L., Foran, J.M., Traer, E., Stuart, R.K., Arellano, M.L., Slack, J.L., Sekeres, M.A., Willekens, C., Choe, S., Wang, H., Zhang, V., Yen, K.E., Kapsalis, S.M., Yang, H., Dai, D., Fan, B., Goldwasser, M., Liu, H., Agresta, S., Wu, B., Attar, E.C., Tallman, M.S., Stone, R.M., Kantarjian, H.M., 2018. Durable remissions with ivosidenib in *IDH1*-mutated relapsed or refractory AML. N. Engl. J. Med. 378, 2386–2398.

DiNardo, C.D., Jonas, B.A., Pullarkat, V., Thirman, M.J., Garcia, J.S., Wei, A.H., Konopleva, M., Döhner, H., Letai, A., Fenaux, P., Koller, E., Havelange, V., Leber, B., Esteve, J., Wang, J., Pejsa, V., Hájek, R., Porkka, K., Illés, Á., Lavie, D., Lemoli, R.M., Yamamoto, K., Yoon, S.-S., Jang, J.-H., Yeh, S.-P., Turgut, M., Hong, W.-J., Zhou, Y., Potluri, J., Pratz, K.W., 2020a. Azacitidine and venetoclax in previously untreated acute myeloid leukemia. N. Engl. J. Med. 383, 617–629.

DiNardo, C.D., Maiti, A., Rausch, C.R., Pemmaraju, N., Naqvi, K., Daver, N.G., Kadia, T.M., Borthakur, G., Ohanian, M., Alvarado, Y., Issa, G.C., Montalban-Bravo, G., Short, N.J., Yilmaz, M., Bose, P., Jabbour, E.J., Takahashi, K., Burger, J.A., Garcia-Manero, G., Jain, N., Kornblau, S.M., Thompson, P.A., Estrov, Z., Masarova, L., Sasaki, K., Verstovsek, S., Ferrajoli, A., Weirda, W.G., Wang, S.A., Konoplev, S., Chen, Z., Pierce, S.A., Ning, J., Qiao, W., Ravandi, F., Andreeff, M., Welch, J.S., Kantarjian, H.M., Konopleva, M.Y., 2020b. 10-Day decitabine with venetoclax for newly diagnosed intensive chemotherapy ineligible, and relapsed or refractory acute myeloid leukaemia: a single-centre, phase 2 trial. Lancet Haematol. 7, e724–e736.

DiNardo, C.D., Tiong, I.S., Quaglieri, A., Mac-Raild, S., Loghavi, S., Brown, F.C., Thijssen, R., Pomilio, G., Ivey, A., Salmon, J.M., Glytsou, C., Fleming, S.A., Zhang, Q., Ma, H., Patel, K.P., Kornblau, S.M., Xu, Z., Chua, C.C., Chen, X., Blombey, P., Flensburg, C., Cummings, N., Aifantis, I., Kantarjian, H., Huang, D.C.S., Roberts, A.W., Majewski, I.J., Konopleva, M., Wei, A.H., 2020c. Molecular patterns of response and treatment failure after frontline venetoclax combinations in older patients with AML. Blood 135, 791–803.

Dong, L., Yu, W.-M., Zheng, H., Loh, M.L., Bunting, S.T., Pauly, M., Huang, G., Zhou, M., Broxmeyer, H.E., Scadden, D.T., Qu, C.-K., 2016. Leukaemogenic effects of *Ptpn11* activating mutations in the stem cell microenvironment. Nature 539, 304–308.

Dunbar, A.J., Rampal, R.K., Levine, R., 2020. Leukemia secondary to myeloproliferative neoplasms. Blood 136, 61–70.
Dunbar, A., Park, Y., Levine, R., 2021. Epigenetic dysregulation of myeloprolifeartive neoplasms. Hematol. Oncol. Clin. N. Am. 35, 237–251.
Elf, S., Abdelfattah, N.S., Chen, E., Perales-Patón, J., Rosen, E.A., Ko, A., Peisker, F., Florescu, N., Giannini, S., Wolach, O., Morgan, E.A., Tothova, Z., Losman, J.-A., Schneider, R.K., Al-Shahrour, F., Mullally, A., 2016. mutant Calreticulin requires both its mutant C-terminus and the thrombopoietin receptor for oncogenic transformation. Cancer Disc. 6, 368–381.
Figueroa, M.E., Abdel-Wahab, O., Lu, C., Ward, P.S., Patel, J., Shih, A., Li, Y., Bhagwat, N., Vasanthakumer, A., Fernandez, H.F., Tallman, M.S., Sun, Z., Wolniak, K., Peeters, J.K., Liu, W., Choe, S.E., Fantin, V.R., Paietta, E., Löwenberg, B., Licht, J.D., Godley, L.A., Delwel, R., Vaik, P.J.M., Thompson, C.B., Levine, R., Melnick, A., 2010. Leukemic *IDH1* and *IDH2* mutations result in a hypermethylation phenotype, disrupt TET2 function, and impair hematopoietic differentiation. Cancer Cell 18, 553–567.
Fiskus, W., Mill, C.P., Nabet, B., Perera, D., Birdwell, C., Manshouri, T., Lara, B., Kadia, T.M., DiNardo, C., Takahashi, K., Daver, N., Bose, P., Masarova, L., Pemmaraju, N., Kornblau, S., Borthakur, G., Montalban-Bravo, G., Garcia Manero, G., Sharma, S., Stubbs, M., Su, X., Green, M.R., Coarfa, C., Verstovsek, S., Khoury, J.D., Vakoc, C.R., Bhalla, K.N., 2021. Superior efficacy of co-targeting GFI1/KDM1A and BRD4 against AML and post-MPN secondary AML cells. Blood Cancer J. 11, 98.
Gangat, N., Morsia, E., Foran, J.M., Palmer, J.M., Elliott, M.A., Tefferi, A., 2020. Venetoclax plus hypomethylating agent in blast-phase myeloproliferative neoplasm: preliminary experience with 12 patients. Br. J. Haematol. 191, e120–e124.
Gangat, N., Guglielmelli, P., Szuber, N., Begna, K.H., Patnaik, M.M., Litzow, M.R., Al-Kali, A., Foran, J.M., Palmer, J.M., Alkhateeb, H., Elliott, M.A., Hanson, C.A., Pardanani, A., Mannelli, F., Vannucchi, A.M., Tefferi, A., 2021. Venetoclax with azacitidine or decitabine in blast-phase myeloproliferative neoplasm: A multicenter series of 32 consecutive cases. Am. J. Hematol. 96, 781–789.
Garcia-Horton, A., Maze, D., McNamara, C.J., Sibai, H., Gupta, V., Murphy, T., 2021. Azacitidine and venetoclax for the treatment of accelerated and blast phase myeloproliferative neoplasms and chronic myelomonocytic leukemia: a case series. Leuk. Lymphoma 62, 1525–1527.
Genovese, G., Kähler, A.K., Handsaker, R.E., Lindberg, J., Rose, S.A., Bakhoum, S.F., Chambert, K., Mick, E., Neale, B.M., Fromer, M., Purcell, S.M., Svantesson, O., Landén, M., Höglund, M., Lehmann, S., Gabriel, S.B., Moran, J.L., Lander, E.S., Sullivan, P.F., Sklar, P., Grönberg, H., Hultman, C.M., McCarroll, S.A., 2014. Clonal hematopoiesis and blood-cancer risk inferred from blood DNA sequence. N. Engl. J. Med. 371, 2477–2487.
Green, A., Beer, P., 2010. Somatic mutations of *IDH1* and *IDH2* in the leukemic transformation of myeloproliferative neoplasms. N. Engl. J. Med. 362, 369–370.
Grinfeld, J., Nangalia, J., Green, A.R., 2017. Molecular determinants of pathogenesis and clinical phenotype in myeloproliferative neoplasms. Haematologica 102, 7–17.
Grinfeld, J., Nangalia, J., Baxter, E.J., Wedge, D.C., Angelopoulos, N., Cantrill, R., Godfrey, A.L., Papaemmanuil, E., Gundem, G., MacLean, C., Cook, J., O'Neil, L., O'Meara, S., Teague, J.W., Butler, A.P., Massie, C.E., Williams, N., Nice, F.L., Andersen, C.L., Hasselbalch, H.C., Guglielmelli, P., McMullin, M.F., Vannucchi, A.M., Harrison, C.N., Gerstung, M., Green, A.R., Campbell, P.J., 2018. Classification and personalized prognosis in myeloproliferative neoplasms. N. Engl. J. Med. 379, 1416–1430.
Gu, Z., Liu, Y., Cai, F., Patrick, M., Zmajkovic, J., Cao, H., Zhang, Y., Tasdogan, A., Chen, M., Qi, L., Liu, X., Li, K., Lyu, J., Dickerson, K.E., Chen, W., Ni, M., Merritt, M.E., Morrison, S.J., Skoda, R.C., Deberardinis, R.J., Xu, J., 2019. Loss of

EZH2 reprograms BCAA metabolism to drive leukemic transformation. Cancer Disc. 9, 1228–1247.

Guglielmelli, P., Barosi, G., Specchia, G., Rambaldi, A., Lo, C.F., Antonioli, E., Pieri, L., Pancrazzi, A., Ponziani, V., Delaini, F., Longo, G., Ammatuna, E., Liso, V., Bosi, A., Barbui, T., Vannuch, A.M., 2009. Identification of patients with poorer survival in primary myelofibrosis based on the burden of *JAK2*V617F mutated allele. Blood 114, 1477–1483.

Guglielmelli, P., Biamonte, F., Score, J., Hidalgo-Curtis, C., Cervantes, F., Maffioli, M., Fanelli, T., Ernst, T., Winkelman, N., Jones, A.V., Zoi, K., Reiter, A., Duncombe, A., Villani, L., Bosi, A., Barosi, G., Cross, N.C.P., Vannuchi, A.M., 2011. *EZH2* mutational status predicts poor survival in myelofibrosis. Blood 118, 5227–5234.

Guglielmelli, P., Lasho, T.L., Rotunno, G., Score, J., Mannarelli, C., Pancrazzi, A., Biamonte, F., Pardanani, A., Zoi, K., Reiter, A., Duncombe, A., Fanelli, T., Pietra, D., Rumi, E., Finke, C., Gangat, N., Ketterling, R.P., Knudson, R.A., Hanson, C.A., Bosi, A., Pereira, A., Manfredini, R., Cervantes, F., Barosi, G., Cazzola, M., Cross, N.C.P., Vannucchi, A.M., Tefferi, A., 2014. The number of prognostically detrimental mutations and prognosis in primary myelofibrosis: an international study of 797 patients. Leukemia 28, 1804–1810.

Guglielmelli, P., Rotunno, G., Fanelli, T., Pacilli, A., Brogi, G., Calabresi, L., Pancrazzi, A., Vannucchi, A.M., on behalf of the Associazione Italiana per la Ricerca sul Cancro Gruppo Italiano Malattie Mieloproliferative (AGIMM), 2015. Validation of the differential prognostic impact of type 1/type 1-like versus type 2/type 2-like *CALR* mutations in myelofibrosis. Blood Cancer J. 5, e360.

Guglielmelli, P., Bartalucci, N., Contini, E., Rotunno, G., Pacilli, A., Romagnoli, S., Mannelli, L., Mannelli, F., Coltro, G., Pancrazzi, A., Ariu, V., Fiaccabrino, S., Semeraro, R., Magi, A., Mannarelli, C., Gesullo, F., Paoli, C., Vannucchi, A.M., 2019. Involvement of RUNX1 pathway is a common event in leukemic transformation of chronic myeloproliferative neoplasms. Blood 134 (Suppl. 1), 2968.

Guo, B.B., Allcock, R.J., Mirzai, B., Malherbe, J.A., Choudry, F.A., Frontini, M., Chuah, H., Liang, J., Kavanagh, S.E., Howman, R., Ouwehand, W.H., Fuller, K.A., Erber, W.N., 2017. Megakaryocytes in myeloproliferative neoplasms have unique somatic mutations. Am. J. Pathol. 187, 1512–1522.

Guo, Y., Zhou, Y., Yamatomo, S., Yang, H., Zhang, P., Chen, S., Nimer, S.D., Zhao, Z.J., Xu, M., Bai, J., Yang, F.-C., 2019. *ASXL1* alteration cooperates with *JAK2* V617F to accelerate myelofibrosis. Leukemia 33, 1287–1291.

Gupta, V., Kennedy, J.A., Capo-Chichi, J.-M., Kim, S., Hu, Z.-H., Alyea, E.P., Popat, U.R., Sobecks, R.M., Scott, B.L., Gerds, A.T., Salit, R.B., Deeg, J., Nakamura, R., Saber, W., 2020. Genetic factors rather than blast reduction determine outcomes of allogeneic HCT in *BCR-ABL*-negative MPN in blast phase. Blood Adv. 4, 5562–5573.

Harrison, C.N., Garcia, J.S., Somervaille, T., Foran, J.M., Verstovsek, S., Jamieson, C., Mesa, R., Ritchie, E.K., Tantravahi, S.K., Vachhani, P., O'Connell, C.L., Komrokji, R.S., Harb, J., Hutti, J.E., Holes, L., Masud, A.A., Potluri, J., Pemmaraju, N., 2021. Navitoclax and ruxolitinib for patients with myelofibrosis and JAK inhibitor experience: response duration in Phase 2 study. In: 2021 European Hematology Association Congress (EHA2021), Abstract EP1076

Harutyunyan, A., Klampfl, T., Cazzola, M., Kralovics, R., 2011. p53 Lesions in leukemic transformation. N. Engl. J. Med. 364, 488–490.

Hautin, M., Mornet, C., Chauveau, A., Bernard, D., Corcos, L., Lippert, E., 2020. Splicing anomalies in myeloproliferative neoplasms: paving the way for new therapeutic venues. Cancers 12, 2216.

Heaney, M.L., Soriano, G., 2013. Acute myeloid leukemia following a myeloproliferative neoplasm: clinical characteristics, genetic features and effects of therapy. Curr. Hematol. Malig. Rep. 8, 116–122.

Hidalgo López, J.E., Carballo-Zarate, A., Verstovsek, S., Wang, S.A., Hu, S., Li, S., Xu, J., Zuo, W., Tang, Z., Yin, C.C., Medeiros, L.J., Bueso-Ramos, C.E., Tang, G., 2018. Bone marrow findings in blast phase of polycythemia vera. Ann. Hematol. 97, 425–434.

Hinds, D.A., Barnholt, K.E., Mesa, R.A., Kiefer, A.K., Do, C.B., Eriksson, N., Mountain, J.L., Francke, U., Tung, J.Y., Nguyen, H.M., Zhang, H., Gojenola, L., Zehnder, J.L., Gotlib, J., 2016. Germ line variants predispose to both *JAK2* V617F clonal hematopoiesis and myeloproliferative neoplasms. Blood 128, 1121–1128.

Issa, G.C., DiNardo, C.D., 2021. Acute myeloid leukemia with *IDH1* and *IDH2* mutations: 2021 treatment algorithm. Blood Cancer J. 11, 107.

Iurlo, A., Cattaneo, D., Gianelli, U., 2019. Blast transformation in myeloproliferative neoplasms: risk factors, biological findings, and targeted therapeutic options. Inter. J. Mol. Sci. 20, 1839.

Jäger, R., Gisslinger, H., Berg, T., Passamonti, F., Cazzola, M., Rumi, E., Pietra, D., Gisslinger, B., Klampfl, T., Harutyunyan, A., Olcaydu, D., Kralovics, R., 2009. Deletions of the transcription factor *Ikaros* in myeloproliferative neoplasms at transformation to acute myeloid leukemia. Blood 114, 435.

Jain, T., Rampal, R.K., 2020. Accelerated and blast phase myeloproliferative neoplasms. Hematol. Oncol. Clin. N. Am. 35, 325–335.

Jaiswal, S., Fontanillas, P., Flannick, J., Manning, A., Grauman, P.V., Mar, B.G., Lindsley, R.C., Mermel, C.H., Burtt, N., Chavez, A., Higgins, J.M., Moltchanov, V., Kuo, F.C., Kluk, M.J., Henderson, B., Kinnunen, L., Koistinen, H.A., Ladenvall, C., Getz, G., Correa, A., Banahan, B.F., Gabriel, S., Kathiresan, S., Stringham, H.M., McCarthy, M.I., Boehnke, M., Tuomilehto, J., Haiman, C., Groop, L., Atzmon, G., Wilson, J.G., Neuberg, D., Altshuler, D., Ebert, B.L., 2014. Age-related clonal hematopoiesis associated with adverse outcomes. N. Engl. J. Med. 371, 2488–2498.

Jamal, R., Bélisle, C., Lessard, M.C., Hébert, J., Roy, D.C., Levine, R., Busque, L., 2008. Evidence suggesting the presence of a stem cell clone anteceding the acquisition of the *JAK2* V617F mutation. Leukemia 22, 1472–1474.

James, C., Ugo, V., Le Couédic, J.-P., Staerk, J., Delhommeau, F., Lacout, C., Garçon, L., Raslova, H., Berger, R., Bennaceur-Griscelli, A., Villeval, J.L., Constantinescu, S.N., Casadevall, N., Vainchenker, W., 2005. A unique clonal *JAK2* mutation leading to constitutive signalling causes polycythaemia vera. Nature 434, 1144–1148.

Jeong, J.J., Gu, X., Nie, J., Sundaravel, S., Liu, H., Kuo, W.-L., Bhagat, T.D., Pradhan, K., Cao, J., Nischal, S., McGraw, K.L., Bhattacharyya, S., Bishop, M.R., Artz, A., Thirman, M.J., Moliterno, A., Ji, P., Levine, R.L., Godley, L.A., Steidl, U., Bieker, J.J., List, A.F., Saunthararajah, Y., He, C., Verma, A., Wickrema, A., 2019. Cytokine-regulated phosphorylation and activation of TET2 by JAK2 in hematopoiesis. Cancer Discov. 9, 778–795.

Kameda, T., Shide, K., Yamaji, T., Kamiunten, A., Sekine, M., Taniguchi, Y., Hidaka, T., Kubuki, Y., Shimoda, H., Marutsuka, K., Sashida, G., Aoyama, K., Yoshimitsu, M., Harada, T., Abe, H., Miike, T., Iwakiri, H., Tahara, Y., Sueta, M., Yamamoto, S., Hasuike, S., Nagata, K., Iwama, A., Kitanaka, A., Shimoda, K., 2015. Loss of *TET2* has dual roles in murine myeloproliferative neoplasms: disease sustainer and disease accelerator. Blood 125, 304–315.

Kantarjian, H., Kadia, T., DiNardo, C., Daver, N., Borthakur, G., Jabbour, E., Garcia-Manero, G., Konopleva, M., Ravandi, F., 2021a. Acute myeloid leukemia: current progress and future directions. Blood Cancer J. 11, 41.

Kantarjian, H., Kadia, T., DiNardo, C., Welch, M.A., Ravandi, F., 2021b. Acute myeloid leukemia: treatment and research outlook for 2021 and the MD Anderson approach. Cancer 127, 1186–1207.

Kishtagari, A., Watts, J., 2017. Biological and clinical implications of telomere disfunction in myeloid malignacies. Ther. Adv. Hematol. 8, 317–326.

Klampfl, T., Gisslinger, H., Harutyunyan, A.S., Nivarthi, H., Rumi, E., Milosevic, J.D., Them, N.C.C., Berg, T., Gisslinger, B., Pietra, D., Chen, D., Vladimer, G.I., Bagienski, K., Milanesi, C., Casetti, I.C., Sant'antonio, E., Ferretti, V., Elena, C., Schischlik, F., Cleary, C., Six, M., Schalling, M., Schönegger, A., Bock, C., Malcovati, L., Pascutto, C., Superti-Furga, G., Cazzola, M., Kralovics, R., 2013. Somatic mutations of calreticulin in myeloproliferative neoplasms. N. Engl. J. Med. 369, 2379–2390.

Kleppe, M., Kwak, M., Koppikar, P., Riester, M., Keller, M., Bastian, L., Hricik, T., Bhagwat, N., McKenney, A.S., Papalexi, E., Abdel-Wahab, O., Rampal, R., Marubayashi, S., Chen, J.J., Romanet, V., Fridman, J.S., Bromberg, J., Teruya-Feldstein, J., Murakami, M., Radimerski, T., Michor, F., Fan, R., Levine, R.L., 2015. JAK-STAT pathway activation in malignant and nonmalignant cells contributes to MPN pathogenesis and therapeutic response. Cancer Disc. 5, 316–331.e7.

Kleppe, M., Koche, R., Zou, L., van Galen, P., Hill, C.E., Dong, L., De Groote, S., Papalexi, E., Hanasoge Somasundara, A.V., Corner, K., Keller, M., Farnoud, N., Medina, J., McGovern, E., Reyes, J., Roberts, J., Witkin, M., Rapaport, F., Teruya-Feldstein, J., Qi, J., Rampal, R., Bernstein, B.B., Bradner, J.E., Levine, R., 2018. Dual targeting of oncogenic activation and inflammatory signaling increases therapeutic efficacy in myeloproliferative neoplasms. Cancer Cell 33, 29–43.

Kodama, T., Hikita, H., Kawaguchi, T., Shigekawa, M., Shimizu, S., Hayashi, Y., Li, W., Miyagi, T., Hosui, A., Tatsumi, T., Kanto, T., Hiramatsu, N., Kiyomizu, K., Tadokoro, S., Tomiyama, Y., Hayashi, N., Takehara, T., 2012. MCL-1 and Bcl-xL regulate Bak/Bax-dependent apoptosis of the megakaryocytic lineage at multistages. Cell Death Differ. 19, 1856–1869.

Konopleva, M., Pollyea, D.A., Potluri, J., Chyla, B., Hogdal, L., Busman, T., McKeegan, E., Salem, A.H., Zhu, M., Ricker, J.L., Blum, W., DiNardo, C.D., Kadia, T., Dunbar, M., Kirby, R., Falotico, N., Leverson, J., Humerickhouse, R., Mabry, M., Stone, R., Kantarjian, H., Letai, A., 2016. Efficacy and biological correlates of response in a phase II study of venetoclax monotherapy in patients with acute myelogenous leukemia. Cancer Discov. 6, 1106–1117.

Konopleva, M., Martinelli, G., Daver, N., Papayiannidis, C., Wei, A., Higgins, B., Ott, M., Mascarenhas, J., Andreeff, M., 2020. MDM2 inhibition: an important step forward in cancer therapy. Leukemia 34, 2858–2874.

Kralovics, R., Passamonti, F., Buser, A.S., Teo, S.-S., Tiedt, R., Passweg, J.R., Tichelli, A., Cazzola, M., Skoda, R.C., 2005. A gain-of-function mutation of *JAK2* in myeloproliferative disorders. N. Engl. J. Med. 352, 1779–1790.

Kremyanskaya, M., Mascarenhas, J., Patriarca, A., Gupta, V., Devos, T., Harrison, C., Passamonti, F., Rampal, R., Mead, A., Scandura, J., Hobbs, G., Talpaz, M., Granacher, N., Somervaille, T., Hoffman, R., Bose, P., Colak, G., Shao, J., Cui, J., Bobba, S., Luptakova, K., Verstovsek, S., Palandri, F., 2021. Clinical benefit of pelabresib (CPI-0610) in combination with ruxolitinib in JAK inhibitor treatment naïve myelofibrosis patients: interim efficacy subgroup analysis from Arm 3 of MANIFEST Phase 2 study. In: 2021 European Hematology Association Congress (EHA2021), Abstract EP1085.

Kubesova, B., Pavlova, S., Malcikova, J., Kabathova, J., Radova, L., Tom, N., Tichy, B., Plevova, K., Kantorova, B., Fiedorova, K., Slavikova, M., Bystry, V., Kissova, J., Gisslinger, B., Gisslinger, H., Penka, M., Mayer, J., Kralovics, R., Pospisilova, S., Doubek, M., 2018. Low-burden *TP53* mutations in chronic phase of myeloproliferative neoplasms: association with age, hydroxyurea administration, disease type and JAK2 mutational status. Leukemia 32, 450–461.

Lachowiez, C., Borthakur, G., Loghavi, S., Zeng, Z., Kadia, T., Masarova, L., Takahashi, K., Tippett, G., Smith, S., Garcia, J., Bose, P., Jabbour, E., Ravandi, F., Daver, N., Garcia-Manero, G., Stoilova, B., Vyas, P., Kantarjian, H., Konopleva, M., DiNardo, C., 2021. A Phase Ib/IIb study of ivosidenib with venetoclax +/- azacitidine in *IDH1*-mutated myeloid malignancies. In: 2021 European Hematology Association Congress (EHA2021). Abstract S136.

Lancman, G., Brunner, A., Hoffman, R., Mascarenhas, J., Hobbs, G., 2018. Outcomes and predictors of survival in blast phase myeloproliferative neoplasms. Leuk. Res. 70, 49–55.

Lasho, T.L., Mudireddy, M., Finke, C.M., Hanson, C.A., Ketterling, R.P., Szuber, N., Begna, K.H., Patnaik, M.M., Gangat, N., Pardanani, A., Tefferi, A., 2018. Targeted next-generation sequencing in blast phase myeloproliferative neoplasms. Blood Adv. 2, 370–380.

Lee, H.J., Daver, N., Kantarjian, H.M., Verstovsek, S., Ravandi, F., 2012. The role of JAK pathway dysregulation in the pathogenesis and treatment of acute myeloid leukemia. Clin. Cancer Res. 19, 327–335.

Levine, R.L., Wadleigh, M., Cools, J., Ebert, B.L., Wernig, G., Huntly, B.J.P., Boggon, T.J., Wlodarska, I., Clark, J.J., Moore, S., Adelsperger, J., Koo, S., Lee, J.C., Gabriel, S., Mercher, T., D'andrea, A., Fröhling, S., Döhner, K., Marynen, P., Vandenberghe, P., Mesa, R.A., Tefferi, A., Griffin, J.D., Eck, M.J., Sellers, W.R., Meyerson, M., Golub, T.R., Lee, S.J., Gilliland, D.G., 2005. Activating mutation in the tyrosine kinase JAK2 in polycythemia vera, essential thrombocythemia, and myeloid metaplasia with myelofibrosis. Cancer Cell 7, 387–397.

Ley, T.J., Ding, L., Walter, M.J., McLellan, M.D., Lamprecht, T., Larson, D.E., Kandoth, C., Payton, J.E., Baty, J., Welch, J., Harris, C.C., Lichti, C.F., Townsend, R.R., Fulton, R.S., Dooling, D.J., Koboldt, D.C., Schmidt, H., Zhang, Q., Osborne, J.R., Lin, L., O'Laughlin, M., McMichael, J., Delehaunty, K.D., McGrath, S.D., Fulton, L.A., Magrini, V.J., Vickery, T.L., Hundal, J., Cook, L.L., Conyers, J.J., Swift, G.W., Reed, J.P., Alldredge, P.A., Wylie, T., Walker, J., Kalicki, J., Watson, M.A., Heath, S., Shannon, W.D., Varghese, N., Nagarajan, R., Westervelt, P., Tomasson, M.H., Link, D.C., Graubert, T.A., DiPersio, J.F., Mardis, E.R., Wilson, R.K., 2010. *DNMT3A* mutations in acute myeloid leukemia. N. Engl. J. Med. 363, 2424–2433.

Li, S., Kralovics, R., De Libero, G., Theocharides, A., Gisslinger, H., Skoda, R.C., 2008. Clonal heterogeneity in polycythemia vera patients with *JAK2* exon12 and *JAK2* V617F mutations. Blood 111, 3863–3866.

Li, B., Mascarenhas, J.O., Rampal, R.K., 2018. Leukemic transformation of myeloproliferative neoplasms: therapeutic and genomic considerations. Curr. Hematol. Malig. Rep. 13, 588–595.

Lin, K.H., Winter, P.S., Xie, A., Roth, C., Martz, C.A., Stein, E.M., Anderson, G.R., Tingley, J.P., Wood, K.C., 2016. Targeting MCL-1/BCL-X_L forestalls the acquisition of resistance to ABT-199 in acute myeloid leukemia. Sci. Rep. 6, 27696.

Lindsley, R.C., Mar, B.G., Mazzola, E., Grauman, P.V., Shareef, S., Allen, S.L., Pigneux, A., Wetzler, M., Stuart, R.K., Erba, H.P., Damon, L.E., Powell, B.L., Lindeman, N., Steensma, D.P., Wadleigh, M., Deangelo, D.J., Neuberg, D., Stone, R.M., Ebert, B.L., 2015. Acute myeloid leukemia ontogeny is defined by distinct somatic mutations. Blood 125, 1367–1376.

Lundberg, P., Karow, A., Nienhold, R., Looser, R., Hao-Shen, H., Nissen, I., Girsberger, S., Lehmann, T., Passweg, J., Stern, M., Beisel, C., Kralovics, R., Skoda, R.C., 2014. Clonal evolution and clinical correlates of somatic mutations in myeloproliferative neoplasms. Blood 123, 2220–2228.

Malherbe, J.A., Fuller, K.A., Arshad, A., Nangalia, J., Romeo, G., Hall, S.L., Meehan, K.S., Guo, B., Howman, R., Erber, W.N., 2016. Megakaryocytic hyperplasia in myeloproliferative neoplasms is driven by disordered proliferative, apoptotic and epigenetic mechanisms. J. Clin. Pathol. 69, 155–163.

Mannelli, F., 2021. Acute myeloid leukemia evolving from myeloproliferative neoplasms: many sides of a challenging disease. J Clin. Med. 10, 436.
Marcellino, B., Mascarenhas, J., 2019. Management of advanced phase myeloproliferative neoplasms. Clin. Adv. Hematol. Oncol. 17, 405–411.
Marcellino, B.K., Hoffman, R., Tripodi, J., Lu, M., Kosiorek, H., Mascarenhas, J., Rampal, R.K., Dueck, A., Najfeld, V., 2018. Advanced forms of MPNs are accompanied by chromosomal abnormalities that lead to dysregulation of TP53. Blood Adv. 2, 3581–3589.
Marinaccio, C., Suraneni, P., Celik, H., Volk, A., Wen, Q.J., Ling, T., Bulic, M., Lasho, T., Koche, R.P., Famulare, C.A., Farnoud, N., Stein, B., Schieber, M., Gurbuxani, S., Root, D.E., Younger, S.T., Hoffman, R., Gangat, N., Ntziachristos, P., Chandel, N.S., Levine, R.L., Rampal, R.K., Challen, G.A., Tefferi, A., Crispino, J.D., 2021. LKB1/*STK11* is a tumor suppressor in the progression of myeloproliferative neoplasms. Cancer Discov. 11, 1398–1410.
Marty, C., Pecquet, C., Nivarthi, H., El-Khoury, M., Chachoua, I., Tulliez, M., Villeval, J.-L., Raslova, H., Kralovics, R., Constantinescu, S.N., Plo, I., Vainchenker, W., 2016. Calreticulin mutants in mice induce an MPL-dependent thrombocytosis with frequent progression to myelofibrosis. Blood 127, 1317–1324.
Masarova, L., Verstovsek, S., Hidalgo-Lopez, J.E., Pemmaraju, N., Bose, P., Estrov, Z., Jabbour, E.J., Ravandi-Kashani, F., Takahashi, K., Cortes, J.E., Ning, J., Ohanian, M., Alvarado, Y., Zhou, L., Pierce, S., Gergis, R., Patel, K.P., Luthra, R., Kadia, T.M., DiNardo, C.D., Borthakur, G., Bhalla, K., Garcia-Manero, G., Bueso-Ramos, C., Kantarjian, H.M., Daver, N., 2018. A phase 2 study of ruxolitinib in combination with azacitidine in patients with myelofibrosis. Blood 132, 1664–1674.
Masarova, L., Bose, P., Pemmaraju, N., Daver, N.G., Zhou, L., Pierce, S., Sasaki, K., Kantarjian, H.M., Estrov, Z., Verstovsek, S., 2020. Prognostic value of blasts in peripheral blood in myelofibrosis in the ruxolitinib era. Cancer 126, 4322–4331.
Masarova, L., DiNardo, C.D., Bose, P., Pemmaraju, N., Daver, N., Kadia, T.M., Chifotides, H.T., Zhou, L., Borthakur, G., Estrov, Z., Konopleva, M., Verstovsek, S., 2021. Single-center experience with venetoclax combinations in patients with newly diagnosed and relapsed AML evolving from MPNs. Blood Adv. 5, 2156–2164.
Mascarenhas, J., Harrison, C., Kiladjian, J.-J., Mesa, R., Komrokji, R.S., Koschmieder, S., Vannuchi, A.M., Berry, T., Sherman, L., Dougherty, S., Sun, L., Huang, F., Wan, Y., Feller, F., Rizo, A., Verstovsek, S., 2020a. A randomized open-label, phase 3 study to evaluate imetelstat versus best available therapy (BAT) in patients with intermediate-2 or high-risk myelofibrosis (MF) refractory to Janus kinase inhibitor. Blood 136 (Suppl. 1), 43–44.
Mascarenhas, J., Harrison, C., Patriarca, A., Devos, T., Palandri, F., Rampal, R., Mead, A., Kremyanskaya, M., Somervaille, T., Wondergem, M., Hoffman, R., Kabir, S., Mertz, J., Colak, G., Shao, J., Bobba, S., Trojer, P., Senderowicz, A., Verstovsek, S., Gupta, V., 2020b. CPI-0610, a Bromodomain and Extraterminal Domain Protein (BET) inhibitor, in combination with ruxolitinib, in JAK inhibitor treatment naïve myelofibrosis patients: update from MANIFEST phase 2 study. HemaSphere 4, 1084.
Mascarenhas, J., Rampal, R.K., Kosiorek, H.E., Bhave, R., Hexner, E., Wang, E.S., Gerds, A., Abboud, C.N., Kremyanskaya, M., Berenzon, D., Odenike, O., Farnoud, N., Krishnan, A., Weinberg, R.S., McGovern, E., Salama, M.E., Najfeld, V., Medina-Martinez, J.S., Arango Ossa, J.E., Levine, M.F., Zhou, Y., Sandy, L., Heaney, M.L., Levine, R.L., Mesa, R.A., Dueck, A.C., Hoffman, R., 2020c. Phase 2 study of ruxolitinib and decitabine in patients with myeloproliferative neoplasm in accelerated and blast phase. Blood Adv. 4, 5246–5256.
Mascarenhas, J., Vannucchi, A.M., Mead, A.J., Garcia-Delgado, R., Pluta, A., Qamoos, H., Uyei, A., Gandhi Laurent, D., Al-Ali, H.K., 2020d. An open-label, global, multicenter phase 1b/2 study of KRT-232, a first-in-class, oral small-molecule inhibitor of murine

double minute 2 (MDM2), combined with ruxolitinib in patients who have myelofibrosis and a suboptimal response to ruxolitinib. Blood 136 (Suppl. 1), 44–45.

Mascarenhas, J., Komrokji, R.S., Palandri, F., Martino, B., Niederwieser, D., Reiter, A., Scott, B.L., Baer, M.R., Hoffman, R., Odenike, O., Vannucchi, A.M., Bussolari, J., Zhu, E., Rose, E., Sherman, L., Dougherty, S., Sun, L., Huang, F., Wan, Y., Feller, F.M., Rizo, A., Kiladjian, J.-J., 2021. Randomized, single-blind, multicenter phase II study of two doses of imetelstat in relapsed or refractory myelofibrosis. J. Clin. Oncol. 39 (26), 2881–2892.

McKenney, A.S., Lau, A.N., Somasundara, A.V.H., Spitzer, B., Intlekofer, A.M., Ahn, J., Shank, K., Rapaport, F.T., Patel, M.A., Papalexi, E., Shih, A.H., Chiu, A., Freinkman, E., Akbay, E.A., Steadman, M., Nagaraja, R., Yen, K., Teruya-Feldstein, J., Wong, K.-K., Rampal, R., Vander Heiden, M.G., Thompson, C.B., Levine, R.L., 2018. *JAK2/IDH*-mutant-driven myeloproliferative neoplasm is sensitive to combined targeted inhibition. J. Clin. Invest. 128, 789–804.

McKerrell, T., Park, N., Moreno, T., Grove, C.S., Ponstingl, H., Stephens, J., Understanding Society Scientific, G., Crawley, C., Craig, J., Scott, M.A., Hodkinson, C., Baxter, J., Rad, R., Forsyth, D.R., Quail, M.A., Zeggini, E., Ouwehand, W., Varela, I., Vassiliou, G.S., 2015. Leukemia-associated somatic mutations drive distinct patterns of age-related clonal hemopoiesis. Cell Rep. 10, 1239–1245.

McKerrell, T., Park, N., Chi, J., Collord, G., Moreno, T., Ponstingl, H., Dias, J., Gerasimou, P., Melanthiou, K., Prokopiou, C., Antoniades, M., Varela, I., Costeas, P.A., Vassiliou, G.S., 2017. *JAK2* V617F hematopoietic clones are present several years prior to MPN diagnosis and follow different expansion kinetics. Blood Adv. 1, 968–971.

McNamara, C.J., Panzarella, T., Kennedy, J.A., Arruda, A., Claudio, J.O., Daher-Reyes, G., Ho, J., Siddiq, N., Devlin, R., Tsui, H., Su, J., Stockley, T., Sukhai, M., Kanwar, N., Chan, S., Maze, D., Schimmer, A., Schuh, A., Sibai, H., Viswabandya, A., Yee, K., Minden, M.D., Kamel-Reid, S., Gupta, V., 2018. The mutational landscape of accelerated- and blast-phase myeloproliferative neoplasms impacts patient outcomes. Blood Adv. 2, 2658–2671.

Medeiros, B.C., Fathi, A.T., DiNardo, C.D., Pollyea, D.A., Chan, S.M., Swords, R., 2017. Isocitrate dehydrogenase mutations in myeloid malignancies. Leukemia 31, 272–281.

Meggendorfer, M., Walter, W., Nadarajah, C., Haferlach, C., Kern, W., Haferlach, T., 2020. Exome sequencing of paired MPN and blast phase shows an accumulation of splicing and chromatin modifying gene mutations, clonal evolution and gain of RAS pathway mutations during progression. In: 2020 European Hematology Association Congress (EHA2020). Abstract S213.

Mesa, R.A., Li, C.-Y., Ketterling, R.P., Schroeder, G.S., Knudson, R.A., Tefferi, A., 2005. Leukemic transformation in myelofibrosis with myeloid metaplasia: a single-institution experience with 91 cases. Blood 105, 973–977.

Mesa, R.A., Verstovsek, S., Cervantes, F., Barosi, G., Reilly, J.T., Dupriez, B., Levine, R., Le Bousse-Kerdiles, M.-C., Wadleigh, M., Campbell, P.J., Silver, R.T., Vannucchi, A.M., Deeg, H.J., Thomas, D., Odenike, O., Soldberg, L.A., Gotlib, J., Hexner, E., Nimer, S.D., Kantarjian, H., Orazi, A., Vardiman, J.W., Thiele, J., Tefferi, A., on behalf of the IWG-MRT, 2007. Primary myelofibrosis (PMF), post polycythemia vera myelofibrosis (post-PV MF), post essential thrombocythemia myelofibrosis (post-ET MF), blast phase PMF (PMF-BP): Consensus on terminology by the international working group for myelofibrosis research and treatment (IWG-MRT). Leuk. Res. 31, 737–740.

Milosevic, J.D., Kralovics, R., 2013. Genetic and epigenetic alterations of myeloproliferative neoplasms. Int. J. Hematol. 97, 183–197.

Milosevic, J.D., Puda, A., Malcovati, L., Berg, T., Hofbauer, M., Stukalov, A., Klampfl, T., Harutyunyan, A.S., Gisslinger, H., Gisslinger, B., Burjanivova, T., Rumi, E., Pietra, D., Elena, C., Vannucchi, A.M., Doubek, M., Dvorakova, D., Robesova, B., Wieser, R., Koller, E., Suvajdzic, N., Tomin, D., Tosic, N., Colinge, J., Racil, Z., Steurer, M., Pavlovic, S., Cazzola, M., Kralovics, R., 2012. Clinical significance of genetic aberrations in secondary acute myeloid leukemia. Am. J. Hematol. 87, 1010–1016.

Mosoyan, G., Kraus, T., Ye, F., Eng, K., Crispino, J.D., Hoffman, R., Iancu-Rubin, C., 2017. Imetelstat, a telomerase inhibitor, differentially affects normal and malignant megakaryopoiesis. Leukemia 31, 2458–2467.

Mudireddy, M., Gangat, N., Hanson, C.A., Ketterling, R.P., Pardanani, A., Tefferi, A., 2018. Validation of the WHO-defined 20% circulating blasts threshold for diagnosis of leukemic transformation in primary myelofibrosis. Blood Cancer J. 8, 57.

Mullally, A., Lane, S.W., Ball, B., Megerdichian, C., Okabe, R., Al-Shahrour, F., Paktinat, M., Haydu, J.E., Housman, E., Lord, A.M., Wernig, G., Kharas, M.G., Mercher, T., Kutok, J.L., Gilliland, D.G., Ebert, B.L., 2010. Physiological *JAK2*V617F expression causes a lethal myeloproliferative neoplasm with differential effects on hematopoietic stem and progenitor cells. Cancer Cell 17, 584–596.

Mullally, A., Poveromo, L., Schneider, R.K., Al-Shahrour, F., Lane, S.W., Ebert, B.L., 2012. Distinct roles for long-term hematopoietic stem cells and erythroid precursor cells in a murine model of *JAK2* V617F-mediated polycythemia vera. Blood 120, 166–172.

Mylonas, E., Yoshida, K., Frick, M., Hoyer, K., Christen, F., Kaeda, J., Obenaus, M., Noerenberg, D., Nennch, C., Chan, W., Ochi, Y., Shiraishi, Y., Shiozawa, Y., Zenz, T., Oakes, C.C., Sawitzki, B., Schawtz, M., Bullinger, L., le Coutre, P., Rose-Zerilli, M.J.-J., Ogawa, S., Damm, F., 2020. Nat. Commun. 11, 73.

Nakatake, M., Monte-Mor, B., Debili, N., Casadevall, N., Ribrag, V., Solary, E., Vainchenker, W., Plo, I., 2012. *JAK2* V617F negatively regulates p53 stabilization by enhancing MDM2 via La expression in myeloproliferative neoplasms. Oncogene 31, 1323–1333.

Nangalia, J., Massie, C.E., Baxter, E.J., Nice, F.L., Gundem, G., Wedge, D.C., Avezov, E., Li, J., Kollmann, K., Kent, D.G., Aziz, A., Godfrey, A.L., Hinton, J., Martincorena, I., Van Loo, P., Jones, A.V., Guglielmelli, P., Tarpey, P., Harding, H.P., Fitzpatrick, J.D., Goudie, C.T., Ortmann, C.A., Loughran, S.J., Raine, K., Jones, D.R., Butler, A.P., Teague, J.W., O'Meara, S., McLaren, S., Bianchi, M., Silber, Y., Dimitropoulou, D., Bloxham, D., Mudie, L., Maddison, M., Robinson, B., Keohane, C., Maclean, C., Hill, K., Orchard, K., Tauro, S., Du, M.-Q., Greaves, M., Bowen, D., Huntly, B.J.P., Harrison, C.N., Cross, N.C.P., Ron, D., Vannucchi, A.M., Papaemmanuil, E., Campbell, P.J., Green, A.R., 2013. Somatic *CALR* mutations in myeloproliferative neoplasms with nonmutated *JAK2*. N. Engl. J. Med. 369, 2391–2405.

Nangalia, J., Nice, F.L., Wedge, D.C., Godfrey, A.L., Grinfeld, J., Thakker, C., Massie, C. E., Baxter, J., Sewell, D., Silber, Y., Campbell, P.J., Green, A.R., 2015. *DNMT3A* mutations occur early or late in patients with myeloproliferative neoplasms and mutation order influences phenotype. Haematologica 100, e438–e442.

Newberry, K.J., Patel, K., Masarova, L., Luthra, R., Manshouri, T., Jabbour, E., Bose, P., Daver, N., Cortes, J., Kantarjian, H., Verstovsek, S., 2017. Clonal evolution and outcomes in myelofibrosis after ruxolitinib discontinuation. Blood 130, 1125–1131.

Nivarthi, H., Chen, D., Cleary, C., Kubesova, B., Jäger, R., Bogner, E., Marty, C., Pecquet, C., Vainchenker, W., Constantinescu, S.N., Kralovics, R., 2016. Thrombopoietin receptor is required for the oncogenic function of CALR mutants. Leukemia 30, 1759–1763.

Odenike, O., 2018. How I treat the blast phase of Philadelphia chromosome–negative myeloproliferative neoplasms. Blood 132, 2339–2350.

Ortmann, C.A., Kent, D.G., Nangalia, J., Silber, Y., Wedge, D.C., Grinfeld, J., Baxter, E.J., Massie, C.E., Papaemmanuil, E., Menon, S., Godfrey, A.L., Dimitropoulou, D., Guglielmelli, P., Bellosillo, B., Besses, C., Döhner, K., Harrison, C.N.,

Vassiliou, G.S., Vannucchi, A., Campbell, P.J., Green, A.R., 2015. Effect of mutation order on myeloproliferative neoplasms. N. Engl. J. Med. 372, 601–612.
Palandri, F., Breccia, M., Tiribelli, M., Bonifacio, M., Benevolo, G., Iurlo, A., Elli, E.M., Binotto, G., Tieghi, A., Polverelli, N., Martino, B., Abruzzese, E., Bergamaschi, M., Heidel, F.H., Cavazzini, F., Crugnola, M., Bosi, C., Isidori, A., Auteri, G., Forte, D., Latagliata, R., Griguolo, D., Cattaneo, D., Trawinska, M., Bartoletti, D., Krampera, M., Vianelli, N., Cavo, M., Palumbo, G.A., 2020a. Risk factors for progression to blast phase and outcome in 589 patients with myelofibrosis treated with ruxolitinib: real-world data. Hematol. Oncol. 38, 372–380.
Palandri, F., Breccia, M., Bonifacio, M., Polverelli, N., Elli, E.M., Benevolo, G., Tiribelli, M., Abruzzese, E., Iurlo, A., Heidel, F.H., Bergamaschi, M., Tieghi, A., Crugnola, M., Cavazzini, F., Binotto, G., Isidori, A., Sgherza, N., Bosi, C., Martino, B., Latagliata, R., Auteri, G., Scaffidi, L., Griguolo, D., Trawinska, M., Cattaneo, D., Catani, L., Krampera, M., Lemoli, R.M., Cuneo, A., Semenzato, G., Foà, R., Di Raimondo, F., Bartoletti, D., Cavo, M., Palumbo, G.A., Vianelli, N., 2020b. Life after ruxolitinib: reasons for discontinuation, impact of disease phase, and outcomes in 218 patients with myelofibrosis. 126, 1243–1252.
Pan, R., Hogdal, L., Benito, J.M., Bucci, D., Han, L., Borthakur, G., Cortes, J., DeAngelo, D.J., Debose, L., Mu, H., Döhner, H., Gaidzik, V.I., Galinsky, I., Golfman, L.S., Haferlach, T., Harutyunyan, K.G., Hu, J., Leverson, J.D., Marcucci, G., Müschen, M., Newman, R., Park, E., Ruvolo, P.P., Ruvolo, V., Ryan, J., Schindela, S., Zweidler-McKay, P., Stone, R.M., Kantarjian, H., Andreeff, M., Konopleva, M., Letai, A.G., 2014. Selective BCL-2 inhibition by ABT-199 causes on-target cell death in acute myeloid leukemia. Cancer Disc. 4, 362–375.
Pan, D., Rampal, R., Mascarenhas, J., 2020. Clinical developments in epigenetic-directed therapies in acute myeloid leukemia. Blood Adv. 4, 970–982.
Pardanani, A., Lasho, T.L., Finke, C.M., Mai, M., McClure, R.F., Tefferi, A., 2010. *IDH1* and *IDH2* mutation analysis in chronic- and blast-phase myeloproliferative neoplasms. Leukemia 24, 1146–1151.
Passamonti, F., Cervantes, F., Vannucchi, A.M., Morra, E., Rumi, E., Pereira, A., Guglielmelli, P., Pungolino, E., Caramella, M., Maffioli, M., Pascutto, C., Lazzarino, M., Cazzola, M., Tefferi, A., 2010. A dynamic prognostic model to predict survival in primary myelofibrosis: a study by the IWG-MRT (International Working Group for Myeloproliferative Neoplasms Research and Treatment). Blood 115, 1703–1708.
Patel, A.A., Cahill, K., Charnot-Katsikas, A., Liu, H., Gurbuxani, S., Thirman, M., Kosuri, S., Artz, A.S., Larson, R.A., Stock, W., Segal, J., Odenike, O., 2020. Clinical outcomes of *IDH2*-mutated advanced-phase Ph-negative myeloproliferative neoplasms treated with enasidenib. Br. J Haematol. 190, e48–e51.
Patel, K.P., Newberry, K.J., Luthra, R., Jabbour, E., Pierce, S., Cortes, J., Singh, R., Mehrotra, M, Routbort, M.J., Luthra, M., Manshouri, T., Santos, F.P., Kantarjian, H., Verstovsek, S., 2015. Correlation of mutation profile and response in patients with myelofibrosis treated with ruxolitinib. Blood 126, 790–797.
Paz, D.L., Riou, J., Verger, E., Cassinat, B., Chauveau, A., Ianotto, J.C., Dupriez, B., Boyer, F., Renard, M., Mansier, O., Murati, A., Rey, J., Etienne, G., Mansat-De Mas, V., Tavitian, S., Nibourel, O., Girault, S., Le Bris, Y., Girodon, F., Ranta D., Chomel, J.-C., Cony-Makhoul, P., Sujobert, P., Robles, M., Abdelali, R.B., Kosmider, O., Cottin, L., Roy, L., Sloma, I., Vacheret, F., Wemeau, M., Mossuz, P., Slama, B., Cussac, V., Denis, G., Walter-Petrich, A., Burroni, B., Jézéquel, N., Giraudier, G., Lippert, E., Socié, G., Kiladjian, J.-J., Ugo, V., 2021. Genomic analysis of primary and secondary myelofibrosis redefines the prognostic impact of ASXL1 mutations: a FIM study. Blood Adv. 5, 1443–1451.

Pérez, C., Pascual, M., Martín-Subero, J.I., Bellosillo, B., Segura, V., Delabesse, E., Álvarez, S., Larrayoz, M.J., Rifón, J., Cigudosa, J.C., Besses, C., Calasanz, M.J., Cross, N.C.P., Prósper, F., Agirre, X., 2013. Aberrant DNA methylation profile of chronic and transformed classic Philadelphia-negative myeloproliferative neoplasms. Haematologica 98, 1414–1420.

Petiti, J., Lo Iacono, M., Rosso, V., Andreani, G., Jovanovski, A., Podestà, M., Lame, D., De Gobbi, M., Fava, C., Saglio, G., Frassoni, F., Cilloni, D., 2020. Bcl-xL represents a therapeutic target in Philadelphia negative myeloproliferative neoplasms. J. Cell Mol. Med. 24, 10978–10986.

Pikman, Y., Lee, B.H., Mercher, T., McDowell, E., Ebert, B.L., Gozo, M., Cuker, A., Wernig, G., Moore, S., Galinsky, I., Deangelo, D.J., Clark, J.J., Lee, S.J., Golub, T.R., Wadleigh, M., Gilliland, D.G., Levine, R.L., 2006. *MPL*W515L is a novel somatic activating mutation in myelofibrosis with myeloid metaplasia. PLoS Med. 3, e270.

Plo, I., Nakatake, M., Malivert, L., De Villartay, J.-P., Giraudier, S., Villeval, J.-L., Wiesmuller, L., Vainchenker, W., 2008. JAK2 stimulates homologous recombination and genetic instability: potential implication in the heterogeneity of myeloproliferative disorders. Blood 112, 1402–1412.

Pollyea, D.A., Amaya, M., Strati, P., Konopleva, M.Y., 2019a. Venetoclax for AML: changing the treatment paradigm. Blood Adv. 3, 4326–4335.

Pollyea, D.A., Tallman, M.S., De Botton, S., Kantarjian, H.M., Collins, R., Stein, A.S., Frattini, M.G., Xu, Q., Tosolini, A., See, W.L., Macbeth, K.J., Agresta, S.V., Attar, E.C., DiNardo, C.D., Stein, E.M., 2019b. Enasidenib, an inhibitor of mutant *IDH2* proteins, induces durable remissions in older patients with newly diagnosed acute myeloid leukemia. Leukemia 33, 2575–2584.

Przychodzen, B., Jerez, A., Guinta, K., Sekeres, M.A., Padgett, R., Maciejewski, J.P., Makishima, H., 2013. Patterns of missplicing due to somatic *U2AF1* mutations in myeloid neoplasms. Blood 122, 999–1006.

Puda, A., Milosevic, J.D., Berg, T., Klampfl, T., Harutyunyan, A.S., Gisslinger, B., Rumi, E., Pietra, D., Malcovati, L., Elena, C., Doubek, M., Steurer, M., Tosic, N., Pavlovic, S., Guglielmelli, P., Pieri, L., Vannucchi, A.M., Gisslinger, H., Cazzola, M., Kralovics, R., 2012. Frequent deletions of *JARID2* in leukemic transformation of chronic myeloid malignancies. Am. J. Hematol. 87, 245–250.

Quintás-Cardama, A., Verstovsek, S., 2013. Molecular pathways: JAK/STAT pathway: mutations, inhibitors, and resistance. Clinical Cancer Res. 19, 1933–1940.

Quintás-Cardama, A., Kantarjian, H., Cortes, J., Verstovsek, S., 2011. Janus kinase inhibitors for the treatment of myeloproliferative neoplasias and beyond. Nat. Rev. Drug Discov. 10, 127–140.

Rampal, R., Mascarenhas, J., 2014. Pathogenesis and management of acute myeloid leukemia that has evolved from a myeloproliferative neoplasm. Curr. Opin. Hematol. 21, 65–71.

Rampal, R., Ahn, J., Abdel-Wahab, O., Nahas, M., Wang, K., Lipson, D., Otto, G.A., Yelensky, R., Hricik, T., McKenney, A.S., Chiosis, G., Chung, Y.R., Pandey, S., Van Den Brink, M.R.M., Armstrong, S.A., Dogan, A., Intlekofer, A., Manshouri, T., Park, C.Y., Verstovsek, S., Rapaport, F., Stephens, P.J., Miller, V.A., Levine, R.L., 2014a. Genomic and functional analysis of leukemic transformation of myeloproliferative neoplasms. Proc. Natl. Acad. Sci U. S. A. 111, E5401–E5410.

Rampal, R., Al-Shahrour, F., Abdel-Wahab, O., Patel, J.P., Brunel, J.-P., Mermel, C.H., Bass, A.J., Pretz, J., Ahn, J., Hricik, T., Kilpivaara, O., Wadleigh, M., Busque, L., Gilliland, D.G., Golub, T.R., Ebert, B.L., Levine, R.L., 2014b. Integrated genomic analysis illustrates the central role of JAK-STAT pathway activation in myeloproliferative neoplasm pathogenesis. Blood 123, e123–e133.

Rampal, R.K., Mascarenhas, J.O., Kosiorek, H.E., Price, L., Berenzon, D., Hexner, E., Abboud, C.N., Kremyanskaya, M., Weinberg, R.S., Salama, M.E., Menghrajani, K.,

Najfeld, V., Sandy, L., Heaney, M.L., Levine, R.L., Mesa, R.A., Dueck, A.C., Goldberg, J.D., Hoffman, R., 2018. Safety and efficacy of combined ruxolitinib and decitabine in accelerated and blast-phase myeloproliferative neoplasms. Blood Adv. 2, 3572–3580.

Rampal, R.K., McGrath, J.P., Krishnan, A., Li, B., Xiao, W., Nikom, D., Wang, J., Levine, R.L., Trojer, P., 2020. LSD1 inhibitor CPI-482 shows efficacy and prolongs survival in mouse models of AML and post-MPN AML in the context of constitutive JAK-STAT pathway activation. Blood 136 (Suppl. 1), 50–51.

Roboz, G.J., DiNardo, C.D., Stein, E.M., De Botton, S., Mims, A.S., Prince, G.T., Altman, J.K., Arellano, M.L., Donnellan, W., Erba, H.P., Mannis, G.N., Pollyea, D.A., Stein, A.S., Uy, G.L., Watts, J.M., Fathi, A.T., Kantarjian, H.M., Tallman, M.S., Choe, S., Dai, D., Fan, B., Wang, H., Zhang, V., Yen, K.E., Kapsalis, S.M., Hickman, D., Liu, H., Agresta, S.V., Wu, B., Attar, E.C., Stone, R.M., 2020. Ivosidenib induces deep durable remissions in patients with newly diagnosed *IDH1*-mutant acute myeloid leukemia. Blood 135, 463–471.

Rozovski, U., Verstovsek, S., Manshouri, T., Dembitz, V., Bozinovic, K., Newberry, K., Zhang, Y., Bove, J.E.T., Pierce, S., Kantarjian, H., Estrov, Z., 2017. An accurate, simple prognostic model consisting of age, *JAK2*, *CALR*, and *MPL* mutation status for patients with primary myelofibrosis. Haematologica 102, 79–84.

Rumi, E., Pietra, D., Pascutto, C., Guglielmelli, P., Martínez-Trillos, A., Casetti, I., Colomer, D., Pieri, L., Pratcorona, M., Rotunno, G., Sant'antonio, E., Bellini, M., Cavalloni, C., Mannarelli, C., Milanesi, C., Boveri, E., Ferretti, V., Astori, C., Rosti, V., Cervantes, F., Barosi, G., Vannucchi, A.M., Cazzola, M., Associazione Italiana Per La Ricerca Sul Cancro Gruppo Italiano Malattie Mieloproliferative I, 2014. Clinical effect of driver mutations of *JAK2*, *CALR*, or *MPL* in primary myelofibrosis. Blood 124, 1062–1069.

Saenz, D.T., Fiskus, W., Manshouri, T., Rajapakshe, K., Krieger, S., Sun, B., Mill, C.P., DiNardo, C., Pemmaraju, N., Kadia, T., Parmar, S., Sharma, S., Coarfa, C., Qiu, P., Verstovsek, S., Bhalla, K.N., 2017a. BET protein bromodomain inhibitor-based combinations are highly active against post-myeloproliferative neoplasm secondary AML cells. Leukemia 31, 678–687.

Saenz, D.T., Fiskus, W., Qian, Y., Manshouri, T., Rajapakshe, K., Raina, K., Coleman, K.G., Crew, A.P., Shen, A., Mill, C.P., Qiu, P., Kadia, T.M., Pemmaraju, N., DiNardo, C., Kim, M.-S., Npwak, A.J., Coarfa, C., Crews, C.M., Verstovsek, S., Bhalla, K.N., 2017b. Novel BET protein proteolysis-targeting chimera exerts superior lethal activity than bromodomain inhibitor (BETi) against post-myeloproliferative neoplasm secondary (s)AML cells. Leukemia 31, 1951–1961.

Saenz, D.T., Fiskus, W., Manshouri, T., Mill, C.P., Qian, Y., Raina, K., Rajapakshe, K., Coarfa, C., Soldi, R., Bose, P., Borthakur, G., Kadia, T.M., Khoury, J.D., Masarova, L., Nowak, A.J., Sun, B., Saenz, D.N., Kornblau, S.M., Horrigan, S., Sharma, S., Qui, P., Crews, C.M., Verstovsek, S., 2019. Targeting nuclear β-catenin as therapy for post-myeloproliferative neoplasm secondary AML. Leukemia 33, 1373–1386.

Saenz, D.T., Fiskus, W., Mill, C.P., Perera, D., Manshouri, T., Lara, B.H., Karkhanis, V., Sharma, S., Horrigan, S.K., Bose, P., Kadia, T.M., Masarova, L., DiNardo, C.D., Borthakur, G., Khoury, J.D., Takahashi, K., Bhaskara, S., Lin, C.Y., Green, M.R., Coarfa, C., Crews, C.M., Verstovsek, S., Bhalla, K.N., 2020. Mechanistic basis and efficacy of targeting β-catenin-TCF7L2-JMJD6-c-Myc axis to overcome resistance to BET inhibitors. Blood 135, 1255–1269.

Santos, F.P.S., Getta, B., Masarova, L., Famulare, C., Schulman, J., Datoguia, T.S., Puga, R.D., Alves Paiva, R.D.M., Arcila, M.E., Hamerschlak, N., Kantarjian, H.M., Levine, R.L., Campregher, P.V., Rampal, R.K., Verstovsek, S., 2020. Prognostic impact of RAS-pathway mutations in patients with myelofibrosis. Leukemia 34, 799–810.

Sashida, G., Wang, C., Tomioka, T., Oshima, M., Aoyama, K., Kanai, A., Mochizuki-Kashio, M., Harada, H., Shimoda, K., Iwama, A., 2016. The loss of Ezh2 drives the pathogenesis of myelofibrosis and sensitizes tumor-initiating cells to bromodomain inhibition. J. Exper. Med. 213, 1459–1477.

Scherber, R.M., Mesa, R., 2020. Management of challenging myelofibrosis after JAK inhibitor failure and/or progression. Blood Rev. 42, 100716.

Schieber, M., Crispino, J.D., Stein, B., 2019. Myelofibrosis in 2019: moving beyond JAK2 inhibition. Blood Cancer J. 9, 74.

Senín, A., Fernández-Rodríguez, C., Bellosillo, B., Camacho, L., Longarón, R., Angona, A., Besses, C., Álvarez-Larrán, A., 2018. Non-driver mutations in patients with *JAK2*V617F-mutated polycythemia vera or essential thrombocythemia with long-term molecular follow-up. Ann. Hematol. 97, 443–451.

Shah, M.T., Saliba, R.M., Verma, A., Ciurea, S.O., Oran, B., Olson, A., Bose, P., Bashir, Q., Masarova, L., Alousi, A.M., Srour, S., Mehta, R.S., Daver, N., Pemmaraju, N., Verstovsek, S., Champlin, R.E., Popat, U.R., 2021. Allogeneic stem cell transplant for patients with myeloproliferative neoplasms in blast phase: improving outcomes in the recent era. Br. J. Haematol. 193, 1004–1008.

Shahin, O.A., Chifotides, H.T., Bose, P., Masarova, L., Verstovsek, S., 2021. Accelerated phase of myeloproliferative neoplasms. Acta Haematol. 144, 484–499.

Shi, Y., Wang, X., Zhuang, Y., Jiang, Y., Melcher, K., Xu, H.E., 2017. Structure of the PRC2 complex and application to drug discovery. Acta Pharmacol. Sin. 38, 963–976.

Shimizu, T., Kubovcakova, L., Nienhold, R., Zmajkovic, J., Meyer, S.C., Hao-Shen, H., Geier, F., Dirnhofer, S., Guglielmelli, P., Vannucchi, A.M., Feenstra, J.D.M., Kralovics, R., Orkin, S.H., Skoda, R.C., 2016. Loss of *Ezh2* synergizes with *JAK2* V617F in initiating myeloproliferative neoplasms and promoting myelofibrosis. J. Exper. Med. 213, 1479–1496.

Short, N.J., Konopleva, M., Kadia, T.M., Borthakur, G., Ravandi, F., DiNardo, C.D., Daver, N., 2020. Advances in the treatment of acute myeloid leukemia: new drugs and new challenges. Cancer Disc. 10, 506–525.

Silva, M., Richard, C., Benito, A., Ollala, I., Fernández-Luna, J.L., 1998. Expression of Bcl-x in erythroid precursors from patients with polycythemia vera. N. Engl. J. Med. 338, 564–571.

Skoda, R.C., Duek, A., Grisouard, J., 2015. Pathogenesis of myeloproliferative neoplasms. Exper. Hematol. 43, 599–608.

Skov, V., 2020. Next generation sequencing in MPNs. Lessons from the past and prospects for use as predictors of prognosis and treatment responses. Cancers 12, 2194.

Spanoudakis, E., Bazdiara, I., Pantelidou, D., Kotsianidis, I., Papadopoulos, V., Margaritis, D., Xanthopoulidis, G., Moustakidis, E., Mantzourani, S., Bourikas, G., Tsatalas, C., 2011. Dynamics of telomere's length and telomerase activity in Philadelphia chromosome-negative myeloproliferative neoplasms. Leuk. Res. 35, 459–464.

Stegelmann, F., Bullinger, L., Schlenk, R.F., Paschka, P., Griesshammer, M., Blersch, C., Kuhn, S., Schauer, S., Döhner, H., Döhner, K., 2011. *DNMT3A* mutations in myeloproliferative neoplasms. Leukemia 25, 1217–1219.

Stein, E.M., DiNardo, C.D., Pollyea, D.A., Fathi, A.T., Roboz, G.J., Altman, J.K., Stone, R.M., DeAngelo, D.J., Levine, R.L., Flinn, I.W., Kantarjian, H.M., Collins, R., Patel, M.R., Frankel, A.E., Stein, A., Sekeres, M.A., Swords, R.T., Medeiros, B.C., Willekens, C., Vyas, P., Tosolini, A., Xu, Q., Knight, R.D., Yen, K.E., Agresta, S., de Botton, S., Tallman, M.S., 2017. Enasidenib in mutant *IDH2* relapsed or refractory acute myeloid leukemia. Blood 130, 722–731.

Stein, E.M., DiNardo, C.D., Fathi, A.T., Pollyea, D.A., Stone, R.M., Altman, J.K., Roboz, G.J., Patel, M.R., Collins, R., Flinn, I.W., Sekeres, M.A., Stein, A.S., Kantarjian, H.M., Levine, R.L., Vyas, P., Macbeth, K.J., Tosolini, A., Vanoostendorp, J., Xu, Q., Gupta, I., Lila, T., Risueno, A., Yen, K.E., Wu, B., Attar, E.C.,

Tallman, M.S., De Botton, S., 2019. Molecular remission and response patterns in patients with mutant-*IDH2* acute myeloid leukemia treated with enasidenib. Blood 133, 676–687.

Takagi, S., Masuoka, K., Uchida, N., Kurokawa, M., Nakamae, H., Imada, K., Iwato, K., Ichinohe, T., Atsuta, Y., Takami, A., Yano, S., 2016. Allogeneic hematopoietic cell transplantation for leukemic transformation preceded by Philadelphia chromosome–negative myeloproliferative neoplasms: a nationwide survey by the Adult Acute Myeloid Leukemia Working Group of the Japan Society for Hematopoietic Cell Transplantation. Biol. Blood Marrow Transplant. 22, 2208–2213.

Tam, C.S., Nussenzveig, R.M., Popat, U., Bueso-Ramos, C.E., Thomas, D.A., Cortes, J.A., Champlin, R.E., Ciurea, S.E., Manshouri, T., Pierce, S.M., Kantarjian, H.M., Verstovsek, S., 2008. The natural history and treatment outcome of blast phase BCR-ABL- myeloproliferative neoplasms. Blood 112, 1628–1637.

Tam, C.S., Kantarjian, H., Cortes, J., Lynn, A., Pierce, S., Zhou, L., Keating, M.J., Thomas, D.A., Verstovsek, S., 2009. Dynamic model for predicting death within 12 months in patients with primary or post-polycythemia vera/essential thrombocythemia myelofibrosis. J. Clin. Oncol. 27, 5587–5593.

Taylor, J., Lee, S.C., 2019. Mutations in spliceosome genes and therapeutic opportunities in myeloid malignancies. Genes, Chromosomes Cancer 58, 889–902.

Tefferi, A., 2010. Novel mutations and their functional and clinical relevance in myeloproliferative neoplasms: *JAK2*, *MPL*, *TET2*, *ASXL1*, *CBL*, *IDH* and *IKZF1*. Leukemia 24, 1128–1138.

Tefferi, A., 2011. Mutations galore in myeloproliferative neoplasms: would the real Spartacus please stand up? Leukemia 25, 1059–1063.

Tefferi, A., Pardanani, A., 2015. Myeloproliferative neoplasms. JAMA Oncol. 1, 97–105.

Tefferi, A., Lasho, T.L., Huang, J., Finke, C., Mesa, R.A., Li, C.Y., Hanson, C.A., Pardanani, A., 2008. Low *JAK2*V617F allele burden in primary myelofibrosis, compared to either a higher allele burden or unmutated status, is associated with inferior overall and leukemia-free survival. Leukemia 22, 756–761.

Tefferi, A., Lasho, T.L., Abdel-Wahab, O., Guglielmelli, P., Patel, J., Caramazza, D., Pieri, L., Finke, C.M., Kilpivaara, O., Wadleigh, M., Mai, M., McClure, R.F., Gilliland, D.G., Levine, R.L., Pardanani, A., Vannucchi, A.M., 2010. *IDH1* and *IDH2* mutation studies in 1473 patients with chronic-, fibrotic- or blast-phase essential thrombocythemia, polycythemia vera or myelofibrosis. Leukemia 24, 1302–1309.

Tefferi, A., Jimma, T., Sulai, N.H., Lasho, T.L., Finke, C.M., Knudson, R.A., McClure, R.F., Pardanani, A., 2012. IDH mutations in primary myelofibrosis predict leukemic transformation and shortened survival: clinical evidence for leukemogenic collaboration with *JAK2*V617F. Leukemia 26, 475–480.

Tefferi, A., Guglielmelli, P., Larson, D.R., Finke, C., Wassie, E.A., Pieri, L., Gangat, N., Fjerza, R., Belachew, A.A., Lasho, T.L., Ketterling, R.P., Hanson, C.A., Rambaldi, A., Finazzi, G., Thiele, J., Barbui, T., Pardanani, A., Vannucchi, A.M., 2014a. Long-term survival and blast transformation in molecularly annotated essential thrombocythemia, polycythemia vera, and myelofibrosis. Blood 124, 2507–2513.

Tefferi, A., Guglielmelli, P., Lasho, T.L., Rotunno, G., Finke, C., Mannarelli, C., Belachew, A.A., Pancrazzi, A., Wassie, E.A., Ketterling, R.P., Hanson, C.A., Pardanani, A., Vannucchi, A.M., 2014b. *CALR* and *ASXL1* mutations-based molecular prognostication in primary myelofibrosis: an international study of 570 patients. Leukemia 28, 1494–1500.

Tefferi, A., Lasho, T.L., Finke, C.M., Knudson, R.A., Ketterling, R., Hanson, C.H., Maffioli, M., Caramazza, D., Passamonti, F., Pardanani, A., 2014c. *CALR*- vs *JAK2*- vs *MPL*-mutated or triple-negative myelofibrosis: clinical, cytogenetic and molecular comparisons. Leukemia 28, 1472–1477.

Tefferi, A., Lasho, T.L., Begna, K.H., Patnaik, M.M., Zblewski, D.L., Finke, C.M., Laborde, R.R., Wassie, E., Schimek, L., Hanson, C.A., Gangat, N., Wang, X.,

Pardanani, A., 2015. A pilot study of the telomerase inhibitor imetelstat for myelofibrosis. N. Engl. J. Med. 373, 908–919.

Tefferi, A., Lasho, T.L., Finke, C.M., Elala, Y., Hanson, C.A., Ketterling, R.P., Gangat, N., Pardanani, A., 2016a. Targeted deep sequencing in primary myelofibrosis. Blood Adv. 1, 105–111.

Tefferi, A., Lasho, T.L., Guglielmelli, P., Finke, C.M., Rotunno, G., Elala, Y., Pacilli, A., Hanson, C.A., Pancrazzi, A., Ketterling, R.P., Mannarelli, C., Barraco, D., Fanelli, T., Pardanani, A., Gangat, N., Vannucchi, A.M., 2016b. Targeted deep sequencing in polycythemia vera and essential thrombocythemia. Blood Adv. 1, 21–30.

Tefferi, A., Nicolosi, M., Mudireddy, M., Szuber, N., Finke, C.M., Lasho, T.L., Hanson, C.A., Ketterling, R.P., Pardanani, A., Gangat, N., Mannarelli, C., Fanelli, T., Guglielmelli, P., Vannucchi, A.M., 2018a. Driver mutations and prognosis in primary myelofibrosis: Mayo-Careggi MPN Alliance study of 1,095 patients. Am. J. Hematol. 93, 348–355.

Tefferi, A., Finke, C.M., Lasho, T.L., Hanson, C.A., Ketterling, R.P., Gangat, N., Pardanani, A., 2018b. *U2AF1* mutation types in primary myelofibrosis: phenotypic and prognostic distinctions. Leukemia 32, 2274–2278.

Tefferi, A., Guglielmelli, P., Lasho, T.L., Gangat, N., Ketterling, R.P., Pardanani, A., Vannucchi, A.M., 2018c. MIPSS70+ Version 2.0: Mutation and karyotype-enhanced international prognostic scoring system for primary myelofibrosis. J. Clin. Oncol. 36, 1769–1770.

Tefferi, A., Guglielmelli, P., Nicolosi, M., Mannelli, F., Mudireddy, M., Bartalucci, N., Finke, C.M., Lasho, T.L., Hanson, C.A., Ketterling, R.P., Begna, K.H., Naseema, G., Pardanani, A., Vannucchi, A.M., 2018d. GIPSS: Genetically inspired prognostic scoring system for primary myelofibrosis. Leukemia 32, 1631–1642.

Tefferi, A., Mudireddy, M., Mannelli, F., Begna, K.H., Patnaik, M.M., Hanson, C.A., Ketterling, R.P., Gangat, N., Yogarajah, M., De Stefano, V., Passamonti, F., Rosti, V., Finazzi, M.C., Rambaldi, A., Bosi, A., Guglielmelli, P., Pardanani, A., Vannucchi, A.M., 2018e. Blast phase myeloproliferative neoplasm: Mayo-AGIMM study of 410 patients from two separate cohorts. Leukemia 32, 1200–1210.

Tefferi, A., Guglielmelli, P., Lasho, T.L., Coltro, G., Finke, C.M., Loscocco, G.G., Sordi, B., Szuber, N., Rotunno, G., Pacilli, A., Hanson, C.A., Ketterling, R.P., Pardanani, A., Gangat, N., Vannucchi, A.M., 2020. Mutation-enhanced international prognostic systems for essential thrombocythaemia and polycythaemia vera. Br. J. Haematol. 189, 291–302.

Theocharides, A., Boissinot, M., Girodon, F., Garand, R., Teo, S.-S., Lippert, E., Talmant, P., Tichelli, A., Hermouet, S., Skoda, R.C., 2007. Leukemic blasts in transformed *JAK2* V617F–positive myeloproliferative disorders are frequently negative for the *JAK2* V617F mutation. Blood 110, 375–379.

Tremblay, D., Feld, J., Dougherty, M., Czaplinska, T., Snachez, G., Kremyanskaya, M., Bar-Natan, M., Shih, A.H., Keyzner, A., Mascarenhas, J., 2020. Venetoclax and hypomethylating agent combination therapy in acute myeloid leukemia secondary to a myeloproliferative neoplasm. Leuk. Res. 98, 106456.

Triviai, I., Zeschke, S., Rentel, J., Spanakis, M., Scherer, T., Gabdoulline, R., Panagiota, V., Thol, F., Heuser, M., Stocking, C., Kröger, N., 2019. *ASXL1/EZH2* mutations promote clonal expansion of neoplastic HSC and impair erythropoiesis in PMF. Leukemia 33, 99–109.

Tsuruta-Kishino, T., Koya, J., Kataoka, K., Narukawa, K., Sumitomo, Y., Kobayashi, H., Sato, T., Kurokawa, M., 2017. Loss of p53 induces leukemic transformation in a murine model of *JAK2* V617F-driven polycythemia vera. Oncogene 36, 3300–3311.

Vainchenker, W., Kralovics, R., 2017. Genetic basis and molecular pathophysiology of classical myeloproliferative neoplasms. Blood 129, 667–679.

Vallapureddy, R.R., Mudireddy, M., Penna, D., Lasho, T.L., Finke, C.M., Hanson, C.A., Ketterling, R.P., Begna, K.H., Gangat, N., Pardanani, A., Tefferi, A., 2019. Leukemic

transformation among 1306 patients with primary myelofibrosis: risk factors and development of a predictive model. Blood Cancer J. 9, 12.
Vannucchi, A.M., Lasho, T.L., Guglielmelli, P., Biamonte, F., Pardanani, A., Pereira, A., Finke, C., Score, J., Gangat, N., Mannarelli, C., Ketterling, R.P., Rotunno, G., Knudson, R.A., Susini, M.C., Laborde, R.R., Spolverini, A., Pancrazzi, A., Pieri, L., Manfredini, R., Tagliafico, E., Zini, R., Jones, A., Zoi, K., Reiter, A., Duncombe, A., Pietra, D., Rumi, E., Cervantes, F., Barosi, G., Cazzola, M., Cross, N.C.P., Tefferi, A., 2013. Mutations and prognosis in primary myelofibrosis. Leukemia 27, 1861–1869.
Vannucchi, A.M., Rotunno, G., Bartalucci, N., Raugei, G., Carrai, V., Balliu, M., Mannarelli, C., Pacilli, A., Calabresi, L., Fjerza, R., Pieri, L., Bosi, A., Manfredini, R., Guglielmelli, P., 2014. Calreticulin mutation-specific immunostaining in myeloproliferative neoplasms: pathogenetic insight and diagnostic value. Leukemia 28, 1811–1818.
Venton, G., Courtier, F., Charbonnier, A., D'incan, E., Saillard, C., Mohty, B., Mozziconacci, M.-J., Birnbaum, D., Murati, A., Vey, N., Rey, J., 2018. Impact of gene mutations on treatment response and prognosis of acute myeloid leukemia secondary to myeloproliferative neoplasms. Am. J. Hematol. 93, 330–338.
Verstovsek, S., Mascarenhas, J., Kremyanskaya, M., Hoffman, R., Rampal, R., Gupta, V., Talpaz, M., Granacher, N., Leber, B., Kiladjian, J.-J., Vannuchi, A., Bose, P., McMullin, M., Sirhan, S., Ribrag, V., Kabir, S., Mertz, J., Colak, G., Shao, J., Bobba, S., Trojer, P., Senderowicz, A., Harrison, C., 2020. CPI-0610, Bromodomain and extraterminal domain protein (BET) inhibitor, as 'add-on' to ruxolitinib (RUX), in advanced myelofibrosis patients with suboptimal response: update of MANIFEST Phase 2 study. HemaSphere 4, 1083.
Verstovsek, S., Haifa, K., Mascarenhas, J., Mead, A.J., Perkins, A., Vannucchi, A.M., Uyei, A., Rothbaum, W.P., Greivy, J., Kiladjian, J.-J., 2021a. BOREAS: a global phase 3 study of KRT-232, a first-in-class murine double minute 2 (MDM2) inhibitor in $TP53^{WT}$ relapsed/refractory (R/R) myelofibrosis (MF). J. Clin. Oncol. 39 (Suppl. 15), TP7057.
Verstovsek, S., Kremyanskaya, M., Mascarenhas, J., Talpaz, M., Harrison, C., Rampal, R., Patriarca, A., Gupta, V., Granacher, N., Somervaille, T., Schiller, G., Drummond, M., Foltz, L., Lambert, J., Prejner, W., Colak, G., Keller, P., Shao, J., Luptakova, K., Hoffman, R., Bose, P., Vannucchi, A., 2021b. Pelabresib (CPI-0610) improved anemia associated with myelofibrosis: interim results from MANIFEST Phase 2 study. In: 2021 European Hematology Association Congress (EHA2021). Abstract EP1077.
Waibel, M., Solomon, V.S., Knight, D.A., Ralli, R.A., Kim, S.-K., Banks, K.-M., Vidacs, E., Virely, C., Sia, K.C.S., Bracken, L.S., Collins-Underwood, R., Drenberg, C., Ramsey, L.B., Meyer, S.C., Takiguchi, M., Dickins, R.A., Levine, R., Ghysdael, J., Dawson, M.A., Lock, R.B., Mullighan, C.G., Johnstone, R.W., 2013. Combined targeting of JAK2 and Bcl-2/Bcl-xL to cure mutant *JAK2*-driven malignancies and overcome acquired resistance to JAK2 inhibitors. Cell Rep. 5, 1047–1059.
Wang, X., Hu, C.S., Petersen, B., Qiu, J., Ye, F., Houldsworth, J., Eng, K., Huang, F., Hoffman, R., 2018. Imetelstat, a telomerase inhibitor, is capable of depleting myelofibrosis stem and progenitor cells. Blood Adv. 2, 2378–2388.
Williams, N., Lee, J., Moore, L., Baxter, J.E., Hewinson, J., Dawson, K.J., Menzies, A., Godfrey, A.L., Green, A.R., Campbell, P.J., Nangalia, J., 2020. Driver mutation acquisition *in utero* and childhood followed by lifelong clonal evolution underlie myeloproliferative neoplasms. Blood 136 (Suppl. 2). LBA-1.
Woods, B., Chen, W., Chiu, S., Marinaccio, C., Fu, C., Gu, L., Bulic, M., Yang, Q., Zouak, A., Jia, S., Suraneni, P.K., Xu, K., Levine, R.L., Crispino, J.D., Wen, Q.J., 2019. Activation of JAK/STAT signaling in megakaryocytes sustains myeloproliferation *in vivo*. Clin. Cancer Res. 25, 5901–5912.

Xie, M., Lu, C., Wang, J., McLellan, M.D., Johnson, K.J., Wendl, M.C., McMichael, J.F., Schmidt, H.K., Yellapantula, V., Miller, C.A., Ozenberger, B.A., Welch, J.S., Link, D.C., Walter, M.J., Mardis, E.R., Dipersio, J.F., Chen, F., Wilson, R.K., Ley, T.J., Ding, L., 2014. Age-related mutations associated with clonal hematopoietic expansion and malignancies. Nat. Med. 20, 1472–1478.

Yamada, O., Kawauchi, K., 2013. The role of the JAK-STAT pathway and related signal cascades in telomerase activation during the development of hematologic malignancies. JAK-STAT 2, e25256.

Yang, H., Kurtenbach, S., Guo, Y., Lohse, I., Durante, M.A., Li, J., Li, Z., Al-Ali, H., Li, L., Chen, Z., Field, M.G., Zhang, P., Chen, S., Yamamoto, S., Li, Z., Zhou, Y., Nimer, S.D., Harbour, J.W., Wahlestedt, C., Xu, M., Yang, F.-C., 2018. Gain of function of ASXL1 truncating protein in the pathogenesis of myeloid malignancies. Blood 131, 328–341.

Yogarajah, M., Tefferi, A., 2017. Leukemic transformation in myeloproliferative neoplasms. Mayo Clin. Proc. 92, 1118–1128.

Young, A.L., Challen, G.A., Birmann, B.M., Druley, T.E., 2016. Clonal haematopoiesis harbouring AML-associated mutations is ubiquitous in healthy adults. Nat. Commun. 7, 12484.

Zhan, H., Ma, Y., Lin, C.H.S., Kaushansky, K., 2016. *JAK2*(V617F)-mutant megakaryocytes contribute to hematopoietic stem/progenitor cell expansion in a model of murine myeloproliferation. Leukemia 30, 2332–2341.

Zhang, Y., Lin, C.H.S., Kaushansky, K., Zhan, H., 2018. *JAK2* V617F megakaryocytes promote hematopoietic stem/progenitor cell expansion in mice through thrombopoietin/MPL signaling. Stem Cells 36, 1676–1684.

Zhang, S.-J., Rampal, R., Manshouri, T., Patel, J., Mensah, N., Kayserian, A., Hricik, T., Heguy, A., Hedvat, C., Gönen, M., Kantarjian, H., Levine, R.L., Abdel-Wahab, O., Verstovsek, S., 2012. Genetic analysis of patients with leukemic transformation of myeloproliferative neoplasms shows recurrent *SRSF2* mutations that are associated with adverse outcome. Blood 119, 4480–4485.

Zhao, W., Du, Y., Ho, W.T., Fu, X., Zhao, Z.J., 2012. *JAK2* V617F and *p53* mutations coexist in erythroleukemia and megakaryoblastic leukemic cell lines. Exper. Hematol. Oncol. 1, 15.

CHAPTER FIVE

Lessons from mouse models of MPN

Camelia Benlabiod[a,b,c,†], Tracy Dagher[a,b,c,†], Caroline Marty[a,b,c,*], and Jean-Luc Villeval[a,b,c,*]
[a]INSERM, UMR 1287, Gustave Roussy, Villejuif, France
[b]Université Paris-Saclay, UMR 1287, Gustave Roussy, Villejuif, France
[c]Gustave Roussy, UMR 1287, Villejuif, France
*Corresponding authors: e-mail address: caroline.marty@gustaveroussy.fr; jean-luc.villeval@gustaveroussy.fr

Contents

Abstract

Over the past decades, a variety of MPN mouse models have been developed to express in HSC the main mutations identified in patients: *JAK2*^{V617F}, *CALRdel52* or *ins5* and *MPL*^{W515L}. These models mimic quite faithfully human PV or ET with their natural evolutions into MF and their hemostasis complications, demonstrating the driver function of these mutations in MPN. Here, we review these models and show how they have

† These authors contributed equally.

International Review of Cell and Molecular Biology, Volume 366
ISSN 1937-6448
https://doi.org/10.1016/bs.ircmb.2021.02.009

improved our general understanding of MPN regarding (1) the mechanisms of fibrosis, thrombosis/hemorrhages and disease initiation, (2) the roles of additional mutations and signaling pathways in disease progression and (3) the preclinical development of novel therapies. We also address controversial results between these models and remind how these models may differ from human MPN onset and also how basically mice are not humans, encouraging caution when one draw lessons from mice to humans. Furthermore, the contribution of germline genetic predisposition, HSC and niche aging, metabolic, oxidative, replicative or genotoxic stress, inflammation, immune escape and additional mutations need to be considered in further investigations to encompass the full complexity of human MPN in mice.

1. "Of mice and Men"

Before starting this review, it seems necessary to remember that "mice are not small people," as Robert Weinberg said (Rangarajan and Weinberg, 2003). Mouse models have proven extraordinarily useful in understanding various aspects of human tumors, including myeloproliferative neoplasms (MPNs), but the two species have accumulated many genetic differences since their most recent common ancestor 91 million years ago (Table 1). These differences may explain some discrepancies between patients and mouse models expressing the same mutation.

Mice have a shorter genome than humans (2.5 compared to 2.9×10^9 letters) with only 1% of unique genes. Mice tend to form cancers in different tissues than human (mesenchymal compared to epithelial) and develop lymphoma, leukemia, sarcoma and less frequently carcinoma or melanoma. They have a short lifespan (2.5 years) with fewer cell divisions than humans (10^{11} versus 10^{16}). Thus, mice undergo about 10^5 fewer cell divisions in a lifetime than humans that should theoretically lead to less risk of genetic damage and cancer. Yet, the lifetime risk of developing cancer is comparable in both species (30%). Therefore, millions of years of evolution have led to a decrease in cancer susceptibility and increased anti-neoplastic mechanisms in humans. Among these mechanisms are the regulation of oxidative stress and telomere length and the usage of different tumor repressor pathways. The high oxidative metabolism rate of mice (sevenfold higher compared to humans) increases the levels of endogenous oxidants and other mutagens causing frequent DNA damages (mice excrete 18-fold more DNA breakdown products per kg of body weight than humans do). Telomerase, protecting the chromosome ends from degradation, is highly active in mice and barely active in humans, explaining the long mouse compared to human

Table 1 Summary of the main differences between mouse and human biology, cancer risk and immuno/hematology.

Characteristics/Species	Mouse	Human
Gene sharing (%)	99	99
Direct gene alignment (%)	40	40
Genome length (letter)	2.5×10^9	2.9×10^9
Weight (kg)	0.035	70
Life span (years)	2.5	79
Division (lifetime)	10^{11}	10^{16}
Oxidative stress metabolism (ratio)	7	1
Risk of developing cancer	30%	30%
DNA breakdown products (/kg of BW)	18	1
Telomere length (kb)	50	10
Telomerase activity	high	Low
Dominant cell-cycle arrest pathway	p53	RB
Dominant age-related cancers	Mesenchymal-related: lymphoma, sarcoma, leukemia, …	Epithelial-related: skin, breast, prostate, lung carcinoma, …
Bone marrow cellularity	High	low
Stress-related spleen hematopoietic activity	High	low

Continued

Table 1 Summary of the main differences between mouse and human biology, cancer risk and immuno/hematology.—cont'd

Characteristics/Species	Mouse	Human
RBC life span (day), size (fL) and number (/μL)	$45/45/9 \times 10^6$	$120/90/4.5 \times 10^6$
Platelet life (day) size (fL) and number (/μL)	$5/5/10^6$	$10/10/0.3 \times 10^6$
Dominant leukocyte in blood	Lymphocytes	Granulocytes
Human immune genes absent in mice	–	TLR10, caspase 10, some chemokines
HSC phenotype/Identity	c-Kit$^+$/*ScaI*$^+$/CD150$^+$/CD48$^-$/CD34$^-$	c-Kit$^+$/CD34$^+$/CD38$^-$/CD90$^+$/CD49f$^+$
HSC replication rate (per week)	2.5	40
Active/quiescent HSC (ratio)	1.46	0.116

These are caused by 180 Mya of cumulating time of independent evolution in distinct environments and pressures of natural selection. For the mice, the C57/Bl6 strain was chosen as a reference, differences exist between strains. RBC, red blood cell; HSC, hematopoietic stem cell; Rb, retinoblastoma.

telomere ($\approx$50 kb versus 10 kb). Therefore, murine cells easily undergo spontaneous immortalization leading to cancer, whereas human cells undergo telomere erosion/shortening with senescence and "crisis"/death as cancer protection, unless de-repressing telomerase expression. Finally, regarding two pathways associated with senescence and cell-cycle arrest, p53 and the retinoblastoma (RB) are respectively dominant in mouse and in humans, possibly explaining that *p53*-deficient mice shift from "murine-like" mesenchymal to "human-like" epithelial cancers (Rangarajan and Weinberg, 2003).

The hematology of mice or humans has also some specific features. In particular, bone marrow (BM) cellularity is elevated in mice (filling 90% of the medullary space) in contrast to the paucicellular marrow of humans. As a consequence, mouse BM cannot extend in response to stress. Instead, the spleen, characterized by the dominant lymphoid white pulp, can become, in the red pulp, an active site for an exuberant trilineage myeloid hematopoiesis (granulocytic, erythroid and megakaryocytic) in response to stress such as anemia, infections and neoplastic hyperproliferation such as MPNs. Therefore, the extramedullary role of the spleen in mice should be considered in addition to the BM site to fully characterize a hematologic phenotype (O'Connell et al., 2015). Indeed, a 4–5-fold increase in spleen weight represents the same cellularity found in the total BM of a mouse (based on one femur representing 6% of total BM and 1 g of spleen representing 10^9 cells).

Mice also have small red blood cells (RBCs) and platelets with a short lifespan compared to human (Table 1). Lymphocytes are the predominant blood leukocytes in C57BL/6 mice. In contrast, in C3H mice granulocytes are the blood dominant cells, similarly to humans. Immunology is different between mice and humans. Indeed, mice are lacking some critical human immune genes such as neutrophil defensins, TLR10, FcγRI and FcγRIIA, caspase 10 and several chemokines (IL-8, NAP-2, ITAC, MCP-4, HCC-1, HCC-2, MPIF-1, PARC, eotaxin-2/3) (Mestas and Hughes, 2004). Furthermore, the phenotypic and functional characteristics of monocytes, dendritic cells and natural killers are also different (Rongvaux et al., 2013).

Considering hematopoietic stem cells (HSCs), the specific signatures of mouse and human repopulating cells are divergent. For example, while CD34 expression is present on human HSC, its expression is minimal on mouse HSC and marks more differentiated progenitors. Sca-1 is very useful in mouse HSC isolation but no such molecule exists in human. Also, human HSCs replicate slowly, once every 40 weeks on average instead of once every 2.5 weeks for mice. Consequently, the steady-state ratio of active clones contributing to hematopoiesis compared with quiescent clones is approximately 1.46 for mice but only 0.116 for humans (Sykes and Scadden, 2013).

Overall, these differences have to be kept in mind when data from mouse models are analyzed and discussed. Having said that, the mouse is and will remain the model of choice for human disease thanks to the knowledge that has been accumulated over the years on this animal, as we try to summarize it. Moreover, mice present practical advantages such as small size, reliable breeding and short lifespan. We will show here how precious murine models have been in the understanding of MPN etiology, disease initiation and evolution, the identification of the mechanisms of fibrosis and hemostasis disorders, the role of additional mutations and the development of some preclinical approaches.

2. MPN driver mutations via the JAK2/STAT signaling pathway

Classical non **BCR-ABL** MPNs are a group of hematological malignancies characterized by the over-production of one or several mature and functional cells from the myeloid lineages: red blood cells with a variable granulocytic/megakaryocytic hyperplasia in polycythemia vera (PV), megakaryocytes (MKs)/platelets in essential thrombocythemia (ET) and granulocytes plus dysplastic MKs with a BM invalidating deposition of collagen fibers in primary myelofibrosis (PMF). The three disorders form a continuum with frequent evolution from ET to PV, from PV to secondary myelofibrosis (MF) and from ET to MF. PMF is the most severe and heterogeneous MPN and usually presents other clinical symptoms such as splenomegaly and/or hepatomegaly due to extramedullary hematopoiesis. MPNs are prone to hemostasis complications such as thrombosis and hemorrhage, and to transformation to acute myeloid leukemia (AML).

MPNs are clonal disorders due to the acquisition of gain-of-function mutations at the HSC level in three main genes encoding: JAK2, the thrombopoietin (TPO) receptor (MPL) and calreticulin (CALR) (James et al., 2005; Klampfl et al., 2013; Nangalia et al., 2013; Pikman et al., 2006; Scott et al., 2007). *JAK2*^{V617F} is the most prevalent mutation before *CALRdel52* (type 1) and *CALRins5* (type 2), *MPL*$^{W515L/K/N}$ and *JAK2 exon12* mutation. These mutations are detected in patients with various phenotypes (*JAK2*^{V617F} in all three MPNs, mutations of *CALR* and *MPL* exclusively in ET and MF, *JAK2 exon12* in PV/MF). They lead to the constitutive activation of type I homodimeric cytokine receptors via the JAK2/STAT signaling pathway: the erythropoietin (EPO), TPO or/and granulocyte colony-stimulating factor (G-CSF) receptors. Based on mouse modeling, these gain-of-function mutations have been characterized as phenotypic drivers of the diseases. Heterogeneity of MPNs, in particular of PMF, is explained by the acquisition of additional non-phenotypic mutations targeting genes encoding epigenetic regulators (DNA methylation and histone modifiers), components of the splicing machinery and transcriptional factors that are frequently found in myelodysplastic syndromes and AML.

Anomalies found in patients have been modeled using different approaches in mice. Most, but not all of them, faithfully mimic human MPNs and have been useful, as we will detail it in the following paragraphs, in understanding the physiopathology of MPN.

3. Modeling MPN in mice—An up-to-date description

Several strategies have been used to model the different recurrent mutations that were identified in MPN patients. They allow characterizing a mutation as driver of the phenotype.

The **retroviral (RV)** strategy will limit the expression of the gene of interest to the hematopoietic system via BM transplantation. This is a fast process that possesses the major drawback of uncontrolled both copy numbers and RV random sites of integration that might cause the undesirable expression of surrounding genes. It is usually characterized by an overexpression of the transgene compared to its physiological level. Moreover, engraftment exposes HSCs to high levels of cytokines, replication and oxidative stress causing possible genetic and physiologic artifacts.

The **transgenic (TG)** technology also uses random gene integration, except when targeting into the ubiquitous constitutively expressed *Rosa26* locus (not used in MPN models). The level and tissue-specific expression of the transgene depends on the promoter used and differs from the physiological level. Compared to the RV technology, mice present the advantage of germline transmission by breeding.

In **knock-in (KI)** models, the modified gene is targeted into its own locus via homologous recombination and its expression is thus dependent on the endogenous gene promoter. This model is the most faithful to the human disease. Constitutive KI models mimic the effect of inherited mutations but a BM graft may limit the expression of the mutated gene to the hematopoietic tissue. The Cre-loxP system has allowed the development of conditional KI mice in which expression of the KI is dependent on the expression of a Cre recombinase that can be controlled (1) by a tissue-specific promoter and (2) in a timely manner using either a modified inducible Cre (tamoxifen-inducible CreER^T, upon nucleus translocation) or an inducible promoter such as the interferon-inducible *Mx1* promoter. As mutations in sporadic MPNs are acquired in the HSCs, conditional inducible KI models efficiently mimic the human pathology using hematopoietic promoters such as the *Vav*-Cre (Croker et al., 2004) (embryonic) and the HSC-*Scl*-CreER^T (Göthert et al., 2005), with the caveats that these promoters are also expressed in some endothelial cells. CRISPR-Cas9-mediated KI models present similar interests and disadvantages than constitutive KI, being present in all cell types of the organism instead of being limited to HSCs and their progenies.

3.1 Pioneer mouse models of MPNs

Before the discovery of *JAK2*^{V617F} and other driver mutations, several models were described encompassing many features of MPN. The early ones were obtained by RV long-term exposure to cytokines, especially by the team of Gregory Johnson (Metcalf, 1998), including an **EPO RV model** leading to an erythrocytosis (Villeval et al., 1992).

Of particular interest for this review, the so-called **TPO^{High} RV model** has the merit to recapitulate the main features of the **evolution of human PMF with osteosclerosis** and is still used for investigating the mechanistic and therapeutic approaches of human PMF (see Section 4.1). This model was developed to determine the side effects associated with long-term exposure to high doses of the recently discovered TPO (Wendling et al., 1994). Several models were described, including the simplest approach by repeated systemic administration of a truncated version of recombinant TPO associated with a polyethylene glycol tail (PEG-rh-MDGF) (Ulich et al., 1996) and more sophisticated approaches by which TPO overexpression in mice was achieved by adenovirus or retrovirus vectors and engraftment or transgenesis (Cannizzo et al., 1997; Frey et al., 1998; Kakumitsu et al., 2005; Ohwada et al., 1996; Villeval et al., 1997; Yan et al., 1995, 1996). Except for one model of TG mice (Zhou et al., 1997), all the other models described the occurrence of BM fibrosis in mice. Of those, the RV model is the one that has been studied the most extensively to investigate the cells and mechanisms involved in MF. The procedure consists in transplanting BM cells transduced with a virus expressing the mouse TPO into lethally irradiated recipient mice, to sustain high levels of TPO production by the hematopoietic system throughout the life of the transplanted mouse. The result was a two-step disease. The first "plethoric" phase of the disease (2 months after BM transplantation, BMT) is characterized by thrombocytosis, leukocytosis and anemia, a striking elevation of progenitor cell numbers except for CFU-E, massive hyperplasia of both MKs and granulocytes in BM and spleen with hypoplasia of erythroblasts in the BM. The second phase ("spent" or "burnt-out" phase) is characterized by severe pancytopenia in blood and BM and extramedullary hematopoiesis in the spleen, liver, kidney, lymph nodes, and lungs. Histological examination revealed severe fibrosis in the BM and spleen and osteosclerosis in BM. All animals died within 10 months after transplantation with severe cytopenia. Notably and unique within the MPN animal models, around 5% of mice died prematurely from undifferentiated acute leukemia. Thus, this fatal myeloproliferative disorder strikingly mimics most of the features attached to the evolution of human PMF. The TPO^{High} model mimics the constitutive overstimulation of MPL and is therefore close to the *MPL*^{W515L} RV model developed by Levine's group (Pikman et al., 2006). However, the *MPL*^{W515L} RV mice involved the overstimulation of other lineages that do not terminally express MPL resulting into polycythemia at the opposite of human *MPL*^{W515L} PMF. Both models have the combined disadvantage of being extremely severe and the advantage of being useful to rapidly test drugs and study the mechanisms involved in MF. In contrast, most *JAK2*- and *CALR*-mutated models slowly develop MF after several months. A possibly re-appraised model may be the one obtained through repeated administration of the TPO peptidomimetic Romiplostim, in use for the treatment of idiopathic thrombocytopenic purpura and some other thrombocytopenia.

The **GATA-1^{low} model** derived from mice (*Gata1*^{neodHS}) with a deletion in a regulatory sequence of GATA1 expressing a deleted version of GATA-1 was developed by Orkin's group with the original aim to investigate the role of this enhancer. As a result of this deletion, mice were both anemic due to a defect in erythroid cell maturation because of a fivefold decrease in GATA-1 expression, and severely thrombocytopenic because of the almost complete lack of GATA-1 expression in MKs that remain immature and release a few abnormally large platelets. Although rapidly lethal in C57BL/6 mice, Vannucchi and Migliaccio reported that in the

CD-1 strain the deletion was viable (Martelli et al., 2005; Vannucchi et al., 2002). Mice slowly recovered from the anemia but remained thrombocytopenic and developed splenomegaly due to the expansion of the MK lineage. After a year, MF and osteosclerosis developed in the BM with splenomegaly associated to extramedullary hematopoiesis. In contrast to the TPO^{High} model, they do not die prematurely. Interestingly, the deletion introduced in the DBA/2 strain induced a severe anemia but no full myelofibrotic phenotype.

Overall, these early models provided evidence that **hyperplasia and/or dysplasia and/or impairment of the MK lineage** were grounds for the development of excess fibrosis in the marrow. Similar observations have been made in non-MPN models such as the ***hBACH1* TG** mouse resulting in MK impairment (Toki et al., 2005), the ***Ts65Dn*** mouse model of Down syndrome (Kirsammer et al., 2008), an ET model with MF caused by the loss-of-function $\boldsymbol{c\text{-}Myb^{E308G/E308G}}$ (Papathanasiou et al., 2017), the chronic induction of TPO in the $\boldsymbol{SMRT^{mRID}}$ **KI** mice (rescue by retinoid antagonists) (Hong et al., 2013) and the lack of alpha-granules in the $\boldsymbol{Nbeal2^{-/-}}$ murine model of gray platelet syndrome (GPS) (Guerrero et al., 2014).

3.2 Engineered *JAK2*-mutated mouse models

$JAK2^{V617F}$ mutation is present in the majority of PV patients (95%) and nearly half (50–60%) of ET and PMF patients. Several mouse models using different strategies expressing either murine (mu) $Jak2^{V617F}$ or human (hu) $JAK2^{V617F}$ have been generated in different genetic backgrounds (Fig. 1A). These models, mimicking the human MPN continuum, have been crucial in the understanding of the molecular pathogenesis of $JAK2^{V617F}$-driven MPNs.

From 2005, four **RV models** using BM cells transduced with a retrovirus expressing mu$Jak2^{V617F}$ were generated in BALB/c or C57BL/6 inbred strains (Bumm et al., 2006; James et al., 2005; Lacout et al., 2006; Wernig et al., 2006; Zaleskas et al., 2006). All these mice developed a disease mimicking human PV progressing or not to MF. Interestingly, leukocytosis, splenomegaly and BM fibrosis were more pronounced in BALB/c compared to C57BL/6 mice, highlighting **mouse strain-specific differences** in disease phenotype probably due to host modifiers (Bumm et al., 2006; Wernig et al., 2006). Bumm et al. (2006) observed a transient non-transplantable disease suggesting no self-renewal capacity and requirement of additional factors such as TNFα levels. Altogether, the RV models

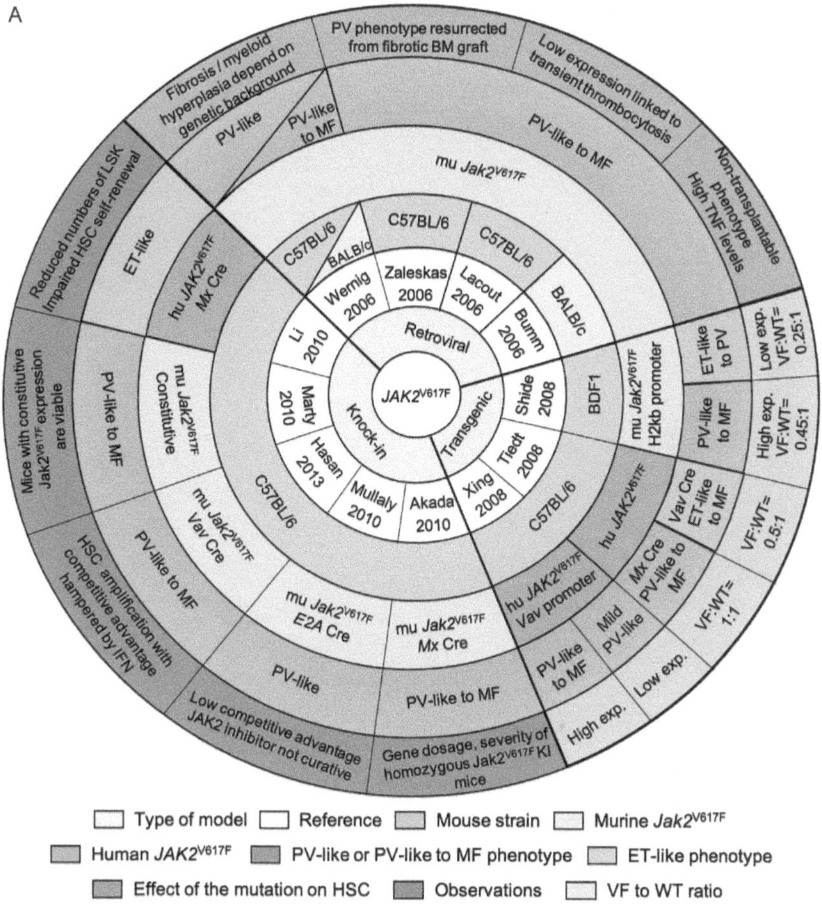

Fig. 1 (A) *JAK2*^{V617F} mouse models. Retroviral, transgenic and knock-in models have been successively developed leading to PV, ET and MF human-like diseases. Starting from the center to the outside are listed in order the mouse model references, mouse strains, the species (murine or human) mutation, the recorded phenotype and some observations related to the phenotype, the expression level or the *JAK2*^{V617F}:*JAK2*^{WT} gene expression ratio (in RV and TG models) or the effect on the stem cell compartment (KI model). (B) *CALRm* mouse models. Retroviral, transgenic (TG) and knock-in models have been successively developed. Starting from the center to the outside are listed in order the mouse model references, the mouse strain, types of KI model, the type of *CALRm*, the phenotypes obtained and in the last circle some observations for each model.

B

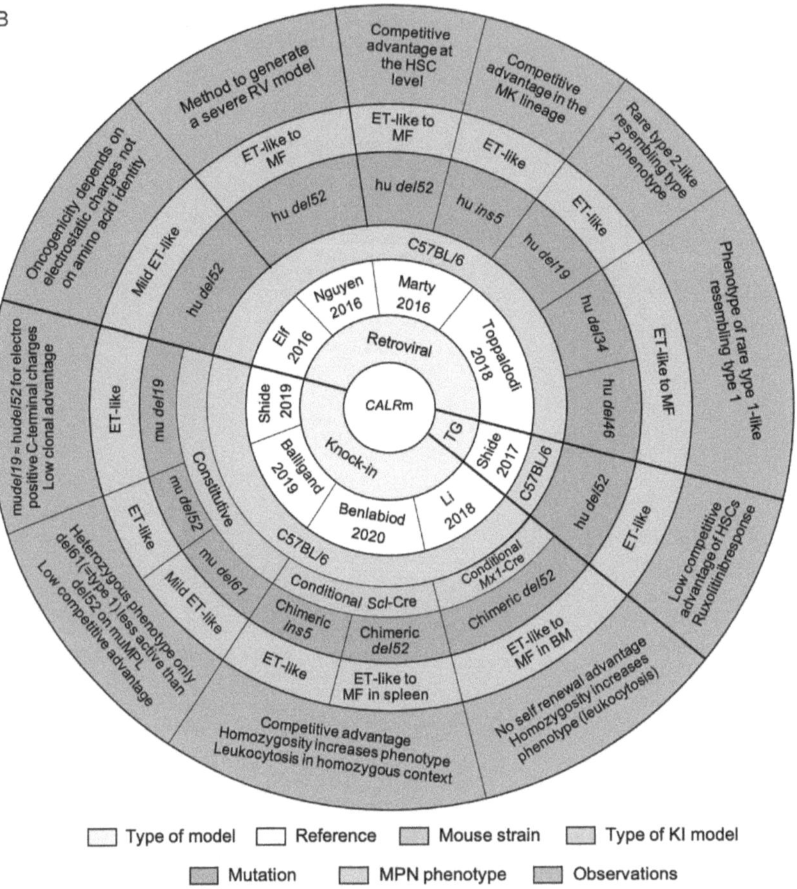

Fig. 1—Cont'd

demonstrated that the expression of **JAK2^{V617F} alone is sufficient to induce a MPN** in mice demonstrating it as a driver of the disease (James et al., 2005).

From 2008, three teams developed **TG models** by transferring mu or hu*JAK2*^{V617F} cDNA under the control of different hematopoietic and non-hematopoietic promoters into mouse oocytes to generate stable transgenic lines. In the TG model generated by Shide et al. (2008), mu*Jak2*^{V617F} was stably expressed under the control of the *H2Kb* promoter. Two lines were established in the BDF1 strain, characterized by different expression levels of JAK2^{V617F}. High expression (VF:WT = 0.45:1) of the mutant generated polycythemia progressing into MF, whereas low JAK2^{V617F}

expression (VF:WT = 0.25:1) induced an inter-individual variability (19% of mice developed PV-like features and 35% developed ET-like features) with no progression to MF. Tiedt et al. (2008) generated an original conditional TG mouse expressing several copies of the hu*JAK2*^{V617F} in inverse orientation flanked by antiparallel LoxP sites ("Flip-Flop" transgene) under the control of a minimal hu*JAK2* promoter. These mice were crossed with TG mice expressing Cre under the *Vav* promoter or the *Mx1* promoter with different doses of pIpC (IFN inducer) to generate first increasing numbers of active and second increasing deletion of hu*JAK2*^{V617F} copies. Few doses of pIpC induced high active hu*JAK2*^{V617F} gene copies (VF: WT = 1:1) leading to a PV phenotype as opposed to repeated doses of pIpC or the strong *Vav* promoter inducing the deletions of active hu*JAK2*^{V617F} gene copies (VF:WT = 0.5:1) leading to an ET phenotype. These data demonstrated that the **JAK2^{V617F}/JAK2^{WT} ratio was critical** for the phenotype manifestation. Interestingly, an equal expression of the hu*JAK2*^{V617F} and the mu*Jak2*^{WT} gene resulted in a PV and not an ET phenotype in contrast to human studies predicting an ET phenotype for a heterozygous expression of JAK2^{V617F}. Finally, Xing et al. (2008) generated TG mice using hu*JAK2*^{V617F} cDNA under the control of the *Vav* promoter also showing a correlation between gene expression levels and disease phenotype. Recently, Chapeau et al. developed a special conditional inducible TG mouse model using the *Scl-tTA-2S* tet-off system to initiate the disease but also reverse the disease hallmarks (Chapeau et al., 2019). Interestingly, this is the only one to report of **gender differences** in the manifestation of the disease phenotype where, in an unexpected way, because at the opposite of human tendency studies, female developed a PV and male developed an ET. Taken together, TG mouse models demonstrated for the first time the importance of the *JAK2*^{V617F}-to-*JAK2*^{WT} ratio and thus the dose-dependent role of JAK2 signaling in the MPN phenotype, as expected from human studies.

From 2010, **KI mouse models** were generated via homologous recombination in ES cells allowing expression of JAK2^{V617F} from its endogenous promoter. The four teams that generated a KI model used the same mouse genetic background (C57BL/6/129Sv) but different approaches. Akada et al., Mullaly et al. and Marty et al. used the mu*Jak2*^{V617F} gene, generating mice with a PV-like phenotype progressing into MF (except for the model developed by Mullaly et al.). Akada et al. (2010) generated an inducible *Mx1*-Cre/mu*Jak2*^{V617F} mouse model with an exon13 mutated cDNA and a BM JAK2^{V617F}:JAK2^{WT} 0.53:1 mRNA ratio. Both, heterozygous and homozygous mice developed a PV but in a more severe manner in the homozygous mice with a progression to MF. Marty et al. generated a constitutive heterozygous (VF:WT mRNA = 1:1) KI mouse model (homozygous mice could not be generated due to the sterility of heterozygous KI females) that developed a severe PV phenotype with a progression to MF (Marty et al., 2010). Mullaly et al. reported *E2A*-Cre/mu*Jak2*^{V617F} KI mice that developed a PV-like phenotype but did not progress to MF (Mullally et al., 2010). In 2013, using the same flexed (Flip-Excision) technology as Mullaly et al., Hasan and colleagues generated a conditional *Vav*-Cre/*Jak2*^{V617F} heterozygous KI mouse model that developed a PV-like phenotype with progression to MF after 6 months (Hasan et al., 2013). Finally, Li et al. developed a *Mx1*-Cre/huJAK2^{V617F} KI mice using the entire hu*JAK2*^{V617F} cDNA resulting in equal expression of murine and human *JAK2* transcripts. Unlike other models, KI mice preferentially developed an ET-like phenotype (10% of PV/MF phenotype) that was not transplantable, revealing a subtle competitive disadvantage relative to normal HSC in the stress associated with BMT (Li et al., 2010). This feature was not observed with mu*Jak2*^{V617F} KI mice asking the question of why the KI model with the mildest phenotype shows the strongest reduction in HSC self-renewal potential (Skoda, 2010).

Overall, several factors including the mouse strain and conditioning (irradiated recipient or native model), gene dosage, gene species (human or mouse) and the method (conditional/inducible models versus constitutive) determine the disease phenotype in engineered mice. KI mice closely recapitulated human MPNs although heterozygosity is always associated with PV, except in Green's model, and heterozygosity is associated with ET in humans (Tiedt et al., 2008). As we will see later, to design more faithful human models and afford a better comprehension of MPN disease initiation, strategies were developed to induce (conditional models) or engraft a **limiting number of mutated HSCs** within the overwhelming WT hematopoiesis, considering that in mouse models the disease is triggered as polyclonal rather than monoclonal or oligoclonal in patients.

In 2007, Scott and colleagues conducted a study on *JAK2*^{V617F}-negative PV and idiopathic erythrocytosis patients and identified four somatic mutations in the **exon 12 of the *JAK2* gene** (Scott et al., 2007). They developed a RV model in BALB/c mice to express the *Jak2*^{K539L} mutation. These mice displayed an elevated hematocrit level but lower leukocytosis and thrombocytosis than the RV *Jak2*^{V617F} mice, suggesting the preferential expansion of the erythroid lineage as observed in patients who preferentially develop an erythrocytosis. In 2016, Grisouard et al. also studied the *JAK2 exon 12* mutation but in TG mice expressing hu*JAK2-N542-E543del exon 12* mutation. They aimed to study the hematologic phenotype of these mice but also the specific changes caused by this mutation on iron metabolism and STATs (developed in chapter 5 in Grisouard et al., 2016). Briefly, mice had a more severe erythrocytosis compared to the model of Scott et al. possibly explained by the differences of strategies (Tefferi, 2016).

3.3 Engineered *MPL*-mutated mouse models

The somatic $MPL^{W515L/K}$ mutation is found in 1–5% of ET and PMF patients (Pikman et al., 2006). This group modeled MPL^{W515L} into RV mice to induce a rapid MPN disorder with high platelet counts, leukocytosis, marked splenomegaly due to extramedullary hematopoiesis, evolving into a lethal MF. Notably, both MPL^{WT} and MPL^{W515L} RV mice developed a marked increased hematocrit supported by a significant increase in BM and spleen erythroid cell populations. This characteristic is at the opposite of the anemia observed in human MF and probably due to the expression of mutant MPL in erythroid cells. Up to now, no MPL^{W515L} KI has been developed. As previously discussed, this model has been used for numerous preclinical MF studies due to the speed of symptom upset and the short survival allowing a rapid appraisal.

3.4 Engineered *CALR*-mutated mouse models (Fig. 1B)

In patients, the two major types of *CALR* mutations (*CALRm*) located in exon 9 of the gene, *del52* (type I) and *ins5* (type II), are closely found in ET while *CALRdel52* is largely predominant in PMF (70% of all *CALRm*) (Cabagnols et al., 2015). ET patients carrying the *CALRins5* mutation were reported with slightly higher platelet counts than *CALRdel52* patients (Pietra et al., 2016; Tefferi et al., 2014).

The first *CALRm* **RV models** were generated by Marty et al. to express hu*CALRdel52* and hu*CALRins5* into C57BL/6 recipient mice (Benlabiod et al., 2020). The *CALRdel52* mice developed stronger thrombocytosis than *CALRins5* mice. Mice expressing a *CALR* lacking the entire exon 9 did not develop the disease indicating that the **specific mutated C-terminal tail sequence** rather than the absence of the WT C-terminal sequence plays a role in the development of the disease. Six months after transplantation, *CALRdel52* mice, and in rare cases *CALRins5* mice, showed a decrease in platelet counts, anemia, BM hypocellularity and splenomegaly due to extramedullary hematopoiesis including MK hyperplasia, associated with the development of MF and osteogenesis. Shortly after, Elf et al. generated a hu*CALRdel52* RV model that developed a mild thrombocytosis explained by BM MK hyperplasia (Elf et al., 2016). No evidence of fibrosis was shown. Afterward, a third hu*CALRdel52* RV mouse model was evidenced to develop marked thrombocytosis progressing to MF, like Marty et al. (Nguyen et al., 2016). Toppaldoddi et al. modeled rare type 1-like and spi2;type-2 like *CALRm* in mice (Toppaldoddi et al., 2019). Mice expressing type 1-like *CALRdel34* and *CALRdel46* displayed a phenotype resembling the *CALRdel52* mice. Similarly, the mild thrombocytosis developed by the type 2-like *CALRdel19* RV mice was close to that of *CALRins5* mice, demonstrating that **type 1-like and type 2-like CALR mutants** are structurally and functionally similar to CALRdel52 and CALRins5 mutants, respectively.

In 2017, Shide et al. generated the only *CALRm* **TG model** (Shide et al., 2017). They expressed hu*CALRdel52* under the control of murine *H-2Kb* promoter. The phenotype developed by this TG mouse was surprisingly milder than the RV models with a moderate thrombocytosis, no significant splenomegaly and no MF.

Mu*CALRm* at the homologous position to the hu*CALRm* does not generate the same amino acid sequence and thus, the same mutated C-terminal tail. Therefore, in 2019, Li et al. generated conditional *CALRdel52* **KI mice** expressing a mu*Calr* with the human mutated exon 9 that were crossed with inducible *Mx1*-Cre TG mice (Li et al., 2018). The phenotype, including thrombocytosis, was more severe in homozygous compared to heterozygous KI mice. Homozygous *CALRdel52* KI mice also developed leukocytosis and displayed a mild decrease in hematocrit at 4 months post-induction when moderate fibrosis was observed in the BM, accompanied by MK hyperplasia and splenomegaly associated with an increase in HSCs, megakaryocytic (MkPs) and other myeloid progenitors. Thus, these KI mice recapitulate an ET-like phenotype progressing toward MF. It was later demonstrated that the global charge of the mutated C-terminal tail rather than the amino acid sequence is important in disease development (Balligand et al., 2016; Shide et al., 2019). As it was shown that the oncogenic effect of human CALR was more the consequence of the positive charge of exon 9 than the amino acid sequence itself. Thus, it was possible to generate KI mice with the mu*Calr*. Using the CRISPR-Cas9 technology, Balligand et al. generated constitutive heterozygous mu*Calrdel52* and mu*Calrdel61* KI mice, two type 1 mutations that differ by the absence of three amino acids at the outset of the new positively charged C-terminal tail of *Calrdel61*

(Balligand et al., 2020). Mu*Calrdel61* KI mice displayed a very similar phenotype than control littermates, with the exception of a 15% increase in platelet count. In comparison, mu*Calrdel52* KI mice showed significant thrombocytosis with mild MK hyperplasia in the BM suggesting that a small difference in the amino acid sequence may be important for the oncogenicity of the mutant protein and that not all type 1 CALR mutants behave the same. The same year, Shide et al. published their work on constitutive mu*CALRdel19* KI mice using the CRISPR/Cas9 system in C57BL/6 ES cells (Shide et al., 2019). The mu*Calrdel19* generates a very similar C-terminal tail than the hu*CALRdel52*. These mice developed a mild increase in platelets and BM MKs with no evidence of fibrosis. Splenomegaly was accompanied by extramedullary hematopoiesis (MKs and myeloid progenitors). Lastly, Benlabiod et al. generated heterozygous and homozygous inducible conditional KI mice expressing mu*Calrdel52* and *Calrins5* with the human mutated C-terminal tail (Benlabiod et al., 2020). They showed that *CALRdel52* mice presented a more severe phenotype compared to *CALRins5* mice including thrombocytosis, leukocytosis, BM hypocellularity compensated by extramedullary hematopoiesis leading to splenomegaly. The phenotype was more severe in the homozygous context.

Most *CALRm* mouse models recapitulated an ET-like disorder. It seems that type-1 and type-1 like mutations induce an ET-like phenotype progressing toward mild MF for certain but not all models, when type-2 and type-2 like mutations induce a milder ET-like phenotype with barely any MF progression. These observations recapitulate what is observed in human MPN patients, except for the difference in platelet counts that is higher in *CALRdel52* than in *CALRins5* KI mice while the opposite is true in ET patients. To explain these differences, multiple hypotheses are to be considered including maybe the role of the homozygous clones in the severity of the disease that will be further discussed in Section 4.3. Moreover, the general phenotype induced by *CALRm* observed in mouse is moderate compared to patients, possibly due to the difference in the activation of the muMPL versus the huMPL, which will be developed in Section 4.5.

4. Lessons from MPN mouse modeling: What it means for patients

4.1 Fibrosis (Fig. 2)

Fibrosis, or scarring, is part of the normal process of tissue repair following an injury. It is defined by the accumulation of disorganized fibers or

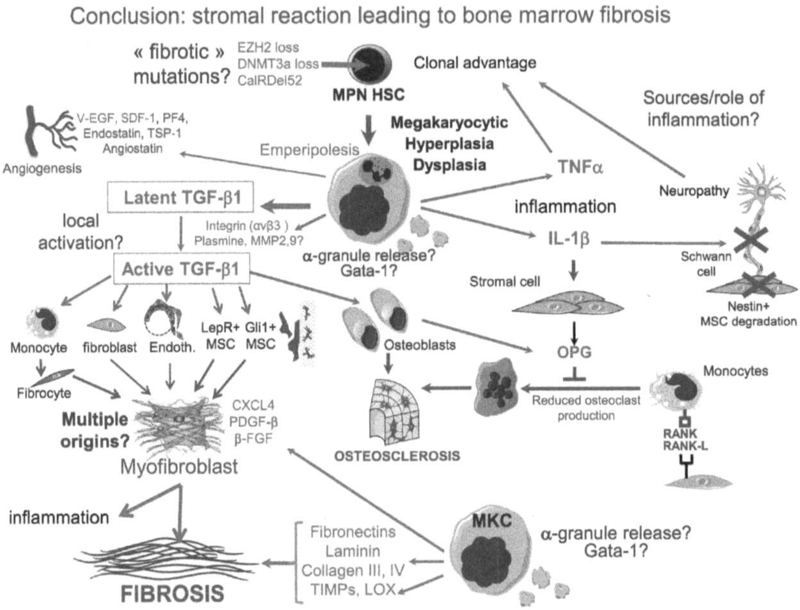

Fig. 2 Main players of the stromal reaction leading to MF identified in MPN mouse models. MKs are the central player in MF. Most MK-derived proteins are found in excess in MF recapitulating the described "cytokine storm" (IL-1, TNF, PDGF, V-EGF, FGF, PF4, ...). It is also the major cellular source of **TGF-β1** in BM. TGF-β1 is required for fibrosis and plays a pivotal role in the physiology of MF. Still, two questions remain related to TGF-β1: (1) its release: TGF-β1 is contained in MK α-granules and should not be released unless a process of activation, degradation or dysplasia (apoptosis/necrosis/senescence) is triggered. What kind of process releases TGF-β1 in MF? Hypotheses are "dysplasia" through gene deregulation (*EZH2, DNMT3*) or defective granule storage possibly due to low GATA-1 levels, abnormal MKC activation induces by JAK/STAT constitutive activation or neutrophil activation in emperipolesis through P-selectin expression; (2) its activation: TGF-β1 is always released as an inactive latent form and sequestered in the ECM. What kind of TGF-β1 activation mechanism occurs in MF? Possible trigger candidates are MK αvβ3 integrin or nonmyelinating Schwann cell integrin αvβ8 also involved for TGF-β1-induced HSC dormancy. Similarly, **a-SMA myofibroblasts** are also key players in MF being the producer of main ECM components (collagen, glycoprotein, proteoglycan, ...) and inflammatory mediators (TGF-β1, IL-1, IL-8, M-CSF and GM-CSF, ...). They are regulated mainly by TGF-β1 but also TNFα, IL-1β or IL13. However, the mechanism of production and **origin(s) of myofibroblasts** or other ECM cell producers are largely questioned. The candidates are (1) monocytes-derived fibrocytes, (2) vascular endothelial cells through an endothelial-to-mesenchymal transition (EndMT), (3) Gli1-positive and/or (4) LepR1-positive mesenchymal stromal cells. Genetic modifiers of MF have been suspected in C57BL/6 versus BALB/c and some mutations favor or accelerate MF occurrence as *EZH2, DNMT3a*. **Osteosclerosis**, as the final MF grade, involves **OPG** through reduced osteoclast production and disruption of bone resorption (osteoclastogenesis) as well as increase bone formation by osteoblasts. Besides this, it is clear that length of **exposition to the neoplastic clone(s)** (an injury by itself) but also with the **JAK2^{V617F}:JAK2^{WT} ratio** determines the slow or rapid apparition of fibrosis in the marrow.

extracellular matrix (ECM) components, mainly collagen, in place of the normal functional tissue. If the wound quickly heals, minimum collagen deposition occurs and no scar is formed. If the injury is prolonged or repeated, extensive deposition of collagen occurs and a pathological/chronic fibrosis takes place. This is observed in virtually all organs: marrow, liver, kidney, heart, lung and gut. MF is observed in end-stage MPNs but also many other hematologic diseases and non-hematologic disorders often associated with inflammation. In MPN, fibrosis appears as a natural evolution and secondary to ET or PV (post-ET or -PV MF), as these diseases cause injury to the marrow. Such a "transformation" occurs in around 20% of patients around 20 years after the initial ET or PV diagnosis. It may also occur without apparent cause or antecedent diagnosis of ET or PV as primary MF (PMF) (previously called idiopathic MF for that reason).

MPN animal models encompass PMF, post-PV MF (PPMF) and post-ET MF (PTMF). The PV or ET caused by *JAK2*^{V617F} expression in HSC leads to secondary MF in most animals if the disease was prolonged long enough (≥6 months). In RV models, PPMF was severe (Lacout et al., 2006), moderate or absent in C57/Bl6 mice (Wernig et al., 2006) and less severe compared to BALB/c mice (Bumm et al., 2006; Wernig et al., 2006), suggesting **genetic modifiers** for fibrosis. In all models, a recurrent finding was that occurrence, speed and grading of fibrosis were correlated not only with the **length of exposition** to the neoplastic cells (injury) but also with the **JAK2^{V617F}:JAK2^{WT} ratio** (Tiedt et al., 2008; Xing et al., 2008). Interestingly, transplantation of BM from a primary recipient in the fibrotic phase resurrected a PV phenotype in secondary mice showing that the fibrosis was not due to a genetic transformation of the "seed" itself (the *JAK2*^{V617F} HSC) but rather to the progressive **seed-induced "modification" of the "soil"** (stromal cells) that does not carry a transplantable driver mutation (Zaleskas et al., 2006). Related to the *CALR*m, it has been shown that PTMF is a relatively late event (≈6 months in the RV model) and is more frequent and severe in decreasing order with the **homozygous *CALRdel52*, the heterozygous *CALRdel52* and the heterozygous *CALRins5***. A discrepancy was also observed in the location of fibrosis in the BM for one model (Li et al., 2018) and in the spleen for another (Benlabiod et al., 2020).

The **TPO^{High} model** was and remains the ultimate model of PMF or PTMF (see Section 3.1). Because of its precocity, rapidity and simplicity, this model has provided numerous information related to the physiopathology of MF. The first lesson and merit of this model were to show that a prolonged amplification and chronic stimulation of the megakaryocytic lineage could by itself generate a disease fully mimicking PMF, making the **MK the central player in PMF**. The second lesson was to demonstrate the **pivotal role of TGF-β** in the physiopathology of MF in showing that no fibrosis (and reduced osteosclerosis) is produced in this RV model from TGF-β deficient hematopoietic donor stem cells (Chagraoui et al., 2002). This result was confirmed using TGF-β inhibitors in the TPO^{High} but also the $JAK2^{V617F}$, MPL^{W515} and $GATA\text{-}1^{Low}$ models (Gastinne et al., 2007; Mascarenhas et al., 2014; Yue et al., 2017; Zingariello et al., 2013). MK is the major source of TGF-β in the BM and two questions remain to be answered: (1) first, as TGF-β is contained in the alpha-granules and should not be released by MKs unless a process of activation or degradation is triggered, we are left to wonder what mechanism may induce the release of TGF-β in PMF and (2) second, TGF-β is released as an inactive latent form and sequestered in the ECM, thus questioning on what might be the TGF-β activation mechanism in PMF. Thrombospondin appeared as an obvious candidate for TGF-β activation but fibrosis is normally developed from thrombospondin-deficient hematopoietic donor stem cells in the TPO^{High} model. Possible other candidates remain such as MK αvβ3 integrin or nonmyelinating Schwann cells αvβ8 integrin involved in TGF-β-induced HSC dormancy (Yamazaki et al., 2011). Interestingly, using the TPO^{High} model with **osteoprotegerin** (OPG) KO recipients, it was shown that OPG is responsible for osteosclerosis appearing at the final MF grade, through disruption of BM resorption (osteoclastogenesis) (Chagraoui et al., 2003).

TGF-β1 is the master regulator of fibrosis in particular as the main regulator of fibrosis-driving **stromal cells**. For many years the precise identity, origin and mechanism of production of these stromal cells, critical for BM fibrosis, remained an enigma. Using different MPN models, different teams described three origins of these cells. In the MPL^{W515L} RV model, neoplastic collagen- and fibronectin-producing monocyte-derived **fibrocytes** were shown to be major contributors to fibrosis, and the specific fibrocyte inhibitor Serum Amyloid P slowed down the development of fibrosis in MPL^{W515L} RV mice and increase survival in xenograft mice (Verstovsek et al., 2016). In a $Jak2^{V617F}$ model, vascular endothelial cells showed mesenchymal functional and morphologic changes through a process called **endothelial-to-mesenchymal transition** (EndMT) (Erba et al., 2017). EndMT occurs within MK proximity and MK-conditioned medium including TGF-β. More recently, using TPO-induced PMF and/or $JAK2^{V617F}$ models and genetic tracing, two groups reported the discovery of **myofibroblasts** as fibrosis-producing cells in the BM (Decker et al., 2017; Schneider et al., 2017). Myofibroblasts are large specialized migrating contractile/secretory cells expressing a-smooth muscle actin (a-SMA) and myosin bundles and producing large amounts of collagen, glycoprotein, proteoglycan and inflammatory mediators. They are the main ECM producers. In solid organ fibrosis, these cells are well characterized (Kramann et al., 2013). In particular, they were reported to express the hedgehog (Hh) transcriptional activator Gli1 (Kramann et al., 2015). Schneider et al. showed that **Gli1^{+} cells** are recruited from the endosteal and perivascular niche to become fibrosis-driving SMA myofibroblasts in the BM (Schneider et al., 2017). Ablation of these Gli1^{+} cells using the diphtheria toxin receptor, abolished BM fibrosis and rescued BM failure. They validated the same pathway in human fibrosis and suggested that Gli protein could be a relevant therapeutic target. Decker et al. more recently found that BM leptin receptor (Lepr)-expressing mesenchymal stromal lineage cells expanded extensively and were the fibrogenic cells in PMF (Decker et al., 2017). These stromal cells down-regulated CXCL12 and SCF compromising HSC niche and upregulated ECM genes associated with fibrosis and osteogenesis, indicating fibrogenic conversion. They also demonstrated that **LepR1^{+} cells** express platelet-derived growth factor receptor α (PDGFRα) that is critical in the development of BM fibrosis because LepR1^{+} cell-specific KO of PDGFRα resulted in greatly reduced reticulin fibrosis. If myofibroblast precursors became important contributors to MF, an open question remains whether the reported Gli1^{+} and LepR1^{+} cells overlap or whether multiple distinct myofibroblast precursors exist in the BM (Kramann and Schneider, 2018). Of note, in a *Jak2*^{V617F} TG MPN model, Arranz et al. performed lineage-tracing using Nestin-CreER^{T} to assess the contribution of Nestin^{+} stromal cells (Arranz et al., 2014). No obvious contribution of these stromal cells to BM fibrosis was noted.

Additional mutations are observed in MF patients compared to ET/PV patients and animal models combining mutations with *JAK2*^{V617F} have been useful to determine which mutations are involved in fibrosis progression. It was shown that mutations of two genes encoding epigenetic regulators **EZH2**, involved in histone modification, and **DNMT3a**, involved in DNA methylation, cooperate with JAK2^{V617F} in the fibrosis process. First, three groups independently showed that EZH2 deficiency leading to decrease repressive function of the polycomb repressive complex 2 (PRC2) as attested by a decreased in H3K27me3, caused increased megakaryopoiesis and platelet production at the expense of erythropoiesis in *JAK2*^{V617F} mice followed by a strong myelofibrosis (Sashida et al., 2016; Shimizu et al., 2016; Yang et al., 2016). It is associated with an increased expression of several PRC2 target genes including ***Hmga2***, a gene activated in PMF patient $CD34^+$ cells (Guglielmelli et al., 2007), which forced expression increases megakaryopoiesis in mice (Shimizu et al., 2016). More recently CRISPR-Cas9 disruption of *Dnmt3a* in *Jak2*^{V617F} KI HSCs blocked the accumulation of erythroid elements in BM and spleen and promoted fibrotic infiltration. Mechanistically this deletion partially mimics the loss of EZH2 with a decreased PTC2 activity, increases chromatin accessibility at active enhancers controlling the expression of pro-inflammatory genes such as *TNFα* via the NF-κB pathways (Jacquelin et al., 2018). For complement information, see the following recent review (Jacquelin et al., 2020).

4.2 Thrombo-hemorrhagic complications (Fig. 3)

Thrombosis (≈60% arterial) and hemorrhages (especially in MF) are the major causes of morbidity and mortality in MPN (12–39% in PV and 11–25% in ET) and often the initial event leading to the diagnosis (≈30% patients) so that MPN is often considered as an acquired thrombophilic disorder. The risk factors for patients are known and are the advanced age (>60 years), a prior history of thrombosis, leukocytosis, thrombocytosis (ET/MF), high hematocrit value (>45% in PV) and *JAK2*^{V617F} allele burden (especially in ET) (Barbui and Falanga, 2016). However, due to patient heterogeneity with different risk factors and especially continuous prophylactic treatments, the physiopathology and efficacy of treatment remain difficult to investigate. While examining *JAK2*^{V617F} animal models, were reported a high percentage of unexpected deaths with a sign of cardiac thrombosis and gangrenous bowel (Akada et al., 2010; Mullally et al., 2010). Therefore, MPN animal models became precious tools to investigate the direct role of the suspected players of these hemostasis disorders such as platelets, leukocytes, erythrocytes, endothelial cells and the driver mutations. Nevertheless, depending on the models, there were variable thrombotic and bleeding complications and to determine whether the hemostasis phenotype was associated with BM fibrosis added to the complexity of the analysis.

In accordance with previous observations in a RV model (Zaleskas et al., 2006), several investigations in 2013–2014 showed that most *Jak2*^{V617F} mouse models displayed hemorrhage signs with an increased bleeding time. Lamrani et al. (2014) focused their studies on **platelet** functions and found

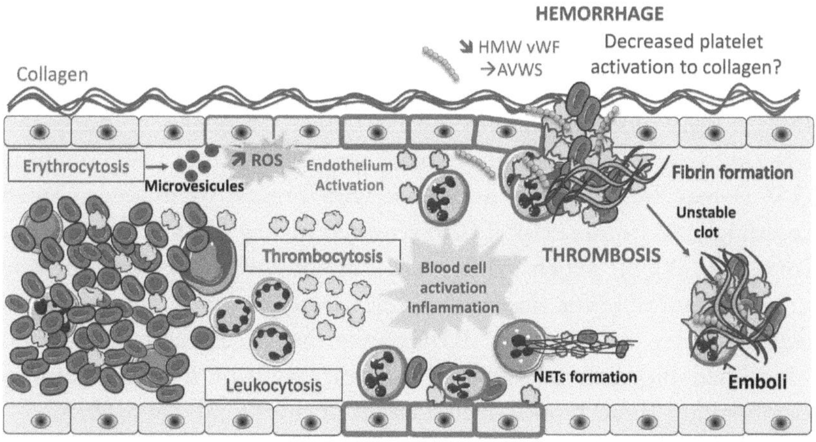

Fig. 3 Hemostasis of MPN animal models. *JAK2*^{V617F} KI animal models experiment frequent sudden and unexpected death with obvious signs of bowel and lung thrombosis or hemorrhages. In most models, an increased bleeding time is observed with an accelerated formation of unstable clots ($FeCl_3$-injury model) and embolization. Studies showed that MPN platelets present minor defects with a decreased reactivity to agonists in most models. However, thrombocytosis, concomitant to enhanced vWF proteolysis, is involved in high molecular weight (HMW) vWF polymer clearance leading to acquired vWF syndrome (AVWS). Comparison with other models showed that most hemostasis disorders are not directly caused by the *JAK2*^{V617F} mutation but rather by the full disease development with possible fibrosis but especially inflammation. Endothelial dysfunctions have been implicated with a pro-inflammatory and pro-adhesive leukocyte phenotype attributed to *JAK2*^{V617F} mutated cells, pro-inflammatory cytokines and *JAK2*^{V617F} erythrocyte-derived microvesicles increasing endothelial oxidative stress, NO pathway inhibition and vascular arterial spasms. Finally, increased propensity for NET formation, possibly due to increased neutrophils with an activated phenotype, has been shown in *Jak2*^{V617F} KI mice and linked to JAK2^{V617F} expression and thrombosis. HMW: high molecular weight, vWF: von Willebrand factor, AVWS: acquired vWF syndrome. The membrane of activated cells is highlighted in red. This figure was partially created using images from Servier Medical Art Commons Attribution 3.0 Unported License.

that PV-like *Jak2*^{V617F} KI mice (Hasan et al., 2013) are characterized by (1) a mild GPVI deficiency, (2) low platelet activation response to GPVI, CLEC-2 and thrombin receptor agonists, (3) a decreased plasma content in large von Willebrand factor (vWF) multimers known to have increased hemostatic potential and, (4) an accelerated formation of unstable clots in $FeCl_3$-injury model explaining the bleeding tendency. Overall, the study suggested that clot instability was due to enhanced vWF proteolysis with a concomitant high molecular weight (HMW) vWF polymer clearance by platelets otherwise presenting minor defects. GPVI downregulation, hemorrhage or thrombosis phenotype occurred 2 months after tamoxifen induction in the *Scl*-CreER^T model demonstrating that they were not directly caused by the *JAK2*^{V617F} mutation but rather by the full disease development. Indeed, most of these events were also observed in the anemic TPO^{High} PMF model and not in an EPO-induced model of erythrocytosis, suggesting that GPVI deficiency, bleeding and acquired vWF syndrome (AVWS) may be somehow associated with fibrosis or its development. Accordingly, Matsuura et al. (2020) reported in *Vav-hJAK2*^{V617F} TG mice (Xing et al., 2008), which exhibit hallmarks of MF, a deficiency in dense granules responsible for impaired platelet secretion upon activation with collagen. Roles of leukocytes, RBCs, endothelial cells and inflammation in increased bleeding of *Jak2*^{V617F} KI mice remain to be determined. Corroborating these results, Etheridge et al., using *Tie2*-Cre/*JAK2*^{V617F} TG mice (TG mice from Tiedt et al. expressing human JAK2^{V617F} in **endothelial** and hematopoietic cells) displaying an ET/PMF phenotype, also showed attenuated arterial thrombosis after $FeCl_3$ injury and an AVWS (Etheridge et al., 2014). Transplantation experiments showed that delayed occlusion after injury required JAK2^{V617F} expression in both hematopoietic and endothelial cells arguing, in agreement with Lamrani et al. data, against the responsibility of an isolated *JAK2*^{V617F} platelet defect. Indeed, platelet-targeted expression of *Jak2*^{V617F} using the *PF4*-Cre TG mice did not lead to dysfunctional hemostasis (Etheridge et al., 2014). Interestingly, decreased clot stability was observed in *Tie2*-Cre/hu *JAK2*^{V617F}, *Vav*-Cre/hu*JAK2*^{V617F} as in *Vav*-Cre/and *Scl*-CreER^T/mu *Jak2*^{V617F} mice, all models displaying a PMF-like phenotype. In contrast to both studies, Hobbs et al. using the *JAK2*^{V617F} KI mouse model from Li et al., showed increased platelet reactivity to some agonists (collagen and thrombin) and decreased tail bleeding volume (Hobbs et al., 2013). Differences may be attributable to the mild ET phenotype of this model.

More recently, different teams investigated, as Etheridge et al., the role of *JAK2*^{V617F} endothelial cells in MPN hemostatic disorders as the mutation was found in these cells in patients, although at low frequencies and still the subject of controversy (Rosti et al., 2013). Rautou's team studied the role of *JAK2*^{V617F} endothelial cells in the severe presentation of patients with the Budd-Chiari syndrome that is defined as a hepatic venous outflow obstruction with liver injury and fibrosis, MPN being the leading cause (25–50%). Using a model based on *Cadherin5*-CreER^T/*Jak2*^{V617F} KI mice they did not find a role for these mutated endothelial cells (Poisson et al., 2018). However, the same team using the same mouse model, demonstrated that *Jak2*^{V617F} erythrocyte-derived microvesicles overexpressed myeloperoxidase increasing endothelial oxidative stress and NO pathway inhibition and vascular hyperreactivity to vasoconstrictors (Poisson et al., 2020). Arterial spasms were reduced by the antioxidant simvastatin. More recently, Guy et al. used a *Pdgfb*-CreER^{T}/*Jak2*^{V617F} model and demonstrated that expression of the mutation in endothelial cells induced a pro-adhesive **leukocyte** phenotype due to increased expression of P-selectin and favors venous thrombosis (Guy et al., 2019). This suggests that mutation may affect the pro-thrombotic nature of endothelial cells. The phenotype was reversed by hydroxyurea. Castiglione et al. confirmed that *JAK2*^{V617F} mutation (using *Tie2*-Cre/*JAK2*^{V617F} mouse model) promotes endothelial dysfunction (Castiglione et al., 2020). This resulted in a pro-inflammatory and pro-adhesive phenotype through an upregulated expression of pro-inflammatory cytokines (IL-1β) and platelet endothelial cell adhesion molecule (PECAM).

Leukocytosis has been linked to an increased risk of thrombosis in patients and MPN neutrophils exhibit enhanced activation. Upon stimulation, neutrophils can expel extracellular strands of decondensed DNA and produce neutrophil extracellular traps (NETs) implicated in the pathogenesis of autoimmunity and thrombosis. Ebert's team showed an increased propensity for NET formation in *Jak2*^{V617F} KI mice (Wolach et al., 2018). They also demonstrated that reduction of NET formation by a JAK1/2 inhibitor or through PAD4 inhibition reduces in vivo thrombosis in mice. Overall, they linked JAK2^{V617F} expression to NET formation and thrombosis.

Finally, several studies using animal models have implicated additional mutation found in MPNs such as deficiency or inactivating mutations in *TET2* and/or *DNMT3a* in atherosclerosis/heart failure (Fuster et al., 2017; Sano et al., 2018) through **inflammation**-associated mechanisms that may be appropriate targets for therapies, confirming data from patients (Dorsheimer et al., 2019). These results (1) are consistent with the variability of hemostatic disorders observed in MPN patients; (2) emphasize the disease phenotype, including its inflammatory components rather than the sole role of MPN mutations in the hemostatic disorders and (3) offer the opportunity to develop biomarkers and appropriated therapies in MPN to avoid one of its leading cause of morbidity.

4.3 MPN disease initiation/clonal advantage (Fig. 4)

MPNs, like cancers, progress as clonal disorders. The mutated clone(s) expand(s) in a hierarchical structure from a stem cell(s) with preserved differentiation capacities to different stages of differentiation with different

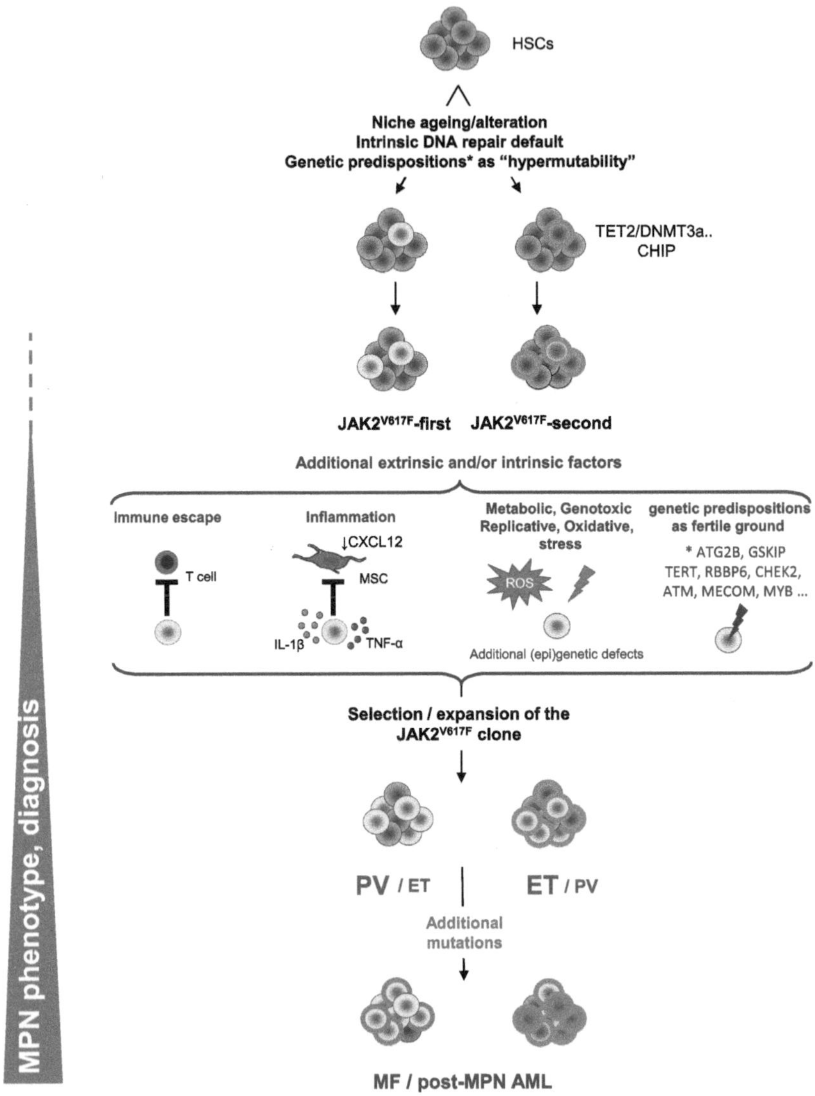

Fig. 4 MPN disease initiation. The initial JAK2^{V617F} mutation may be caused in the HSC by the niche alteration, intrinsic DNA repair defaults or genetic predispositions resulting in "hypermutability" or even induced HSC fitness change "fertile ground" (germline mutations not represented). This mutation is a common age-related mutation in clonal hematopoiesis but other mutations, such as *TET2* or *DNMT3* mutations, are more frequent and may precede the acquisition of JAK2^{V617F}. This order may modify later the MPN phenotype. At this stage, the mutated HSC may be eliminated, kept at low level without MPN diagnosis or amplified into the development of a clonal hematopoiesis of indeterminate potential (CHIP). However, several lines of evidence have shown that *JAK2*^{V617F} occur several years before the MPN manifestation. Extrinsic factors, associated

(Continued)

mechanisms that endow its progeny with competitive advantages (Nguyen et al., 2012). In MPN, especially in ET and PV, the number of somatic mutations is relatively low, with sometimes only one identified isolated driver mutation. Driver mutations activate the cytokine JAK/STAT signaling pathway, explaining the strong amplification of late stages of differentiation in MPN with less amplification of the immature ones (Delhommeau et al., 2007). Therefore, the role of these mutations in clonal amplification has been seriously questioned and the importance of additional molecular events has been evaluated. The case of $JAK2^{V617F}$ is iconic. Some groups noticed in animal models, a modest or no effect on the immature LSK compartment (Mullally et al., 2010) and others even impaired HSC functions with reduced self-renewal capacities (Kent et al., 2013; Li et al., 2010). These data suggested that the acquisition of this sole mutation was insufficient for clonal expansion and disease progression and may even cause a neoplastic HSC exhaustion. Other data showed that $JAK2^{V617F}$ was capable by itself to amplify by around fourfold the pool of early progenitor cells (LSK or SLAM) (Akada et al., 2010; Hasan et al., 2013). However, all these data on clonal expansion might have been misleading because the mutation was expressed in all HSCs and not at the clonal level (TG and KI).

Using the setting of competitive transplantation, Mullally et al. showed that in a 1:4 $JAK2^{V617F}$:$JAK2^{WT}$ ratio, mutated SLAMs (Lin^{-} $c\text{-}Kit^{+}$ $Sca\text{-}1^{+}CD48^{-}CD150^{+}$) out-competed normal SLAM (Mullally et al., 2012) in contrast to another team (Li et al., 2010). We could also show that 17 $JAK2^{V617F}$ SLAMs (corresponding to approximately three HSCs) in a ratio of 1/10 mutated/normal SLAMs could sustain the disease for over 6 months (Hasan et al., 2013). Finally, the group of Skoda used a single HSC (SLAM, $CD34^{-}Flk2^{-}$) transplantation to study whether mono clonal $JAK2^{V617F}$ hematopoiesis was sufficient to induce a MPN in mice (Lundberg et al., 2014). Only a small subset of mice developed long-term reconstitution (6/113) and a MPN with exclusive erythrocytosis or

Fig. 4—Cont'd with the immune escape, inflammation, the metabolic, genotoxic, replicative or oxidative stress and intrinsic factors such as previous or additional mutations or genetic predispositions providing a "fertile ground" contribute to the expansion of the mutated clone. The role of these factors in increased $JAK2^{V617F}$ allele burden and MPN development have been investigated and already demonstrated in animal models, such as high TNF-α and IL-1β levels for inflammation and *Tet2* or *DNMT3a* as primary/secondary mutations. The same factors, including specific mutations involving EZH2, ASXL1, *IDH1/2*, *TP53* as shown in MPN models, are involved into the evolution of PV and ET into myelofibrosis or acute leukemia transformation.

thrombocytosis. Secondary transplantation of BM from the only ET mouse did not induce either a stable reconstitution or a phenotype. This study demonstrates that a **single HSC expressing *JAK2*^{V617F} can initiate a MPN**. The low disease occurrence frequency may be due to the TG model (Tiedt et al., 2008) with an increasing huJAK2^{V617F} expression that might lead to HSC exhaustion in transplanted mice (Lundberg et al., 2014).

Recently a *PF4*-Cre/*Jak2*^{V617F} KI mouse model was described developing a full PV-like disease at 10 weeks of age from a very low number of recombined HSCs and without using deleterious irradiation or transplantation that may introduce a bias in the results (Mansier et al., 2019). A progressive amplification of all immature progenitors was observed without amplification of the HSC pool showing a minor impact of the mutation on the more immature cells but confirming *JAK2*^{V617F} HSC fitness, although possibly low, to give rise alone to MPN.

Discrepancies between results in *JAK2*^{V617F} MPN models may be linked to the fact that the mutation was either expressed as its human or mouse form, overexpressed (RV) or expressed at low levels, and the ratio between the JAK2^{V617F} and JAK2^{WT} proteins remains yet to be evaluated since no specific antibody is available. Nonetheless, the capacity of *JAK2*^{V617F} to initiate a disease by itself maybe clearly amplified or favored by intrinsic or extrinsic factors such as genetic predispositions (Bellanné-Chantelot et al., 2020; Saliba et al., 2015), metabolic reprogramming (obesity, diabetes, metabolic diseases) (Faubert et al., 2020), inflammation/immunity (Greten and Grivennikov, 2019; Grivennikov et al., 2010), additional mutations (*TET2*, *DMNT3A*), microenvironment or aging.

In the first instance, as discussed later, several genetic defects such as ***DNMT3a* loss** (Challen et al., 2011) or ***TET2* loss** (Moran-Crusio et al., 2011), cooperate with *JAK2*^{V617F} to amplify its ability for clonal expansion (Chen et al., 2015; Jacquelin et al., 2018; Kameda et al., 2015). These epigenetic defects involved in HSC fitness are also the most common age-related mutations in clonal hematopoiesis with *JAK2*^{V617F} (Busque et al., 2012; Jaiswal et al., 2014; Xie et al., 2014), comforting the hypothesis that *JAK2*^{V617F} is also an early initiating event for MPN. *DNMT3a* mutations and *TET2* loss often precede but may also follow *JAK2*^{V617F}, this order influencing the phenotype as shown by Green's group for *DNMT3a* (Nangalia et al., 2015) and for *TET2*, to induce when first an ET and when second a PV, in young age with high risks of thrombocytosis (Ortmann et al., 2015). Following *JAK2*^{V617F}, these mutations may have occurred through the genetic instability generated by *JAK2*^{V617F} mutation as it was shown in mice to mediate a decrease in the detoxifying enzyme catalase resulting in an increase of reactive oxygen species (ROS) and oxidative stress leading to DNA damages (Marty et al., 2013).

Extrinsic factors seem to favor MPN disease development. Indeed, it was shown that gene dysregulations leading to PTPN11 phosphatase activation or IκBα or Rb loss, could promote the development of a myeloproliferative disorder (although not a MPN per se) in mice through interaction with the BM microenvironment (Dong et al., 2016; Rupec et al., 2005; Walkley et al., 2007). Arranz et al. demonstrated that BM **Nestin^{+} mesenchymal** stem cells were reduced in *JAK2*^{V617F}-mutated mice due to BM neural damage and Schwann cell death caused by IL-1β produced by mutant HSCs (Arranz et al., 2014). The in vivo depletion of Nestin^{+} cells induced the expansion of mutated HSCs and the progression of the MPN phenotype. Very recently, **IL-1β deficiency** was shown to reduce the MPN phenotype of the engraftment of *JAK2*^{V617F} TG BM cells in competitive experiments with WT cells, demonstrating that IL-1β, produced by the *JAK2* mutated cells, favors early clonal expansion and is required for optimal MPN disease initiation. Rai et al. demonstrated that IL-1β secreted from *JAK2*^{V617F}-mutated cells is involved in early clonal expansion and is necessary for MPN disease initiation (Rai et al., 2019). Moreover, Tyner et al. (2010) observed an increase in tumor **TNFα levels** in a *JAK2*^{V617F} RV model that was confirmed in patients by Fleischman et al. (2011). The authors were able to show that TNFα promotes the clonal expansion of *JAK2*^{V617F} cells in a *JAK2*^{V617F} RV model using $TNF\alpha^{+/+}$ or $TNF\alpha^{-/-}$ donor cells. Hosts grafted with $TNF\alpha^{-/-}$ BM displayed reduced splenomegaly and a loss of the competitive advantage over *JAK2*^{WT} cells in the spleen and BM. Interestingly, *JAK2*^{V617F}-*Dnmt3a*-Cas9 LSKs with increased competitive advantage showed a strong enrichment of inflammatory pathways, and specifically of TNFα (Jacquelin et al., 2018).

Due to its relatively recent discovery, the capacity of ***CALRm*** to provide clonal selection has been less investigated. However, *CALRm* patients generally reach a higher allele burden in granulocytes at diagnosis and develop a MPN disease approximately 10 years earlier than *JAK2*^{V617F} patients, suggesting that *CALRm* induce a stronger clonal advantage than *JAK2*^{V617F} (Rumi et al., 2014). Different studies on RV, TG and KI models showed that *CALRm* induces an increased frequency of HSCs in the BM but also in the spleen (Benlabiod et al., 2020; Li et al., 2018; Marty et al., 2016; Shide et al., 2017; Toppaldoddi et al., 2019). Contrary to *JAK2*^{V617F}, which gives a competitive advantage in later stages of hematopoiesis

(Delhommeau et al., 2007; Scott et al., 2007), *CALRm* seems to give an early competitive advantage at the HSC level especially for type 1 mutations (Marty et al., 2016; Toppaldoddi et al., 2019). Indeed, using engraftment studies, Benlabiod et al. showed that both *CALRdel52* and *CALRins5* induced a competitive advantage over the WT cells but only *CALRdel52* induced an increase in proliferation of HSCs ($Lin^-c\text{-}Kit^+Sca\text{-}1^+ CD48^-CD150^+$) (Benlabiod et al., 2020). However, other competitive engraftments using constitutive KI models revealed a low competitive advantage of mu*CALRdel52* and *CALRdel19* cells that was not preserved in serial transplantation assays (Balligand et al., 2020; Shide et al., 2019). Moreover, Li et al. demonstrated that homozygous *CALRdel52* cells had no competitive advantage over the WT cells in peripheral blood (Li et al., 2018). The difference between these models and Benlabiod et al. study might be explained by different engraftment strategies. Benlabiod et al. grafted floxed-cells expressing a WT phenotype and induced the KI after the engraftment, contrary to Li et al., Balligand et al. and Shide et al. who transplanted already induced BM cells expressing a *CALRm* phenotype. This could suggest that CALRm-expressing cells might display a homing or an HSC function defect. The results from Benlabiod et al. also correlated with what was found in patients by El-Khoury et al. who demonstrated that *CALRm* was associated with clonal dominance at the HSC level (El-Khoury et al., 2020).

In the next part, we will discuss the role of additional mutations in the selection of the mutated clone and the impact on the MPN phenotype.

4.4 Role of additional mutations

MPNs, and in particular MF, are heterogeneous disorders. In mice, as mentioned earlier, *JAK2*^{V617F} models generate ET and/or PV evolving, in most of cases, into MF. No genetic event responsible for this "natural" and possibly clonal evolution has been characterized; as mentioned before a PV phenotype resurrects from a fibrotic BM transplantation (Zaleskas et al., 2006), and also polyclonal hematopoiesis in MF phase was observed as in PV phase (Lacout et al., 2006). Therefore, modifications or wound/scar of the microenvironment by the MPN clone seem to be responsible for the evolution toward MF in MPN animal models. Concerning the evolution of MPN into AML, there is no report describing the transformation of *JAK2*^{V617F} or *CALRdel52* models into a blastic crisis, except in the TPO^{High} model (Villeval et al., 1997). In humans, additional mutations are mainly found in MF and post-MPN leukemic patients, facilitating the progression of MPN into MF or AML.

These additional mutations are also found in other myeloid malignancies, especially MDS and they affect genes encoding **epigenetic regulators** (TET2, ASXL1, DNMT3A, EZH2, IDH1/2), components of the **splicing machinery** (SF3B1, SRSF2, U2AF1, others) and **transcriptional factors** such as p53. In humans, these mutations are acquired before or after the phenotypic drivers and may correspond to an age-related clonal hematopoiesis modifying the fitness of the *JAK2/CALR/MPL*-mutated clone or to a predisposition. They may also modify the disease phenotype by favoring disease progression. Mutations in a set of five genes, namely *ASXL1*, *EZH2*, *SRSF2* and *IDH1/2* have been associated with poor PMF prognosis (reduced overall survival and increased risk of leukemic transformation) (Vannucchi et al., 2013). The most frequent mutations associated with *JAK2*^{V617F} post-MPN AML are *ASXL1*, *IDH2*, *SRSF2* and *TP53* mutations, *ASXL1* and *SRSF2* mutations being also often associated with *JAK2*^{V617F}-negative post-MPN AML (Rampal et al., 2014).

Several mouse models have been developed combining *JAK2*^{V617F} to these mutations in order to assess and demonstrate their contribution to the disease. We will not address mutations of the splicing machinery or transcriptional factors, such as RUNX1 and NF-E2, as compound mice have yet to be generated.

4.4.1 Alteration of the epigenetic landscape of MPN clones (Shih et al., 2012)

4.4.1.1 *EZH2* mutations

Several mutations in the polycomb repressive complex 2 (PRC2) components have been identified in myeloid malignancies but *EZH2* mutations are by far the most frequently found in MPNs. EZH2 is the catalytic subunit of PRC2 (H3K27me3). Both heterozygous and homozygous mutations of *EZH2* have been identified in ET but mostly in MF patients. They have been identified as the first genetic lesion, preceding both *JAK2*^{V617F} and *CALRm*. Most of these *EZH2* mutations result in the truncation of the protein predicting a loss-of-function (Ernst et al., 2010). Three teams investigated the result of conditional loss of *Ezh2* in *JAK2*^{V617F}-driven MPN models. Floxed *Ezh2* KO mice were crossed with conditional murine *Jak2*^{V617F} KI mouse (Yang et al., 2016) or with two different human *JAK2*^{V617F} TG models (Sashida et al., 2016; Shide et al., 2008; Shimizu et al., 2016). Concomitant induction of *Jak2*^{V617F} and *Ezh2* KO by Yang et al. resulted in the promotion of megakaryopoiesis at the expense of erythropoiesis with an accelerated **progression to MF** and an increase **clonal advantage** of HSCs. The modification of the *Jak2*^{V617F}-induced phenotype was dependent on the de-repression of EZH2 target genes involved in megakaryopoiesis (*S100a8*, *S100a9*, *Ifi2712a*, *Hmga2*) due to decreased H3K27me3 levels. Concomitant induction of *Jak2*^{V617F} expression by the group of Skoda with tamoxifen and deletion of one or both *Ezh2* alleles resulted in a more pronounced phenotype evolving to MF with the heterozygous deletion or MF with the homozygous deletion suggesting that the severity of the phenotype was dependent on the amount of residual EZH2. In accordance with the other studies, they showed a **shift of erythropoiesis toward megakaryopoiesis**. Single HSC engraftments suggested a predominant role for *EZH2* loss in the initiation of the *JAK2*^{V617F} disease with a synergistic increase in HSC self-renewal. Gene expression analysis of LT-HSCs and MEPs revealed an increased expression of genes linked to self-renewal (including *Hmga2*), proliferation, megakaryopoiesis and MF due to sub-representation of H3K7me3 marks on their promoters. The group of Iwama used transplantation to restrict to hematopoietic cells the expression of *JAK2*^{V617F} and a tamoxifen-inducible *EZH2* deletion under the control of ubiquitin promoter (Sashida et al., 2016). They obtained very similar results than previously described with transcriptional program modifications including also a de-repressed expression of *Hmga2*. The MF-initiating capacity of *JAK2*$^{V617F/+}$/*Ezh2*$^{-/-}$ cells was inhibited with a bromodomain inhibitor that reduces H3K27ac levels at the promoter regions of PRC2 targets and down-regulated their expression.

Taken together, these studies showed that loss of *EZH2* drives clonal hematopoiesis and promotes the rapid progression of the *JAK2*^{V617F} disease toward MF with **de-repression of *Hmga2***. When upregulated, HMGA2 led to a faster and more severe transplantable *JAK2*^{V617F} MPN with anemia due to apoptosis of erythroblasts in mice (Ueda et al., 2017).

4.4.1.2 *TET2* mutations

In contrast to other epigenetic regulators, mutations of *TET2* are commonly associated with either of the MPNs and not specifically with MF. TET2 is a dioxygenase that catalyzes the conversion of 5-methyl-cytosine (5-mC) associated with low gene expression into 5-hydroxymethylcytosine (hmC), an intermediate toward demethylation of cytosine and gene expression. Heterozygous and homozygous inactivating *TET2* mutations and deletions are frequently associated with PV and MF *JAK2*^{V617F} MPNs and may appear before or after *JAK2*^{V617F} acquisition. *TET2* mutations do not clearly impact MPN prognosis may be owing to when the mutation occurs (early or in a full-bloomed MPN), but have been associated with leukemic transformation. Delhommeau et al. showed in immunodeficient mice that the *TET2*-mutated/JAK2^{V617F} clone was associated with an increase clonal advantage at the early and committed progenitor levels (Delhommeau et al., 2009). *Tet2* KO mice have a marked expansion of myeloid-biased HSCs and progenitor cells and present an impaired myeloid differentiation, an increase of immature lymphoid cells, splenomegaly associated with extramedullary hematopoiesis and developed a CMML-like disease (Moran-Crusio et al., 2011; Quivoron et al., 2011). *Tet2*$^{+/-}$ mice had a similar phenotype than *Tet2*$^{-/-}$ mice. *Tet2* loss was found to accelerate the *JAK2*^{V617F}-driven MPN phenotype and **increased clonal dominance of HSCs** in the *Vav*-Cre compound *Jak2*^{V617F}*Tet2*^{null} mice of Chen et al. (2015). In particular, a HSC self-renewal signature was enriched in the GSEA analysis of both *Jak2*V617F/*Tet2*null and *Tet2*^{null} mice compared to *Jak2*^{V617F} model. Kameda et al. (2015) crossed their *Jak2*^{V617F} TG mice with an engineered *Tet2*$^{trap/trap}$ model that has residual *Tet2* mRNA (Shide et al., 2012). To overcome the early post-natal lethality of *Tet2*$^{trap/trap}$ mice, authors chose to engraft fetal livers in recipient mice that developed a worse MPN (MF) than mice engrafted with *Jak2*^{V617F} fetal livers alone, with poor survival and an expansion of LT-HSCs. In serial transplantations, the progressive disadvantage conferred by *Jak2*^{V617F} was defeated by the reduction of TET2 level that allowed sustaining MPN, suggesting that it restores the self-renewal defect due to JAK2^{V617F} in this model. In all these mouse models, *TET2* loss worsens *JAK2*^{V617F}-induced phenotype and increases the self-renewal capacity of HSCs and their expansion.

4.4.1.3 *DNMT3A* mutations

DNMT3A catalyzes the de novo demethylation of CpG islets leading to gene inactivation. Heterozygous mutations are more associated with PMF ($\approx$15%) than PV and ET ($\approx$5%) (Abdel-Wahab et al., 2011). *Dnmt3a* KO mice present a major expansion of HSC with increased self-renewal (Challen et al., 2011). One model has been generated combining *Jak2*^{V617F} with *Dnmt3a* loss using the CRISPR-Cas9 technology into *Jak2*^{V617F} KI mice (Jacquelin et al., 2018). Both events cooperate to favor a rapid progression of PV to lethal MF accompanied by **increased HSC self-renewal**.

4.4.1.4 *IDH1/2* mutations

IDH1/2 are two isoforms of the Krebs cycle enzyme that catalyzes the decarboxylation of isocitrate to alpha-ketoglutarate. Mutations are associated with PMF and affect the methylation of DNA via the depletion of alpha-ketoglutarate (necessary for TET2 and other demethylases) and predict a poor MPN outcome and reduced leukemia-free survival. With the hope of investigating the role of IDH1/2 to promote the transformation of *JAK2*^{V617F}-MPNs, the group of Levine crossed *Jak2*^{V617F} KI mice shown to display a mild phenotype without MF (Mullally et al., 2010) with conditional hu*IDH1*^{R132H} or mu*Idh2*^{R140Q} with *Mx1*-Cre (McKenney et al., 2018). Six-month-old compound mice were studied and showed a PV-like phenotype with leukocytosis and splenomegaly resembling *Jak2*^{V617F} mice and concomitant elevated 2-hydroxyglutarate compared to *Idh1/2* mutated mice. Mouse MPN progressed to a **pre-leukemic stage** and was transplantable from the LT-HSC compartment that was expanded. Phenotype with *IDH1*^{R132H} was milder than with *Idh2*^{R140Q}. Phenotype (and spleen blasts) regressed with a combination of JAK2 and IDH2 inhibitors. Interestingly, treatment reverses extramedullary hematopoiesis and the leukemic signature that was highlighted in *Vav*-Cre/*Idh1*^{R132H} mice (Sasaki et al., 2012).

4.4.1.5 *ASXL1* mutations

ASXL1 mono-ubiquitinylates the DNA packaging protein histone H2A (H2AK119ub1).

Frameshift and nonsense *ASXL1* mutations are of equally poor prognosis in *JAK2*^{V617F} than in type 2 *CALRm* MPNs. In contrast, when associated with type 1 *CALRm* they have not this poor prognosis. *Asxl1* KO is embryonic lethal peri-birth in a fraction of animals (Fisher et al., 2010; Wang et al., 2014). Survivors show a slight reduction in myeloid and erythroid progenitors and moderate splenomegaly. In contrast, Wang et al. showed also significant features of MDS and a reduced progenitor cell pool. In a recent conditional model using *Mx1*-Cre and *Vav*-Cre, Abdel-Wahab et al. (2013) confirmed these results and further studied the effect of *Asxl1* loss on hematopoiesis after birth. Mice developed a severe and rapid pancytopenia and myelodysplasia due to impaired progenitor differentiation. Mice presented an increased number of HSC and progenitor cells but **impaired self-renewal** characterized by the competitive disadvantage in serial transplantation. When crossing *Asxl1*$^{+/-}$ mice with *JAK2*^{V617F} TG mice (Guo et al., 2019; Xing et al., 2008) disease **progression to MF** was accelerated and a few mice developed myeloid leukemia.

Intriguingly, Inoue et al. demonstrated that similar phenotypes were developed by both *Asxl1* KO and RV overexpression of mutant ASXL1, suggesting that these mutations might rather be dominant-negative and/or gain-of-function (Inoue et al., 2013, 2016). Following these observations, several groups developed *Asxl1*-mutated KI models with various effect on hematopoiesis including a decrease in the number and self-renewal of HSCs and age-dependent myelodysplasia and splenomegaly (Hsu et al., 2017; Uni et al., 2019). Similarly, two other models presented dysplastic features and increased susceptibility for leukemic transformation/age-related MPN, MDS and AML but opposed effects on HSCs with impaired numbers (Nagase et al., 2018) or increased HSC self-renewal and myeloid bias (Yang et al., 2018). Thus, in the future, it will be interesting to combine *JAK2*^{V617F} with *ASXL1* mutations rather than with *ASXL1* deletion to study hematopoiesis.

Taken together, these studies were important to characterize these additional mutations when they are first events as drivers of clonal hematopoiesis due to the increase in HSC self-renewal, maybe at the exception of *ASXL1* mutations. As a second event, they accelerate the degree of malignancy of the already developed MPN and act as phenotypic modifiers. Some studies have documented that *CALRm* are most of the time the first genetic lesion suggesting that additional mutations may rather play a role in disease progression instead of initiation in these cases.

4.4.2 Alterations in transcription factors

4.4.2.1 p53

TP53 is a tumor suppressor gene. Mutations have been associated with transformation into secondary post-MPN AML (50% of cases) as well as de novo AML. At the chronic MPN stage *TP53* mutations are usually mono-allelic and with a very low allele frequency (VAF), not always proved as loss of function and mostly associated with PMF. Transformation to AML is associated with bi-allelic *TP53* mutations and increased VAF. Two groups studied the effects of combining *JAK2*^{V617F} and *TP53* loss in mice using the same approach. They engineered a RV model where p53-null BM was transduced to express *Jak2*^{V617F} and engrafted in either WT or *Tp53*$^{-/-}$ recipient mice (Rampal et al., 2014; Tsuruta-Kishino et al., 2017). In both studies, mice developed a PV shortly followed with **progression to leukemia** and a drastically reduced median survival after BMT between 46.5 and 110 days, respectively. Mice showed an expansion of progenitors (especially of erythroid progenitors), MEPs and HSCs possessing leukemia-initiating capacity as shown by the shortened latency to develop leukemia with serial transplantations. The cooperativity between the two genetic events was also highlighted by the increased self-renewal capacity of committed progenitors in replating assays. Moreover, the group of Kurokawa demonstrated that *Tp53* heterozygosity had no impact on the $JAK2^{V617F}$-induced PV, confirming that the passage from the chronic phase to AML is mostly dependent on the gain of *TP53* homozygosity (Tsuruta-Kishino et al., 2017).

In aggregate, compound mouse models have helped to understand the role of additional mutations in MPN. They show that these additional mutations rarely cause a myeloproliferation by themselves. However, they may increase self-renewal of HSCs helping disease initiation through the expansion of the $JAK2^{V617F}$ clone and probably of the *CALR*-mutated clone (*TET2, EZH2, DNMT3a*). Some promote myelofibrosis (EZH2, DNMT3a, ASXL1) or acute leukemic transformation (*ASXL1, IDH1/2, TP53* loss). Importantly, these mutations can be co-associated suggesting that it might be necessary to generate more complex mouse models to understand the interrelationships between them in MPN.

4.5 Deciphering the mechanisms responsible for the MPN phenotype

Mouse models have been helpful in determining that driver mutations were responsible for the occurrence of MPN and that their oncogenic properties were all dependent on the activation of the JAK2/STAT signaling pathway. Mouse models were further used to decipher the mechanisms contributing to disease pathogenesis, in particular how $JAK2^{V617F}$ could cause different disease phenotypes, by evaluating the contribution of specific organs, partners (cytokines and their receptors) and signaling pathways.

4.5.1 Role of the spleen

As mentioned before, the spleen does not play the same role in mice and men. Therefore, one must be careful in evaluating the role of the spleen in a mouse model of MPN. With this caveat, splenomegaly in patients still results from extramedullary hematopoiesis and is a major clinical manifestation of MPNs. Moreover, the JAK1/2 inhibitor ruxolitinib can be a safer alternative to splenectomy as it significantly reduces spleen size. A study was conducted in 2009 by Mo and colleagues using a RV $Jak2^{V617F}$ PV model demonstrated that splenectomy normalizes the hematocrit in mice (Mo et al., 2009). However, disease burden, fibrosis and osteosclerosis develop normally in the marrow of these mice. Interestingly, the hematocrit was only slightly blunted in an EPO-induced secondary polycythemia model suggesting that the spleen was specifically required for the polycythemia induced by *JAK2*^{V617F}. Obviously, the role of the spleen in human PV is not the same, underlying the differences between the MPN pathogenesis of the two species. Splenectomy in MF patients is a potential therapeutic option. However, this surgery can be associated with a decrease in overall survival (Santos et al., 2014) in addition to transformation to AML (Barosi et al., 1998; Morel et al., 2010).

4.5.2 Importance of MPL

It has been shown in cell lines that the expression of a homodimeric type I cytokine receptor is required for JAK2^{V617F}-mediated transformation and signaling activities (Lu et al., 2005) and MPL is the only one being essential for HSC maintenance making it a good candidate for MPN disease initiation and development. Using *JAK2*^{V617F} TG mice developing ET (Tiedt et al., 2008), Sangkhae et al. demonstrated that *JAK2*^{V617F}/*Mpl*$^{-/-}$ mice remained thrombocytopenic with normal neutrophil counts and spleen weight and a profound HSC pool deficiency while *JAK2*^{V617F}/*Mpl*$^{+/-}$ mice with reduced expression of MPL developed an attenuated MPN with slightly elevated platelet and neutrophil counts and a rescue of the normal HSC pool. Therefore, the presence of MPL but also its level is required for the full development of MPN making MPL a good therapeutic target. They also found that TPO was not crucial for the development of MPN (Sangkhae et al., 2014). Of note, further investigation is still necessary to see if MPL is required for the full development of PV and MF using another mouse model with an inducible *Mpl*$^{-/-}$.

In order to study the implication of MPL in *CALRm*-mediated thrombocytosis in mice, Marty and colleagues used *c-mpl*- and *tpo*-deficient mice to show that MPL, but not TPO, was essential for CALR mutant to induce thrombocytosis in a RV mouse model (Marty et al., 2016). This study demonstrates that like other driver mutations, *CALRm* induces MPN via the activation of the cytokine receptor/JAK2/STAT signaling pathway. Moreover, this experiment confirmed in vitro data demonstrating that *CALRm* tropism (ET/PMF) was linked to the specific activation of MPL by CALR mutants, while JAK2^{V617F} can bind and activate EPOR, MPL and the G-CSFR. Balligand and colleagues also assessed the importance of MPL signaling in the development of *CALRm*-mediated ET. They crossed heterozygous *Calrde52* KI with mice expressing MPL$^{\Delta 60/\Delta 60}$ missing part of the receptor cytoplasmic tail crucial for signaling.

Resulting mice did not develop thrombocytosis, highlighting the importance of MPL signaling in *CALRm*-induced pathogenesis (Balligand et al., 2020). Interestingly, *CALRm* KI mouse models were found to develop an unexpectedly mild phenotype compared to heterozygous human counterparts. They demonstrated that the chimeric CALRdel52 and CALRins5 proteins (murine backbone with human mutated C-terminal part) were less efficient in activating muMPL than huMPL and that treating these mice with the MPL agonist Romiplostim increased the severity of the disease suggesting that the mild phenotype was linked to a sub-optimal stimulation of muMPL in the KI mouse models (Benlabiod et al., 2020).

These results demonstrate how MPN, driven by gain-of-function mutations of the JAK2/STAT pathway relied on JAK2-dependent MPL signaling for HSC maintenance.

4.5.3 Role of zygosity

In MPN patients, *JAK2*^{V617F} heterozygosity is mostly associated with an ET phenotype whereas homozygosity, resulting from mitotic recombination of chromosome 9p (acquired uniparental disomy or aUPD), is more frequent in PV and MF patients (Perrier et al., 2005; Tiedt et al., 2008). **Loss of heterozygosity** has rarely been observed in *Jak2*^{V617F}-mutated mice models. Marty et al. characterized the development of 14% of homozygous KI clones in a heterozygous *Vav*-Cre/*Jak2*^{V617F} KI mouse model due to either homologous recombination as found in MPN patients or to the deletion of the *WT* allele (Marty et al., 2013). The group of Mohi demonstrated that loss of the *Jak2*^{WT} allele in mice increased the severity of the phenotype with an acceleration of MF suggesting that it acts as a **negative regulator** of the *Jak2*^{V617F} allele (Akada et al., 2014). The group of Skoda using an inducible *Jak2* KO model corroborated this result. Deletion of the *WT* allele worsens erythrocytosis (but not leukocytosis and thrombocytosis) and reduced survival of a *Vav*-Cre/*Jak2*^{V617F} TG model, strengthening the link between the WT:V617F JAK2 ratio in the responsiveness of erythropoiesis to JAK2^{V617F} (Kubovcakova et al., 2013). Finally, the group of Green showed that homozygosity in their *JAK2*^{V617F} KI mice resulted in a phenotypic switch from ET to PV and increasing defects in HSC (Li et al., 2014).

Related to *CALRdel52* and *CALRins5* KI mouse models, disease phenotype was also more severe in the homozygous context (Benlabiod et al., 2020; Li et al., 2018). Moreover, to study the role of *CALRins5* homozygosity in the severity of disease development they performed competitive engraftment of 20:80% of homozygous: heterozygous *CALRins5* BM cells into lethally irradiated recipient mice. These mice developed higher thrombocytosis than mice grafted with 100% heterozygous *CALRins5* BM and, interestingly, with 100% heterozygous *CALRdel52* BM cells. Considering that homozygosity is more frequent in *CALRins5* patients, these observations provided a possible explanation for the apparent discrepancy between higher platelet counts found in *CALRins5* compared to *CALRdel52* ET patients and mouse data where thrombocytosis if higher in heterozygous *CALRdel52* compared to heterozygous *CALRins5* KI mice. Six months after KI expression induction, they observed that homozygous *CALRins5* HSCs (SLAM) were able to take the advantage over the heterozygous cells. In contrast, Balligand and colleagues were unable to generate homozygous *Calrdel61* and *Calrdel52* mice using their constitutive KI models as homozygosity reproduced the lethal phenotype of endogenous *CalrWT* KO for heart development (Balligand et al., 2020; Mesaeli et al., 1999). Taken together, gene-dosage hypothesis for $JAK2^{V617F}$ as for *CALRm* linked to the strength of JAK2 signaling seems to be verified, as homozygosity induces in both cases a more severe MPN phenotype.

4.5.4 Role of STATs in $JAK2^{V617F}$-induced disease phenotype heterogeneity

When EPO, TPO and G-CSF ligand bind to their respective receptors, JAK2 is activated by auto and trans-phosphorylation before phosphorylating some key tyrosine residues in the cytoplasmic tails of the receptors. These phosphorylated tyrosine residues serve as anchors to recruit signaling molecules such as the STAT1/3/5 that get phosphorylated by JAK2 and translocate to the nucleus to activate target genes. $JAK2^{V617F}$, *CALRm*, and $MPL^{W515L/K}$ mutations lead to the constitutive activation of the JAK2/STATs pathway. The individual implication of the different STATs in MPN pathogenesis has been evaluated in mice. The hypothesis was that the heterogeneity of $JAK2^{V617F}$-induced MPN phenotype might be due in part to gene dosage and the strength of JAK2 signaling but also to STAT qualitative implications in signaling.

In humans, Teofili et al. showed that different p-STAT3/5 status are linked to different phenotypes: high p-STAT3/5 to PV, high p-STAT3 and low p-STAT5 in ET and low p-STAT3/5 in MF (Teofili et al., 2007). E. Chen and colleagues showed that increased STAT1 activity produced a $JAK2^{V617F}$ ET phenotype when the downregulation of STAT1 activity in ET-progenitors produces a PV phenotype (Chen et al., 2010). In mice, defective megakaryopoiesis of *Gata-1^{low}* mice is rescued by

STAT1 ectopic expression while *Stat1* KO mice display defective MK polyploidization (Huang et al., 2007).

To unravel the contribution of the STAT1/3/5 family members in MPN phenotypes, KO mice of the *Stat* family were crossed with *JAK2*^{V617F} mouse models. Duek and colleagues showed that in *JAK2*^{V617F}/*Stat1*$^{-/-}$ mice thrombocytosis was reduced in favor of erythropoiesis, which led to the conclusion that activation of IFNγ/**STAT1** signaling promotes megakaryopoiesis and is involved in *JAK2*^{V617F}-induced ET phenotype (Duek et al., 2014). However, later, Grisouard et al. crossed *Stat1* KO mice with *JAK2 exon12* TG mice (displaying a PV phenotype) and found that these mice had no change in platelet and red cell parameters (Grisouard et al., 2016) driving the conclusion that STAT1 activation is not an essential mediator of *JAK2 exon12*-dependent erythropoiesis, in contrast to the one of *JAK2*^{V617F}-driven phenotype.

The role of **STAT3** was also investigated using a double *JAK2*^{V617F}/*Stat3*$^{fl/fl}$ mouse model (Grisouard et al., 2015). These mice have an increased platelet count and reduced overall survival. One hypothesis was that the deletion of *Stat3* may trigger the expression/activation of STAT1 due to a compensation mechanism between STAT1 and STAT3 (Regis et al., 2008). In another study, Yan and colleagues used *JAK2*^{V617F}/*Stat3*-deficient mice and found that the deletion of STAT3 increased platelet counts, enhanced the expansion of progenitor and myeloid precursor cells and enhanced the severity of MPN (Yan et al., 2015). The latter observation suggests that STAT3 has a negative role in *JAK2*^{V617F}-induced PV.

The effect of **STAT5** deficiency in the pathogenesis of MPN has been studied by two teams using two different *JAK2*^{V617F} mouse models. D. Yan and colleagues crossing conditional *Jak2*^{V617F} KI and floxed *Stat5*-deficient mice showed an inhibition in the development of PV, normal blood parameters and the absence of splenomegaly with a rescue of the PV phenotype upon re-expression of STAT5 (Yan et al., 2012). Walz et al. showed that *JAK2*^{V617F} RV expression in *Stat5a/b*-deficient mice failed to induce polycythemia but the loss of STAT5 did not prevent the development of MF at the opposite of Yan et al. data (Walz et al., 2012). This shows that STAT5 is crucial for the development of PV and possibly prevent or delay fibrosis transformation. However, unexpectedly giving the importance of Stat5 for *Jak2*^{V617F} MPN, no increase in Stat5 was observed in *JAK2exon12* TG mice (Grisouard et al., 2016).

4.5.5 Metabolic alterations

Only one study was conducted on the role of $JAK2^{V617F}$-dependent metabolic alterations in MPN most notably focusing on the increased need for energy, which is one of cancer hallmarks. To study this, Rao and colleagues compared WT mice and conditional TG mice expressing either *JAK2*^{V617F} or *JAK2 exon12* (Rao et al., 2019). Authors showed that *JAK2* mutations increase glycolysis and oxidative phosphorylation leading to hypoglycemia correlated with the exacerbated erythropoiesis. A high-fat diet in contrast to a high-glucose diet (not correcting hypoglycemia) improved the erythroid MPN phenotype. Chemical suppression of glycolysis normalized glucose level, increased reactive oxygen species (ROS) and decreased hemoglobin and leukocytosis. Thus, metabolic alterations are implicated in worsening the disease, leading to complications such as atrophy of the adipose tissue and a decreased overall survival.

4.5.6 Reactive oxygen species in the pathogenesis of MPN

C. Marty and colleagues were the first to show that *JAK2*^{V617F} induces an accumulation of ROS in mice (Marty et al., 2013). Increase ROS levels were partly due to a decrease in the detoxifying enzyme catalase and were demonstrated to be involved in $JAK2^{V617F}$-induced genetic instability and the loss of heterozygosity. Furthermore, the accumulation of ROS was significantly higher in immature cells (LSK), compared to more engaged cells, suggesting a potential role in clonal advantage.

4.5.7 Changes in iron metabolism caused by JAK2 exon12 mutation

PV patients carrying the *JAK2 exon12* mutation usually present with an isolated erythrocytosis, characterized by a persistent increase of red cell mass and hematocrit. Grisouard and colleagues showed that the inducible TG mice expressing a human *JAK2 exon12* mutation had isolated erythrocytosis due to modifications in the iron metabolism, including decreased expression of hepcidin and increased expression of transferrin receptor-1 and erythroferrone, making iron more available for maximal production of red blood cells (Grisouard et al., 2016).

4.6 Preclinical mouse MPN model studies, two iconic examples

Mouse MPN models have been extensively used in preclinical studies to test numerous drugs, as elegantly reviewed elsewhere (Bartalucci et al., 2013).

Thus, we chose to focus on JAK inhibitors and interferon-alpha as a proof-of-concept. These extensive preclinical studies have demonstrated the validity of MPN mouse models for the evaluation of drug efficacy or inefficacy.

4.6.1 JAK2 inhibitors

The first JAK2 inhibitor tested in mice was Fedratinib (formerly TG101348/SAR302503 and recent trade name Inrebic). Wernig et al. showed that Fedratinib reduced hematocrit and leukocyte counts as well as the splenomegaly in a *Jak2*^{V617F}RV mouse model (Wernig et al., 2008). Debeurme et al. compared the effects of Fedratinib in two models, a *Jak2*^{V617F} KI with a PV phenotype and PPMF and a TPO^{High} mouse model with PTMF. This study showed that Fedratinib reduced the erythroid hyperplasia and splenomegaly without reducing the mutant allele burden like previous studies (Mullally et al., 2010) but also normalized WBC and hematocrit levels and even in the thrombocytopenic PPMF model prevented the development of MF and increased platelet counts in contrast to the severe PTMF model (Debeurme et al., 2015). Fedratinib as Inrebic was FDA approved for the treatment of MF. Ruxolitinib (formerly INCB018424) is a JAK1/2 inhibitor and has been tested in many mouse models (Kubovcakova et al., 2013; Quintás-Cardama et al., 2010). In 2011, Ruxolitinib was the first of the JAK inhibitors to be approved in clinics for high-risk MF and later on for hydroxyurea-refractory PV and steroid-refractory acute graft-versus-host disease. MPN model studies showed that ruxolitinib treatment improved clinical manifestations in mice without eradication of MPN-initiating cells assayed in secondary transplantations.

The type II JAK2 inhibitor CHZ868, reversing the type I JAK inhibitor persistence acquired by MPN cells through reactivation of the JAK/STAT signaling (Koppikar et al., 2012), was more efficient than type I inhibitor in decreasing BM allele burden of *Vav*-Cre/*Jak2*^{V617F} KI mice (Mullally et al., 2010) but was not tested in term of MPN-initiating cells in secondary transplantations. In addition, CHZ868 was also efficient in improving the phenotype and survival of *MPL*^{W515L} RV transduced BALB/c and C57BL/6 mice, slightly reducing allele burden in blood (from 73% to 63% GFP + cells) (Meyer et al., 2015).

Overall, JAK2 inhibitors in MPN animal models rapidly reduced splenomegaly, RBC/neutrophil parameters and circulating levels of inflammatory cytokines without providing a convincing molecular remission. In that sense, MPN model reproduces the effects observed with these molecules in patients and represent good preclinical models, type II JAK2 inhibitors remaining to be tested in humans. They are presently tested in association with other small molecule inhibiting either other signaling molecules or epigenetic regulators.

4.6.2 Interferon-alpha (IFNα)

IFNα is the treatment providing the most frequent molecular remissions in patients (8–20% of patients) (Gowin et al., 2017; Kiladjian et al., 2008; Quintás-Cardama et al., 2011). IFNα treatment of MPN mouse models reproduced quite faithfully the effects observed in patients treated with pegylated-IFNα. Indeed, in contrast to JAK2 inhibitors, IFNα tested in chimeric models with transplanted *Jak2*^{WT} and *Jak2*^{V617F} KI cells, by Mullally et al. (2013) and Hasan et al. (2013), led to complete eradication of the disease-initiating cells assayed in secondary BM transplantation. Yet, the mechanism of action of IFNα remains largely unknown, limiting the improvement of the treatment (Kiladjian et al., 2008). MPN animal models provided good tools to investigate them. Mullally et al. showed that *Jak2*^{V617F} BM cells invalidated for the IFNα receptor (*Ifnar1*$^{-/-}$) transplanted with normal BM cells did not respond to IFNα treatment. This experiment demonstrated that IFNα acts by a mechanism intrinsic to the *Jak2*^{V617F} clones rather than by an immunological rejection of tumor cells provided by T/B cells mostly not invalidated for *Ifnar1* in this chimeric model. This intrinsic mechanism could be a preferential depletion or apoptosis of mutated clones by IFNα. Mullaly and colleagues found that IFNα was more potent in inducing *Jak2*^{V617F} than *Jak2*^{WT} stem cell-cycle activation suggesting a loss of self-renewal of mutated HSC. Hasan et al. demonstrated that IFNα was more potent in inducing spleen *Jak2*^{V617F} than *Jak2*^{WT} cells apoptosis. Recently, Austin et al., using chronic peg-IFN-α administration, showed that IFNα induced a greater STAT1 pathway activation in *Jak2*^{V617F} than in *Jak2*^{WT} HSCs, with increased induction of ROS, DNA damage and a reduction in quiescence (Austin et al., 2020). Related to the role of ROS, this seems in contradiction with preclinical studies using the antioxidant *N*-acetyl-L-cysteine (NAC) that improved *Vav*-Cre/*Jak2*^{V617F} mouse symptoms (reduction of splenomegaly and blood parameters) as well as decreased MPN-initiating cells (Marty et al., 2013) or the overall survival of mice by inhibiting thrombosis (Craver et al., 2020). Of note, Austin et al. demonstrated distinct effects of IFNα and Ruxolitinib on HSC that may be useful for ongoing clinical efforts to combine both. More recently, Dagher and colleagues treated *Jak2*^{V617F} KI mice with IFNα alone or a combination with arsenic trioxide (ATO). They found that IFNα+ATO was more efficient than IFNα alone in reducing disease-initiating cells and relapses after treatment arrest. To better understand these observations, they crossed the *Jak2*^{V617F} mice with *PML* KO mice. Results show that in the absence of the tumor suppressor gene *PML*, IFNα is less effective, predicting potent clinical efficacy of the IFNα + ATO combination in patients and identifying PML as a major effector of therapy (Dagher et al., 2021).

Overall, these preclinical studies in MPN mouse models paved the way to improve curative treatments of MPNs by IFNα.

5. Conclusions and remarks

MPN mouse models have proven extraordinarily useful in understanding the etiology, initiation, transformation, mechanisms of fibrosis/hemostasis, roles of additional mutations and novel clinical approaches of human MPN. Expressions of *JAK2*^{V617F}, *CALR* and *MPL* mutations in mouse HSCs mimic quite faithfully human PV or ET with their natural evolutions into MF, demonstrating that these mutations are drivers in MPNs. These models also identified the role of additional mutations in disease evolution, in particular *TET2* and *DNMT3a* in accelerated clonal progression, *EZH2*, *DNMT3a*, *ASXL1* in MF development, *ASXL1*, *IDH1* or *TP53* loss in secondary acute leukemia transformation. They also largely contributed to the identification of mechanisms implicated in the onset of MF, pointing out the essential roles of dysplastic MKs with TGF-β1 and other factors, the cellular origins of the excess ECM producer myofibroblasts or the contribution of mutations including *DNMT3a* and *EZH2* with its PCR2 target *Hmga2*. Furthermore, animal models, similarly to patients, present high percentages of unexpected deaths due to thrombosis or hemorrhages and have proven to be precious tools to investigate mechanisms and key players involved in these complications, with the fact that they were not preventively treated. In most models, thrombosis with occlusive clots is not prevalent in contrast to hemorrhage with accelerated formations of unstable clots and embolization probably due to an acquired vWF syndrome. Identified players are pro-inflammatory/adhesive leukocytes leading to increased NET formation, microvesicles leading to endothelial oxidative stress and vascular arterial spasms. The therapeutic efforts to target the main protagonists of fibrosis and hemostasis disorders have not yet been successful but no doubt that preclinical animal models will be precious in discovering new drugs able to prevent these two main life-threatening MPN complications. In this regards, extensive data with JAK2 inhibitors and IFNα showed that these models reproduced quite faithfully the effects observed in patients with these drugs.

Still, controversial results remain such as the full implication of *JAK2*^{V617F} in disease initiation and clonal expansion. Indeed, these models cannot fully reproduce the human disease related to the development of the initial clone without severe replicative/oxidative stress caused by BM transplantation or a perfect heterozygous mutated/normal 1/1 gene ratio with appropriate signaling partners. As mentioned in this review, mice themself are not small people and they differ from humans, considering MPN, by the spleen function, the usage of different primary tumor repressor pathways, the steady state of active/quiescent HSCs and different components of cellular immunology/inflammation. Finally, the contribution of germline genetic predisposition, age, HSC niche, metabolic, oxidative, replicative or genotoxic stress, inflammation, immune escape and multiple mutations need to be considered and further investigated to encompass the full complexity of human MPN in mice.

Acknowledgments

This work was supported by grants from MPN research foundation on IFN initiative, INCA Plbio 2014, 2018, Labex GR-Ex. C.B. and T.D. were supported by PhD grants from Université Paris Sud Saclay and Ligue Nationale contre le Cancer, respectively. We thanks Marie Charlotte Bourrienne, from Bichat hospital, Paris, France, for her great contribution to the hemostasis chapter, Dr. Isabelle Plo and Dr. William Vainchenker from INSERM, UMR 1287, Gustave Roussy, Villejuif, France, for critical review and contributions to the manuscript. Authors have declared that no conflict of interest exists.

References

Abdel-Wahab, O., Pardanani, A., Rampal, R., Lasho, T.L., Levine, R.L., Tefferi, A., 2011. DNMT3A mutational analysis in primary myelofibrosis, chronic myelomonocytic leukemia and advanced phases of myeloproliferative neoplasms. Leukemia 25, 1219–1220. https://doi.org/10.1038/leu.2011.82.

Abdel-Wahab, O., Gao, J., Adli, M., Dey, A., Trimarchi, T., Chung, Y.R., Kuscu, C., Hricik, T., Ndiaye-Lobry, D., Lafave, L.M., Koche, R., Shih, A.H., Guryanova, O.A., Kim, E., Li, S., Pandey, S., Shin, J.Y., Telis, L., Liu, J., Bhatt, P.K., Monette, S., Zhao, X., Mason, C.E., Park, C.Y., Bernstein, B.E., Aifantis, I., Levine, R.L., 2013. Deletion of Asxl1 results in myelodysplasia and severe developmental defects in vivo. J. Exp. Med. 210, 2641–2659. https://doi.org/10.1084/jem.20131141.

Akada, H., Yan, D., Zou, H., Fiering, S., Hutchison, R.E., Mohi, M.G., 2010. Conditional expression of heterozygous or homozygous Jak2V617F from its endogenous promoter induces a polycythemia vera-like disease. Blood 115, 3589–3597. https://doi.org/10.1182/blood-2009-04-215848.

Akada, H., Akada, S., Hutchison, R.E., Mohi, G., 2014. Loss of wild-type Jak2 allele enhances myeloid cell expansion and accelerates myelofibrosis in Jak2V617F knock-in mice. Leukemia 28, 1627–1635. https://doi.org/10.1038/leu.2014.52.

Arranz, L., Sánchez-Aguilera, A., Martín-Pérez, D., Isern, J., Langa, X., Tzankov, A., Lundberg, P., Muntión, S., Tzeng, Y.-S., Lai, D.-M., Schwaller, J., Skoda, R.C., Méndez-Ferrer, S., 2014. Neuropathy of haematopoietic stem cell niche is essential for myeloproliferative neoplasms. Nature 512, 78–81. https://doi.org/10.1038/nature13383.

Austin, R.J., Straube, J., Bruedigam, C., Pali, G., Jacquelin, S., Vu, T., Green, J., Gräsel, J., Lansink, L., Cooper, L., Lee, S.-J., Chen, N.-T., Lee, C.-W., Haque, A., Heidel, F.H., D'Andrea, R., Hill, G.R., Mullally, A., Milsom, M.D., Bywater, M., Lane, S.W., 2020. Distinct effects of ruxolitinib and interferon-alpha on murine JAK2V617F myeloproliferative neoplasm hematopoietic stem cell populations. Leukemia 34, 1075–1089. https://doi.org/10.1038/s41375-019-0638-y.

Balligand, T., Achouri, Y., Pecquet, C., Chachoua, I., Nivarthi, H., Marty, C., Vainchenker, W., Plo, I., Kralovics, R., Constantinescu, S.N., 2016. Pathologic activation of thrombopoietin receptor and JAK2-STAT5 pathway by frameshift mutants of mouse calreticulin. Leukemia 30, 1775–1778. https://doi.org/10.1038/leu.2016.47.

Balligand, T., Achouri, Y., Pecquet, C., Gaudray, G., Colau, D., Hug, E., Rahmani, Y., Stroobant, V., Plo, I., Vainchenker, W., Kralovics, R., Van den Eynde, B.J., Defour, J.-P., Constantinescu, S.N., 2020. Knock-in of murine Calr del52 induces essential thrombocythemia with slow-rising dominance in mice and reveals key role of Calr exon 9 in cardiac development. Leukemia 34 (2), 510–521. https://doi.org/10.1038/s41375-019-0538-1.

Barbui, T., Falanga, A., 2016. Molecular biomarkers of thrombosis in myeloproliferative neoplasms. Thromb. Res. 140 (Suppl. 1), S71–S75. https://doi.org/10.1016/S0049-3848(16)30102-5.

Barosi, G., Ambrosetti, A., Centra, A., Falcone, A., Finelli, C., Foa, P., Grossi, A., Guarnone, R., Rupoli, S., Luciano, L., Petti, M.C., Pogliani, E., Russo, D., Ruggeri, M., Quaglini, S., 1998. Splenectomy and risk of blast transformation in myelofibrosis with myeloid metaplasia. Italian cooperative study group on myeloid with myeloid metaplasia. Blood 91, 3630–3636.

Bartalucci, N., Bogani, C., Vannucchi, A.M., 2013. Preclinical models for drug selection in myeloproliferative neoplasms. Curr. Hematol. Malig. Rep. 8, 317–324. https://doi.org/10.1007/s11899-013-0182-1.

Bellanné-Chantelot, C., Rabadan Moraes, G., Schmaltz-Panneau, B., Marty, C., Vainchenker, W., Plo, I., 2020. Germline genetic factors in the pathogenesis of myeloproliferative neoplasms. Blood Rev. 42, 100710. https://doi.org/10.1016/j.blre.2020.100710.

Benlabiod, C., da Cacemiro, M.C., Nédélec, A., Edmond, V., Muller, D., Rameau, P., Touchard, L., Gonin, P., Constantinescu, S.N., Raslova, H., Villeval, J.-L., Vainchenker, W., Plo, I., Marty, C., 2020. Calreticulin del52 and ins5 knock-in mice recapitulate different myeloproliferative phenotypes observed in patients with MPN. Nat. Commun. 11, 4886. https://doi.org/10.1038/s41467-020-18691-3.

Bumm, T.G.P., Elsea, C., Corbin, A.S., Loriaux, M., Sherbenou, D., Wood, L., Deininger, J., Silver, R.T., Druker, B.J., Deininger, M.W.N., 2006. Characterization of murine JAK2V617F-positive myeloproliferative disease. Cancer Res. 66, 11156–11165. https://doi.org/10.1158/0008-5472.CAN-06-2210.

Busque, L., Patel, J.P., Figueroa, M.E., Vasanthakumar, A., Provost, S., Hamilou, Z., Mollica, L., Li, J., Viale, A., Heguy, A., Hassimi, M., Socci, N., Bhatt, P.K., Gonen, M., Mason, C.E., Melnick, A., Godley, L.A., Brennan, C.W.,

Abdel-Wahab, O., Levine, R.L., 2012. Recurrent somatic TET2 mutations in normal elderly individuals with clonal hematopoiesis. Nat. Genet. 44, 1179–1181. https://doi.org/10.1038/ng.2413.

Cabagnols, X., Defour, J.P., Ugo, V., Ianotto, J.C., Mossuz, P., Mondet, J., Girodon, F., Alexandre, J.H., Mansier, O., Viallard, J.F., Lippert, E., Murati, A., Mozziconacci, M.J., Saussoy, P., Vekemans, M.C., Knoops, L., Pasquier, F., Ribrag, V., Solary, E., Plo, I., Constantinescu, S.N., Casadevall, N., Vainchenker, W., Marzac, C., Bluteau, O., 2015. Differential association of calreticulin type 1 and type 2 mutations with myelofibrosis and essential thrombocytemia: relevance for disease evolution. Leukemia 29, 249–252. https://doi.org/10.1038/leu.2014.270.

Cannizzo, S.J., Frey, B.M., Raffi, S., Moore, M.A., Eaton, D., Suzuki, M., Singh, R., Mack, C.A., Crystal, R.G., 1997. Augmentation of blood platelet levels by intratracheal administration of an adenovirus vector encoding human thrombopoietin cDNA. Nat. Biotechnol. 15, 570–573. https://doi.org/10.1038/nbt0697-570.

Castiglione, M., Jiang, Y.-P., Mazzeo, C., Lee, S., Chen, J.-S., Kaushansky, K., Yin, W., Lin, R.Z., Zheng, H., Zhan, H., 2020. Endothelial JAK2V617F mutation leads to thrombosis, vasculopathy, and cardiomyopathy in a murine model of myeloproliferative neoplasm. J. Thromb. Haemost. 18 (12), 3359–3370. https://doi.org/10.1111/jth.15095.

Chagraoui, H., Komura, E., Tulliez, M., Giraudier, S., Vainchenker, W., Wendling, F., 2002. Prominent role of TGF-beta 1 in thrombopoietin-induced myelofibrosis in mice. Blood 100, 3495–3503. https://doi.org/10.1182/blood-2002-04-1133.

Chagraoui, H., Tulliez, M., Smayra, T., Komura, E., Giraudier, S., Yun, T., Lassau, N., Vainchenker, W., Wendling, F., 2003. Stimulation of osteoprotegerin production is responsible for osteosclerosis in mice overexpressing TPO. Blood 101, 2983–2989. https://doi.org/10.1182/blood-2002-09-2839.

Challen, G.A., Sun, D., Jeong, M., Luo, M., Jelinek, J., Berg, J.S., Bock, C., Vasanthakumar, A., Gu, H., Xi, Y., Liang, S., Lu, Y., Darlington, G.J., Meissner, A., Issa, J.-P.J., Godley, L.A., Li, W., Goodell, M.A., 2011. Dnmt3a is essential for hematopoietic stem cell differentiation. Nat. Genet. 44, 23–31. https://doi.org/10.1038/ng.1009.

Chapeau, E.A., Mandon, E., Gill, J., Romanet, V., Ebel, N., Powajbo, V., Andraos-Rey, R., Qian, Z., Kininis, M., Zumstein-Mecker, S., Ito, M., Hynes, N.E., Tiedt, R., Hofmann, F., Eshkind, L., Bockamp, E., Kinzel, B., Mueller, M., Murakami, M., Baffert, F., Radimerski, T., 2019. A conditional inducible JAK2V617F transgenic mouse model reveals myeloproliferative disease that is reversible upon switching off transgene expression. PLoS One 14. https://doi.org/10.1371/journal.pone.0221635, e0221635.

Chen, E., Beer, P.A., Godfrey, A.L., Ortmann, C.A., Li, J., Costa-Pereira, A.P., Ingle, C.E., Dermitzakis, E.T., Campbell, P.J., Green, A.R., 2010. Distinct clinical phenotypes associated with JAK2V617F reflect differential STAT1 signaling. Cancer Cell 18, 524–535. https://doi.org/10.1016/j.ccr.2010.10.013.

Chen, E., Schneider, R.K., Breyfogle, L.J., Rosen, E.A., Poveromo, L., Elf, S., Ko, A., Brumme, K., Levine, R., Ebert, B.L., Mullally, A., 2015. Distinct effects of concomitant Jak2V617F expression and Tet2 loss in mice promote disease progression in myeloproliferative neoplasms. Blood 125, 327–335. https://doi.org/10.1182/blood-2014-04-567024.

Craver, B.M., Ramanathan, G., Hoang, S., Chang, X., Mendez Luque, L.F., Brooks, S., Lai, H.Y., Fleischman, A.G., 2020. N-acetylcysteine inhibits thrombosis in a murine model of myeloproliferative neoplasm. Blood Adv. 4, 312–321. https://doi.org/10.1182/bloodadvances.2019000967.

Croker, B.A., Metcalf, D., Robb, L., Wei, W., Mifsud, S., DiRago, L., Cluse, L.A., Sutherland, K.D., Hartley, L., Williams, E., Zhang, J.-G., Hilton, D.J., Nicola, N.A., Alexander, W.S., Roberts, A.W., 2004. SOCS3 is a critical physiological negative regulator of G-CSF signaling and emergency granulopoiesis. Immunity 20, 153–165. https://doi.org/10.1016/s1074-7613(04)00022-6.

Dagher, T., Maslah, N., Edmond, V., Cassinat, B., Vainchenker, W., Giraudier, S., Pasquier, F., Verger, E., Niwa-Kawakita, M., Lallemand-Breitenbach, V., Plo, I., Kiladjian, J.-J., Villeval, J.-L., de Thé, H., 2021. JAK2V617F myeloproliferative neoplasm eradication by a novel interferon/arsenic therapy involves PML. J. Exp. Med. 218 (2), e20201268. https://doi.org/10.1084/jem.20201268.

Debeurme, F., Lacout, C., Moratal, C., Bagley, R.G., Vainchenker, W., Adrian, F., Villeval, J.-L., 2015. JAK2 inhibition has different therapeutic effects according to myeloproliferative neoplasm development in mice. J. Cell. Mol. Med. 19, 2564–2574. https://doi.org/10.1111/jcmm.12608.

Decker, M., Martinez-Morentin, L., Wang, G., Lee, Y., Liu, Q., Leslie, J., Ding, L., 2017. Leptin-receptor-expressing bone marrow stromal cells are myofibroblasts in primary myelofibrosis. Nat. Cell Biol. 19, 677–688. https://doi.org/10.1038/ncb3530.

Delhommeau, F., Dupont, S., Tonetti, C., Massé, A., Godin, I., Le Couedic, J.-P., Debili, N., Saulnier, P., Casadevall, N., Vainchenker, W., Giraudier, S., 2007. Evidence that the JAK2 G1849T (V617F) mutation occurs in a lymphomyeloid progenitor in polycythemia vera and idiopathic myelofibrosis. Blood 109, 71–77. https://doi.org/10.1182/blood-2006-03-007146.

Delhommeau, F., Dupont, S., Della Valle, V., James, C., Trannoy, S., Massé, A., Kosmider, O., Le Couedic, J.-P., Robert, F., Alberdi, A., Lécluse, Y., Plo, I., Dreyfus, F.J., Marzac, C., Casadevall, N., Lacombe, C., Romana, S.P., Dessen, P., Soulier, J., Viguié, F., Fontenay, M., Vainchenker, W., Bernard, O.A., 2009. Mutation in TET2 in myeloid cancers. N. Engl. J. Med. 360, 2289–2301. https://doi.org/10.1056/NEJMoa0810069.

Dong, L., Yu, W.-M., Zheng, H., Loh, M.L., Bunting, S.T., Pauly, M., Huang, G., Zhou, M., Broxmeyer, H.E., Scadden, D.T., Qu, C.-K., 2016. Leukaemogenic effects of Ptpn11 activating mutations in the stem cell microenvironment. Nature 539, 304–308. https://doi.org/10.1038/nature20131.

Dorsheimer, L., Assmus, B., Rasper, T., Ortmann, C.A., Ecke, A., Abou-El-Ardat, K., Schmid, T., Brüne, B., Wagner, S., Serve, H., Hoffmann, J., Seeger, F., Dimmeler, S., Zeiher, A.M., Rieger, M.A., 2019. Association of mutations contributing to clonal hematopoiesis with prognosis in chronic ischemic heart failure. JAMA Cardiol. 4, 25–33. https://doi.org/10.1001/jamacardio.2018.3965.

Duek, A., Lundberg, P., Shimizu, T., Grisouard, J., Karow, A., Kubovcakova, L., Hao-Shen, H., Dirnhofer, S., Skoda, R.C., 2014. Loss of Stat1 decreases megakaryopoiesis and favors erythropoiesis in a JAK2-V617F-driven mouse model of MPNs. Blood 123, 3943–3950. https://doi.org/10.1182/blood-2013-07-514208.

Elf, S., Abdelfattah, N.S., Chen, E., Perales-Patón, J., Rosen, E.A., Ko, A., Peisker, F., Florescu, N., Giannini, S., Wolach, O., Morgan, E.A., Tothova, Z., Losman, J.-A., Schneider, R.K., Al-Shahrour, F., Mullally, A., 2016. Mutant calreticulin requires both its mutant C-terminus and the thrombopoietin receptor for oncogenic transformation. Cancer Discov. 6, 368–381. https://doi.org/10.1158/2159-8290.CD-15-1434.

El-Khoury, M., Cabagnols, X., Mosca, M., Vertenoeil, G., Marzac, C., Favale, F., Bluteau, O., Lorre, F., Tisserand, A., Rabadan Moraes, G., Ugo, V., Lanotto, J.-C., Rey, J., Solary, E., Roy, L., Rameau, P., Debili, N., Pasquier, F., Casadevall, N., Marty, C., Constantinescu, S.N., Raslova, H., Vainchenker, W., Plo, I., 2020. Different impact of calreticulin mutations on human hematopoiesis in myeloproliferative neoplasms. Oncogene 39 (31), 5323–5337. https://doi.org/10.1038/s41388-020-1368-3.

Erba, B.G., Gruppi, C., Corada, M., Pisati, F., Rosti, V., Bartalucci, N., Villeval, J.-L., Vannucchi, A.M., Barosi, G., Balduini, A., Dejana, E., 2017. Endothelial-to-mesenchymal transition in bone marrow and spleen of primary myelofibrosis. Am. J. Pathol. 187, 1879–1892. https://doi.org/10.1016/j.ajpath.2017.04.006.

Ernst, T., Chase, A.J., Score, J., Hidalgo-Curtis, C.E., Bryant, C., Jones, A.V., Waghorn, K., Zoi, K., Ross, F.M., Reiter, A., Hochhaus, A., Drexler, H.G., Duncombe, A., Cervantes, F., Oscier, D., Boultwood, J., Grand, F.H., Cross, N.C.P., 2010. Inactivating mutations of the histone methyltransferase gene EZH2 in myeloid disorders. Nat. Genet. 42, 722–726. https://doi.org/10.1038/ng.621.

Etheridge, S.L., Roh, M.E., Cosgrove, M.E., Sangkhae, V., Fox, N.E., Chen, J., López, J.A., Kaushansky, K., Hitchcock, I.S., 2014. JAK2V617F-positive endothelial cells contribute to clotting abnormalities in myeloproliferative neoplasms. Proc. Natl. Acad. Sci. U. S. A. 111, 2295–2300. https://doi.org/10.1073/pnas.1312148111.

Faubert, B., Solmonson, A., DeBerardinis, R.J., 2020. Metabolic reprogramming and cancer progression. Science 368 (6487). https://doi.org/10.1126/science.aaw5473, eaaw5473.

Fisher, C.L., Pineault, N., Brookes, C., Helgason, C.D., Ohta, H., Bodner, C., Hess, J.L., Humphries, R.K., Brock, H.W., 2010. Loss-of-function additional sex combs like 1 mutations disrupt hematopoiesis but do not cause severe myelodysplasia or leukemia. Blood 115, 38–46. https://doi.org/10.1182/blood-2009-07-230698.

Fleischman, A.G., Aichberger, K.J., Luty, S.B., Bumm, T.G., Petersen, C.L., Doratotaj, S., Vasudevan, K.B., LaTocha, D.H., Yang, F., Press, R.D., Loriaux, M.M., Pahl, H.L., Silver, R.T., Agarwal, A., O'Hare, T., Druker, B.J., Bagby, G.C., Deininger, M.W., 2011. TNFα facilitates clonal expansion of JAK2V617F positive cells in myeloproliferative neoplasms. Blood 118, 6392–6398. https://doi.org/10.1182/blood-2011-04-348144.

Frey, B.M., Rafii, S., Crystal, R.G., Moore, M.A., 1998. Adenovirus long-term expression of thrombopoietin in vivo: a new model for myeloproliferative syndrome and osteomyelofibrosis. Schweiz. Med. Wochenschr. 128, 1587–1592.

Fuster, J.J., MacLauchlan, S., Zuriaga, M.A., Polackal, M.N., Ostriker, A.C., Chakraborty, R., Wu, C.-L., Sano, S., Muralidharan, S., Rius, C., Vuong, J., Jacob, S., Muralidhar, V., Robertson, A.A.B., Cooper, M.A., Andrés, V., Hirschi, K.K., Martin, K.A., Walsh, K., 2017. Clonal hematopoiesis associated with TET2 deficiency accelerates atherosclerosis development in mice. Science 355, 842–847. https://doi.org/10.1126/science.aag1381.

Gastinne, T., Vigant, F., Lavenu-Bombled, C., Wagner-Ballon, O., Tulliez, M., Chagraoui, H., Villeval, J.-L., Lacout, C., Perricaudet, M., Vainchenker, W., Benihoud, K., Giraudier, S., 2007. Adenoviral-mediated TGF-beta1 inhibition in a mouse model of myelofibrosis inhibit bone marrow fibrosis development. Exp. Hematol. 35, 64–74. https://doi.org/10.1016/j.exphem.2006.08.016.

Göthert, J.R., Gustin, S.E., Hall, M.A., Green, A.R., Göttgens, B., Izon, D.J., Begley, C.G., 2005. In vivo fate-tracing studies using the Scl stem cell enhancer: embryonic hematopoietic stem cells significantly contribute to adult hematopoiesis. Blood 105, 2724–2732. https://doi.org/10.1182/blood-2004-08-3037.

Gowin, K., Jain, T., Kosiorek, H., Tibes, R., Camoriano, J., Palmer, J., Mesa, R., 2017. Pegylated interferon alpha—2a is clinically effective and tolerable in myeloproliferative neoplasm patients treated off clinical trial. Leuk. Res. 54, 73–77. https://doi.org/10.1016/j.leukres.2017.01.006.

Greten, F.R., Grivennikov, S.I., 2019. Inflammation and cancer: triggers, mechanisms, and consequences. Immunity 51, 27–41. https://doi.org/10.1016/j.immuni.2019.06.025.

Grisouard, J., Shimizu, T., Duek, A., Kubovcakova, L., Hao-Shen, H., Dirnhofer, S., Skoda, R.C., 2015. Deletion of Stat3 in hematopoietic cells enhances thrombocytosis and shortens survival in a JAK2-V617F mouse model of MPN. Blood 125, 2131–2140. https://doi.org/10.1182/blood-2014-08-594572.

Grisouard, J., Li, S., Kubovcakova, L., Rao, T.N., Meyer, S.C., Lundberg, P., Hao-Shen, H., Romanet, V., Murakami, M., Radimerski, T., Dirnhofer, S., Skoda, R.C., 2016. JAK2 exon 12 mutant mice display isolated erythrocytosis and changes in iron metabolism favoring increased erythropoiesis. Blood 128, 839–851. https://doi.org/10.1182/blood-2015-12-689216.

Grivennikov, S.I., Greten, F.R., Karin, M., 2010. Immunity, inflammation, and cancer. Cell 140, 883–899. https://doi.org/10.1016/j.cell.2010.01.025.

Guerrero, J.A., Bennett, C., van der Weyden, L., McKinney, H., Chin, M., Nurden, P., McIntyre, Z., Cambridge, E.L., Estabel, J., Wardle-Jones, H., Speak, A.O., Erber, W.N., Rendon, A., Ouwehand, W.H., Ghevaert, C., 2014. Gray platelet syndrome: proinflammatory megakaryocytes and α-granule loss cause myelofibrosis and confer metastasis resistance in mice. Blood 124, 3624–3635. https://doi.org/10.1182/blood-2014-04-566760.

Guglielmelli, P., Zini, R., Bogani, C., Salati, S., Pancrazzi, A., Bianchi, E., Mannelli, F., Ferrari, S., Le Bousse-Kerdilès, M.-C., Bosi, A., Barosi, G., Migliaccio, A.R., Manfredini, R., Vannucchi, A.M., 2007. Molecular profiling of CD34+ cells in idiopathic myelofibrosis identifies a set of disease-associated genes and reveals the clinical significance of Wilms' tumor gene 1 (WT1). Stem Cells 25, 165–173. https://doi.org/10.1634/stemcells.2006-0351.

Guo, Y., Zhou, Y., Yamatomo, S., Yang, H., Zhang, P., Chen, S., Nimer, S.D., Zhao, Z.J., Xu, M., Bai, J., Yang, F.-C., 2019. ASXL1 alteration cooperates with JAK2V617F to accelerate myelofibrosis. Leukemia 33, 1287–1291. https://doi.org/10.1038/s41375-018-0347-y.

Guy, A., Gourdou-Latyszenok, V., Le Lay, N., Peghaire, C., Kilani, B., Dias, J.V., Duplaa, C., Renault, M.-A., Denis, C., Villeval, J.L., Boulaftali, Y., Jandrot-Perrus, M., Couffinhal, T., James, C., 2019. Vascular endothelial cell expression of JAK2V617F is sufficient to promote a pro-thrombotic state due to increased P-selectin expression. Haematologica 104, 70–81. https://doi.org/10.3324/haematol.2018.195321.

Hasan, S., Lacout, C., Marty, C., Cuingnet, M., Solary, E., Vainchenker, W., Villeval, J.-L., 2013. JAK2V617F expression in mice amplifies early hematopoietic cells and gives them a competitive advantage that is hampered by IFNα. Blood 122, 1464–1477. https://doi.org/10.1182/blood-2013-04-498956.

Hobbs, C.M., Manning, H., Bennett, C., Vasquez, L., Severin, S., Brain, L., Mazharian, A., Guerrero, J.A., Li, J., Soranzo, N., Green, A.R., Watson, S.P., Ghevaert, C., 2013. JAK2V617F leads to intrinsic changes in platelet formation and reactivity in a knock-in mouse model of essential thrombocythemia. Blood 122, 3787–3797. https://doi.org/10.1182/blood-2013-06-501452.

Hong, S.-H., Dvorak-Ewell, M., Stevens, H.Y., Barish, G.D., Castro, G.L., Nofsinger, R., Frangos, J.A., Shoback, D., Evans, R.M., 2013. Rescue of a primary myelofibrosis model by retinoid-antagonist therapy. Proc. Natl. Acad. Sci. U. S. A. 110, 18820–18825. https://doi.org/10.1073/pnas.1318974110.

Hsu, Y.-C., Chiu, Y.-C., Lin, C.-C., Kuo, Y.-Y., Hou, H.-A., Tzeng, Y.-S., Kao, C.-J., Chuang, P.-H., Tseng, M.-H., Hsiao, T.-H., Chou, W.-C., Tien, H.-F., 2017. The distinct biological implications of Asxl1 mutation and its roles in leukemogenesis revealed by a knock-in mouse model. J. Hematol. Oncol. 10, 139. https://doi.org/10.1186/s13045-017-0508-x.

Huang, Z., Richmond, T.D., Muntean, A.G., Barber, D.L., Weiss, M.J., Crispino, J.D., 2007. STAT1 promotes megakaryopoiesis downstream of GATA-1 in mice. J. Clin. Invest. 117, 3890–3899. https://doi.org/10.1172/JCI33010.

Inoue, D., Kitaura, J., Togami, K., Nishimura, K., Enomoto, Y., Uchida, T., Kagiyama, Y., Kawabata, K.C., Nakahara, F., Izawa, K., Oki, T., Maehara, A., Isobe, M., Tsuchiya, A.,

Harada, Y., Harada, H., Ochiya, T., Aburatani, H., Kimura, H., Thol, F., Heuser, M., Levine, R.L., Abdel-Wahab, O., Kitamura, T., 2013. Myelodysplastic syndromes are induced by histone methylation–altering ASXL1 mutations. J. Clin. Invest. 123, 4627–4640. https://doi.org/10.1172/JCI70739.

Inoue, D., Matsumoto, M., Nagase, R., Saika, M., Fujino, T., Nakayama, K.I., Kitamura, T., 2016. Truncation mutants of ASXL1 observed in myeloid malignancies are expressed at detectable protein levels. Exp. Hematol. 44, 172–176.e1. https://doi.org/10.1016/j.exphem.2015.11.011.

Jacquelin, S., Straube, J., Cooper, L., Vu, T., Song, A., Bywater, M., Baxter, E., Heidecker, M., Wackrow, B., Porter, A., Ling, V., Green, J., Austin, R., Kazakoff, S., Waddell, N., Hesson, L.B., Pimanda, J.E., Stegelmann, F., Bullinger, L., Döhner, K., Rampal, R.K., Heckl, D., Hill, G.R., Lane, S.W., 2018. Jak2V617F and Dnmt3a loss cooperate to induce myelofibrosis through activated enhancer-driven inflammation. Blood 132, 2707–2721. https://doi.org/10.1182/blood-2018-04-846220.

Jacquelin, S., Kramer, F., Mullally, A., Lane, S.W., 2020. Murine models of myelofibrosis. Cancers 12 (9), 2381. https://doi.org/10.3390/cancers12092381.

Jaiswal, S., Fontanillas, P., Flannick, J., Manning, A., Grauman, P.V., Mar, B.G., Lindsley, R.C., Mermel, C.H., Burtt, N., Chavez, A., Higgins, J.M., Moltchanov, V., Kuo, F.C., Kluk, M.J., Henderson, B., Kinnunen, L., Koistinen, H.A., Ladenvall, C., Getz, G., Correa, A., Banahan, B.F., Gabriel, S., Kathiresan, S., Stringham, H.M., McCarthy, M.I., Boehnke, M., Tuomilehto, J., Haiman, C., Groop, L., Atzmon, G., Wilson, J.G., Neuberg, D., Altshuler, D., Ebert, B.L., 2014. Age-related clonal hematopoiesis associated with adverse outcomes. N. Engl. J. Med. 371, 2488–2498. https://doi.org/10.1056/NEJMoa1408617.

James, C., Ugo, V., Le Couédic, J.-P., Staerk, J., Delhommeau, F., Lacout, C., Garçon, L., Raslova, H., Berger, R., Bennaceur-Griscelli, A., Villeval, J.L., Constantinescu, S.N., Casadevall, N., Vainchenker, W., 2005. A unique clonal JAK2 mutation leading to constitutive signalling causes polycythaemia vera. Nature 434, 1144–1148. https://doi.org/10.1038/nature03546.

Kakumitsu, H., Kamezaki, K., Shimoda, K., Karube, K., Haro, T., Numata, A., Shide, K., Matsuda, T., Oshima, K., Harada, M., 2005. Transgenic mice overexpressing murine thrombopoietin develop myelofibrosis and osteosclerosis. Leuk. Res. 29, 761–769. https://doi.org/10.1016/j.leukres.2004.12.009.

Kameda, T., Shide, K., Yamaji, T., Kamiunten, A., Sekine, M., Taniguchi, Y., Hidaka, T., Kubuki, Y., Shimoda, H., Marutsuka, K., Sashida, G., Aoyama, K., Yoshimitsu, M., Harada, T., Abe, H., Miike, T., Iwakiri, H., Tahara, Y., Sueta, M., Yamamoto, S., Hasuike, S., Nagata, K., Iwama, A., Kitanaka, A., Shimoda, K., 2015. Loss of TET2 has dual roles in murine myeloproliferative neoplasms: disease sustainer and disease accelerator. Blood 125, 304–315. https://doi.org/10.1182/blood-2014-04-555508.

Kent, D.G., Li, J., Tanna, H., Fink, J., Kirschner, K., Pask, D.C., Silber, Y., Hamilton, T.L., Sneade, R., Simons, B.D., Green, A.R., 2013. Self-renewal of single mouse hematopoietic stem cells is reduced by JAK2V617F without compromising progenitor cell expansion. PLoS Biol. 11. https://doi.org/10.1371/journal.pbio.1001576, e1001576.

Kiladjian, J.-J., Cassinat, B., Chevret, S., Turlure, P., Cambier, N., Roussel, M., Bellucci, S., Grandchamp, B., Chomienne, C., Fenaux, P., 2008. Pegylated interferon-alfa-2a induces complete hematologic and molecular responses with low toxicity in polycythemia vera. Blood 112, 3065–3072. https://doi.org/10.1182/blood-2008-03-143537.

Kirsammer, G., Jilani, S., Liu, H., Davis, E., Gurbuxani, S., Le Beau, M.M., Crispino, J.D., 2008. Highly penetrant myeloproliferative disease in the Ts65Dn mouse model of down syndrome. Blood 111, 767–775. https://doi.org/10.1182/blood-2007-04-085670.

Klampfl, T., Gisslinger, H., Harutyunyan, A.S., Nivarthi, H., Rumi, E., Milosevic, J.D., Them, N.C.C., Berg, T., Gisslinger, B., Pietra, D., Chen, D., Vladimer, G.I., Bagienski, K., Milanesi, C., Casetti, I.C., Sant'Antonio, E., Ferretti, V., Elena, C., Schischlik, F., Cleary, C., Six, M., Schalling, M., Schönegger, A., Bock, C., Malcovati, L., Pascutto, C., Superti-Furga, G., Cazzola, M., Kralovics, R., 2013. Somatic mutations of calreticulin in myeloproliferative neoplasms. N. Engl. J. Med. 369, 2379–2390. https://doi.org/10.1056/NEJMoa1311347.

Koppikar, P., Bhagwat, N., Kilpivaara, O., Manshouri, T., Adli, M., Hricik, T., Liu, F., Saunders, L.M., Mullally, A., Abdel-Wahab, O., Leung, L., Weinstein, A., Marubayashi, S., Goel, A., Gönen, M., Estrov, Z., Ebert, B.L., Chiosis, G., Nimer, S.D., Bernstein, B.E., Verstovsek, S., Levine, R.L., 2012. Heterodimeric JAK-STAT activation as a mechanism of persistence to JAK2 inhibitor therapy. Nature 489, 155–159. https://doi.org/10.1038/nature11303.

Kramann, R., Schneider, R.K., 2018. The identification of fibrosis-driving myofibroblast precursors reveals new therapeutic avenues in myelofibrosis. Blood 131, 2111–2119. https://doi.org/10.1182/blood-2018-02-834820.

Kramann, R., DiRocco, D.P., Humphreys, B.D., 2013. Understanding the origin, activation and regulation of matrix-producing myofibroblasts for treatment of fibrotic disease. J. Pathol. 231, 273–289. https://doi.org/10.1002/path.4253.

Kramann, R., Schneider, R.K., DiRocco, D.P., Machado, F., Fleig, S., Bondzie, P.A., Henderson, J.M., Ebert, B.L., Humphreys, B.D., 2015. Perivascular Gli1+ progenitors are key contributors to injury-induced organ fibrosis. Cell Stem Cell 16, 51–66. https://doi.org/10.1016/j.stem.2014.11.004.

Kubovcakova, L., Lundberg, P., Grisouard, J., Hao-Shen, H., Romanet, V., Andraos, R., Murakami, M., Dirnhofer, S., Wagner, K.-U., Radimerski, T., Skoda, R.C., 2013. Differential effects of hydroxyurea and INC424 on mutant allele burden and myeloproliferative phenotype in a JAK2-V617F polycythemia vera mouse model. Blood 121, 1188–1199. https://doi.org/10.1182/blood-2012-03-415646.

Lacout, C., Pisani, D.F., Tulliez, M., Gachelin, F.M., Vainchenker, W., Villeval, J.-L., 2006. JAK2V617F expression in murine hematopoietic cells leads to MPD mimicking human PV with secondary myelofibrosis. Blood 108, 1652–1660. https://doi.org/10.1182/blood-2006-02-002030.

Lamrani, L., Lacout, C., Ollivier, V., Denis, C.V., Gardiner, E., Ho Tin Noe, B., Vainchenker, W., Villeval, J.-L., Jandrot-Perrus, M., 2014. Hemostatic disorders in a JAK2V617F-driven mouse model of myeloproliferative neoplasm. Blood 124, 1136–1145. https://doi.org/10.1182/blood-2013-10-530832.

Li, J., Spensberger, D., Ahn, J.S., Anand, S., Beer, P.A., Ghevaert, C., Chen, E., Forrai, A., Scott, L.M., Ferreira, R., Campbell, P.J., Watson, S.P., Liu, P., Erber, W.N., Huntly, B.J.P., Ottersbach, K., Green, A.R., 2010. JAK2 V617F impairs hematopoietic stem cell function in a conditional knock-in mouse model of JAK2 V617F-positive essential thrombocythemia. Blood 116, 1528–1538. https://doi.org/10.1182/blood-2009-12-259747.

Li, J., Kent, D.G., Godfrey, A.L., Manning, H., Nangalia, J., Aziz, A., Chen, E., Saeb-Parsy, K., Fink, J., Sneade, R., Hamilton, T.L., Pask, D.C., Silber, Y., Zhao, X., Ghevaert, C., Liu, P., Green, A.R., 2014. JAK2V617F homozygosity drives a phenotypic switch in myeloproliferative neoplasms, but is insufficient to sustain disease. Blood 123, 3139–3151. https://doi.org/10.1182/blood-2013-06-510222.

Li, J., Prins, D., Park, H.J., Grinfeld, J., Gonzalez-Arias, C., Loughran, S., Dovey, O.M., Klampfl, T., Bennett, C., Hamilton, T.L., Pask, D.C., Sneade, R., Williams, M., Aungier, J., Ghevaert, C., Vassiliou, G.S., Kent, D.G., Green, A.R., 2018. Mutant calreticulin knockin mice develop thrombocytosis and myelofibrosis without a stem cell self-renewal advantage. Blood 131, 649–661. https://doi.org/10.1182/blood-2017-09-806356.

Lu, X., Levine, R., Tong, W., Wernig, G., Pikman, Y., Zarnegar, S., Gilliland, D.G., Lodish, H., 2005. Expression of a homodimeric type I cytokine receptor is required for JAK2V617F-mediated transformation. Proc. Natl. Acad. Sci. U. S. A. 102, 18962–18967. https://doi.org/10.1073/pnas.0509714102.

Lundberg, P., Takizawa, H., Kubovcakova, L., Guo, G., Hao-Shen, H., Dirnhofer, S., Orkin, S.H., Manz, M.G., Skoda, R.C., 2014. Myeloproliferative neoplasms can be initiated from a single hematopoietic stem cell expressing JAK2-V617F. J. Exp. Med. 211, 2213–2230. https://doi.org/10.1084/jem.20131371.

Mansier, O., Kilani, B., Guitart, A.V., Guy, A., Gourdou-Latyszenok, V., Marty, C., Parrens, M., Plo, I., Vainchenker, W., James, C., 2019. Description of a knock-in mouse model of JAK2V617F MPN emerging from a minority of mutated hematopoietic stem cells. Blood 134, 2383–2387. https://doi.org/10.1182/blood.2019001163.

Martelli, F., Ghinassi, B., Panetta, B., Alfani, E., Gatta, V., Pancrazzi, A., Bogani, C., Vannucchi, A.M., Paoletti, F., Migliaccio, G., Migliaccio, A.R., 2005. Variegation of the phenotype induced by the Gata1low mutation in mice of different genetic backgrounds. Blood 106, 4102–4113. https://doi.org/10.1182/blood-2005-03-1060.

Marty, C., Lacout, C., Martin, A., Hasan, S., Jacquot, S., Birling, M.-C., Vainchenker, W., Villeval, J.-L., 2010. Myeloproliferative neoplasm induced by constitutive expression of JAK2V617F in knock-in mice. Blood 116, 783–787. https://doi.org/10.1182/blood-2009-12-257063.

Marty, C., Lacout, C., Droin, N., Le Couédic, J.-P., Ribrag, V., Solary, E., Vainchenker, W., Villeval, J.-L., Plo, I., 2013. A role for reactive oxygen species in JAK2 V617F myeloproliferative neoplasm progression. Leukemia 27, 2187–2195. https://doi.org/10.1038/leu.2013.102.

Marty, C., Pecquet, C., Nivarthi, H., El-Khoury, M., Chachoua, I., Tulliez, M., Villeval, J.L., Raslova, H., Kralovics, R., Constantinescu, S.N., Plo, I., Vainchenker, W., 2016. Calreticulin mutants in mice induce an MPL-dependent thrombocytosis with frequent progression to myelofibrosis. Blood 127, 1317–1324. https://doi.org/10.1182/blood-2015-11-679571.

Mascarenhas, J., Li, T., Sandy, L., Newsom, C., Petersen, B., Godbold, J., Hoffman, R., 2014. Anti-transforming growth factor-β therapy in patients with myelofibrosis. Leuk. Lymphoma 55, 450–452. https://doi.org/10.3109/10428194.2013.805329.

Matsuura, S., Thompson, C.R., Belghasem, M.E., Bekendam, R.H., Piasecki, A., Leiva, O., Ray, A., Italiano, J., Yang, M., Merill-Skoloff, G., Chitalia, V.C., Flaumenhaft, R., Ravid, K., 2020. Platelet dysfunction and thrombosis in JAK2V617F-mutated primary myelofibrotic mice. Arterioscler. Thromb. Vasc. Biol. 40, e262–e272. https://doi.org/10.1161/ATVBAHA.120.314760.

McKenney, A.S., Lau, A.N., Somasundara, A.V.H., Spitzer, B., Intlekofer, A.M., Ahn, J., Shank, K., Rapaport, F.T., Patel, M.A., Papalexi, E., Shih, A.H., Chiu, A., Freinkman, E., Akbay, E.A., Steadman, M., Nagaraja, R., Yen, K., Teruya-Feldstein, J., Wong, K.-K., Rampal, R., Vander Heiden, M.G., Thompson, C.B., Levine, R.L., 2018. JAK2/IDH-mutant-driven myeloproliferative neoplasm is sensitive to combined targeted inhibition. J. Clin. Invest. 128, 789–804. https://doi.org/10.1172/JCI94516.

Mesaeli, N., Nakamura, K., Zvaritch, E., Dickie, P., Dziak, E., Krause, K.H., Opas, M., MacLennan, D.H., Michalak, M., 1999. Calreticulin is essential for cardiac development. J. Cell Biol. 144, 857–868. https://doi.org/10.1083/jcb.144.5.857.

Mestas, J., Hughes, C.C.W., 2004. Of mice and not men: differences between mouse and human immunology. J. Immunol. 172, 2731–2738. https://doi.org/10.4049/jimmunol.172.5.2731.

Metcalf, D., 1998. The molecular control of hematopoiesis: progress and problems with gene manipulation. Stem Cells 16, 314–321. https://doi.org/10.1002/stem.160314.

Meyer, S.C., Keller, M.D., Chiu, S., Koppikar, P., Guryanova, O.A., Rapaport, F., Xu, K., Manova, K., Pankov, D., O'Reilly, R.J., Kleppe, M., McKenney, A.S., Shih, A.H., Shank, K., Ahn, J., Papalexi, E., Spitzer, B., Socci, N., Viale, A., Mandon, E., Ebel, N., Andraos, R., Rubert, J., Dammassa, E., Romanet, V., Dölemeyer, A., Zender, M., Heinlein, M., Rampal, R., Weinberg, R.S., Hoffman, R., Sellers, W.R., Hofmann, F., Murakami, M., Baffert, F., Gaul, C., Radimerski, T., Levine, R.L., 2015. CHZ868, a type II JAK2 inhibitor, reverses type I JAK inhibitor persistence and demonstrates efficacy in myeloproliferative neoplasms. Cancer Cell 28, 15–28. https://doi.org/10.1016/j.ccell.2015.06.006.

Mo, J.-R., Mathur, A., Angagaw, M., Zhao, S., Wang, Y., Gargano, D., DiBacco, A., Bachman, E.S., 2009. Splenectomy normalizes hematocrit in murine polycythemia vera. PLoS One 4, e7286. https://doi.org/10.1371/journal.pone.0007286.

Moran-Crusio, K., Reavie, L., Shih, A., Abdel-Wahab, O., Ndiaye-Lobry, D., Lobry, C., Figueroa, M.E., Vasanthakumar, A., Patel, J., Zhao, X., Perna, F., Pandey, S., Madzo, J., Song, C., Dai, Q., He, C., Ibrahim, S., Beran, M., Zavadil, J., Nimer, S.D., Melnick, A., Godley, L.A., Aifantis, I., Levine, R.L., 2011. Tet2 loss leads to increased hematopoietic stem cell self-renewal and myeloid transformation. Cancer Cell 20, 11–24. https://doi.org/10.1016/j.ccr.2011.06.001.

Morel, P., Duhamel, A., Hivert, B., Stalniekiewicz, L., Demory, J.-L., Dupriez, B., 2010. Identification during the follow-up of time-dependent prognostic factors for the competing risks of death and blast phase in primary myelofibrosis: a study of 172 patients. Blood 115, 4350–4355. https://doi.org/10.1182/blood-2009-12-255943.

Mullally, A., Lane, S.W., Ball, B., Megerdichian, C., Okabe, R., Al-Shahrour, F., Paktinat, M., Haydu, J.E., Housman, E., Lord, A.M., Wernig, G., Kharas, M.G., Mercher, T., Kutok, J.L., Gilliland, D.G., Ebert, B.L., 2010. Physiological Jak2V617F expression causes a lethal myeloproliferative neoplasm with differential effects on hematopoietic stem and progenitor cells. Cancer Cell 17, 584–596. https://doi.org/10.1016/j.ccr.2010.05.015.

Mullally, A., Poveromo, L., Schneider, R.K., Al-Shahrour, F., Lane, S.W., Ebert, B.L., 2012. Distinct roles for long-term hematopoietic stem cells and erythroid precursor cells in a murine model of Jak2V617F-mediated polycythemia vera. Blood 120, 166–172. https://doi.org/10.1182/blood-2012-01-402396.

Mullally, A., Bruedigam, C., Poveromo, L., Heidel, F.H., Purdon, A., Vu, T., Austin, R., Heckl, D., Breyfogle, L.J., Kuhn, C.P., Kalaitzidis, D., Armstrong, S.A., Williams, D.A., Hill, G.R., Ebert, B.L., Lane, S.W., 2013. Depletion of Jak2V617F myeloproliferative neoplasm-propagating stem cells by interferon-α in a murine model of polycythemia vera. Blood 121, 3692–3702. https://doi.org/10.1182/blood-2012-05-432989.

Nagase, R., Inoue, D., Pastore, A., Fujino, T., Hou, H.-A., Yamasaki, N., Goyama, S., Saika, M., Kanai, A., Sera, Y., Horikawa, S., Ota, Y., Asada, S., Hayashi, Y., Kawabata, K.C., Takeda, R., Tien, H.-F., Honda, H., Abdel-Wahab, O., Kitamura, T., 2018. Expression of mutant Asxl1 perturbs hematopoiesis and promotes susceptibility to leukemic transformation. J. Exp. Med. 215, 1729–1747. https://doi.org/10.1084/jem.20171151.

Nangalia, J., Massie, C.E., Baxter, E.J., Nice, F.L., Gundem, G., Wedge, D.C., Avezov, E., Li, J., Kollmann, K., Kent, D.G., Aziz, A., Godfrey, A.L., Hinton, J., Martincorena, I., Van Loo, P., Jones, A.V., Guglielmelli, P., Tarpey, P., Harding, H.P., Fitzpatrick, J.D., Goudie, C.T., Ortmann, C.A., Loughran, S.J., Raine, K., Jones, D.R., Butler, A.P., Teague, J.W., O'Meara, S., McLaren, S., Bianchi, M., Silber, Y., Dimitropoulou, D., Bloxham, D., Mudie, L., Maddison, M., Robinson, B., Keohane, C., Maclean, C., Hill, K., Orchard, K., Tauro, S., Du, M.-Q., Greaves, M., Bowen, D., Huntly, B.J.P., Harrison, C.N., Cross, N.C.P., Ron, D., Vannucchi, A.M.,

Papaemmanuil, E., Campbell, P.J., Green, A.R., 2013. Somatic CALR mutations in myeloproliferative neoplasms with nonmutated JAK2. N. Engl. J. Med. 369, 2391–2405. https://doi.org/10.1056/NEJMoa1312542.

Nangalia, J., Nice, F.L., Wedge, D.C., Godfrey, A.L., Grinfeld, J., Thakker, C., Massie, C. E., Baxter, J., Sewell, D., Silber, Y., Campbell, P.J., Green, A.R., 2015. DNMT3A mutations occur early or late in patients with myeloproliferative neoplasms and mutation order influences phenotype. Haematologica 100, e438–e442. https://doi.org/10.3324/haematol.2015.129510.

Nguyen, L.V., Vanner, R., Dirks, P., Eaves, C.J., 2012. Cancer stem cells: an evolving concept. Nat. Rev. Cancer 12, 133–143. https://doi.org/10.1038/nrc3184.

Nguyen, T.K., Morse, S.J., Fleischman, A.G., 2016. Transduction-transplantation mouse model of myeloproliferative neoplasm. J. Vis. Exp. 118, 54624. https://doi.org/10.3791/54624.

O'Connell, K.E., Mikkola, A.M., Stepanek, A.M., Vernet, A., Hall, C.D., Sun, C.C., Yildirim, E., Staropoli, J.F., Lee, J.T., Brown, D.E., 2015. Practical murine hematopathology: a comparative review and implications for research. Comp. Med. 65, 96–113.

Ohwada, A., Rafii, S., Moore, M.A., Crystal, R.G., 1996. In vivo adenovirus vector-mediated transfer of the human thrombopoietin cDNA maintains platelet levels during radiation-and chemotherapy-induced bone marrow suppression. Blood 88, 778–784.

Ortmann, C.A., Kent, D.G., Nangalia, J., Silber, Y., Wedge, D.C., Grinfeld, J., Baxter, E.J., Massie, C.E., Papaemmanuil, E., Menon, S., Godfrey, A.L., Dimitropoulou, D., Guglielmelli, P., Bellosillo, B., Besses, C., Döhner, K., Harrison, C.N., Vassiliou, G.S., Vannucchi, A., Campbell, P.J., Green, A.R., 2015. Effect of mutation order on myeloproliferative neoplasms. N. Engl. J. Med. 372, 601–612. https://doi.org/10.1056/NEJMoa1412098.

Papathanasiou, P., Petvises, S., Hey, Y.-Y., Perkins, A.C., O'Neill, H.C., 2017. Impact of the c-MybE308G mutation on mouse myelopoiesis and dendritic cell development. PLoS One 12, e0176345. https://doi.org/10.1371/journal.pone.0176345.

Perrier, A., Roy, P.-M., Sanchez, O., Le Gal, G., Meyer, G., Gourdier, A.-L., Furber, A., Revel, M.-P., Howarth, N., Davido, A., Bounameaux, H., 2005. Multidetector-row computed tomography in suspected pulmonary embolism. N. Engl. J. Med. 352, 1760–1768. https://doi.org/10.1056/NEJMoa042905.

Pietra, D., Rumi, E., Ferretti, V.V., Di Buduo, C.A., Milanesi, C., Cavalloni, C., Sant'Antonio, E., Abbonante, V., Moccia, F., Casetti, I.C., Bellini, M., Renna, M.C., Roncoroni, E., Fugazza, E., Astori, C., Boveri, E., Rosti, V., Barosi, G., Balduini, A., Cazzola, M., 2016. Differential clinical effects of different mutation subtypes in CALR-mutant myeloproliferative neoplasms. Leukemia 30, 431–438. https://doi.org/10.1038/leu.2015.277.

Pikman, Y., Lee, B.H., Mercher, T., McDowell, E., Ebert, B.L., Gozo, M., Cuker, A., Wernig, G., Moore, S., Galinsky, I., DeAngelo, D.J., Clark, J.J., Lee, S.J., Golub, T.R., Wadleigh, M., Gilliland, D.G., Levine, R.L., 2006. MPLW515L is a novel somatic activating mutation in myelofibrosis with myeloid metaplasia. PLoS Med. 3, e270. https://doi.org/10.1371/journal.pmed.0030270.

Poisson, J., Hilscher, M.B., Tanguy, M., Hammoutene, A., Boulanger, C.M., Villeval, J.-L., Simonetto, D.A., Valla, D., Shah, V.H., Rautou, P.-E., 2018. Endothelial JAK2V617F does not enhance liver lesions in mice with Budd-Chiari syndrome. J. Hepatol. 68, 1086–1087. https://doi.org/10.1016/j.jhep.2018.01.010.

Poisson, J., Tanguy, M., Davy, H., Camara, F., El Mdawar, M.-B., Kheloufi, M., Dagher, T., Devue, C., Lasselin, J., Plessier, A., Merchant, S., Blanc-Brude, O., Souyri, M., Mougenot, N., Dingli, F., Loew, D., Hatem, S.N., James, C.,

Villeval, J.-L., Boulanger, C.M., Rautou, P.-E., 2020. Erythrocyte-derived microvesicles induce arterial spasms in JAK2V617F myeloproliferative neoplasm. J. Clin. Invest. 130, 2630–2643. https://doi.org/10.1172/JCI124566.

Quintás-Cardama, A., Vaddi, K., Liu, P., Manshouri, T., Li, J., Scherle, P.A., Caulder, E., Wen, X., Li, Y., Waeltz, P., Rupar, M., Burn, T., Lo, Y., Kelley, J., Covington, M., Shepard, S., Rodgers, J.D., Haley, P., Kantarjian, H., Fridman, J.S., Verstovsek, S., 2010. Preclinical characterization of the selective JAK1/2 inhibitor INCB018424: therapeutic implications for the treatment of myeloproliferative neoplasms. Blood 115, 3109–3117. https://doi.org/10.1182/blood-2009-04-214957.

Quintás-Cardama, A., Kantarjian, H., Cortes, J., Verstovsek, S., 2011. Janus kinase inhibitors for the treatment of myeloproliferative neoplasias and beyond. Nat. Rev. Drug Discov. 10, 127–140. https://doi.org/10.1038/nrd3264.

Quivoron, C., Couronné, L., Della Valle, V., Lopez, C.K., Plo, I., Wagner-Ballon, O., Do Cruzeiro, M., Delhommeau, F., Arnulf, B., Stern, M.-H., Godley, L., Opolon, P., Tilly, H., Solary, E., Duffourd, Y., Dessen, P., Merle-Beral, H., Nguyen-Khac, F., Fontenay, M., Vainchenker, W., Bastard, C., Mercher, T., Bernard, O.A., 2011. TET2 inactivation results in pleiotropic hematopoietic abnormalities in mouse and is a recurrent event during human lymphomagenesis. Cancer Cell 20, 25–38. https://doi.org/10.1016/j.ccr.2011.06.003.

Rai, S., Hansen, N., Hao-Shen, H., Dirnhofer, S., Rao Tata, N., Skoda, R., 2019. IL-1β Secreted from mutant cells carrying JAK2-V617F favors early clonal expansion and promotes MPN disease initiation and progression. Blood 134 (Supplement 1), 307.

Rampal, R., Ahn, J., Abdel-Wahab, O., Nahas, M., Wang, K., Lipson, D., Otto, G.A., Yelensky, R., Hricik, T., McKenney, A.S., Chiosis, G., Chung, Y.R., Pandey, S., van den Brink, M.R.M., Armstrong, S.A., Dogan, A., Intlekofer, A., Manshouri, T., Park, C.Y., Verstovsek, S., Rapaport, F., Stephens, P.J., Miller, V.A., Levine, R.L., 2014. Genomic and functional analysis of leukemic transformation of myeloproliferative neoplasms. Proc. Natl. Acad. Sci. U. S. A. 111, E5401–E5410. https://doi.org/10.1073/pnas.1407792111.

Rangarajan, A., Weinberg, R.A., 2003. Opinion: comparative biology of mouse versus human cells: modelling human cancer in mice. Nat. Rev. Cancer 3, 952–959. https://doi.org/10.1038/nrc1235.

Rao, T.N., Hansen, N., Hilfiker, J., Rai, S., Majewska, J.-M., Leković, D., Gezer, D., Andina, N., Galli, S., Cassel, T., Geier, F., Delezie, J., Nienhold, R., Hao-Shen, H., Beisel, C., Di Palma, S., Dimeloe, S., Trebicka, J., Wolf, D., Gassmann, M., Fan, T.W.-M., Lane, A.N., Handschin, C., Dirnhofer, S., Kröger, N., Hess, C., Radimerski, T., Koschmieder, S., Čokić, V.P., Skoda, R.C., 2019. JAK2-mutant hematopoietic cells display metabolic alterations that can be targeted to treat myeloproliferative neoplasms. Blood 134, 1832–1846. https://doi.org/10.1182/blood.2019000162.

Regis, G., Pensa, S., Boselli, D., Novelli, F., Poli, V., 2008. Ups and downs: the STAT1: STAT3 seesaw of interferon and gp130 receptor signalling. Semin. Cell Dev. Biol. 19, 351–359. https://doi.org/10.1016/j.semcdb.2008.06.004.

Rongvaux, A., Takizawa, H., Strowig, T., Willinger, T., Eynon, E.E., Flavell, R.A., Manz, M.G., 2013. Human hemato-lymphoid system mice: current use and future potential for medicine. Annu. Rev. Immunol. 31, 635–674. https://doi.org/10.1146/annurev-immunol-032712-095921.

Rosti, V., Villani, L., Riboni, R., Poletto, V., Bonetti, E., Tozzi, L., Bergamaschi, G., Catarsi, P., Dallera, E., Novara, F., Massa, M., Campanelli, R., Fois, G., Peruzzi, B., Lucioni, M., Guglielmelli, P., Pancrazzi, A., Fiandrino, G., Zuffardi, O., Magrini, U., Paulli, M., Vannucchi, A.M., Barosi, G., Associazione Italiana per la Ricerca sul Cancro Gruppo Italiano Malattie Mieloproliferative (AGIMM) investigators, 2013. Spleen endothelial cells from patients with myelofibrosis harbor the JAK2V617F mutation. Blood 121, 360–368. https://doi.org/10.1182/blood-2012-01-404889.

Rumi, E., Pietra, D., Ferretti, V., Klampfl, T., Harutyunyan, A.S., Milosevic, J.D., Them, N.C.C., Berg, T., Elena, C., Casetti, I.C., Milanesi, C., Sant'antonio, E., Bellini, M., Fugazza, E., Renna, M.C., Boveri, E., Astori, C., Pascutto, C., Kralovics, R., Cazzola, M., Associazione Italiana per la Ricerca sul Cancro Gruppo Italiano Malattie Mieloproliferative Investigators, 2014. JAK2 or CALR mutation status defines subtypes of essential thrombocythemia with substantially different clinical course and outcomes. Blood 123, 1544–1551. https://doi.org/10.1182/blood-2013-11-539098.

Rupec, R.A., Jundt, F., Rebholz, B., Eckelt, B., Weindl, G., Herzinger, T., Flaig, M.J., Moosmann, S., Plewig, G., Dörken, B., Förster, I., Huss, R., Pfeffer, K., 2005. Stroma-mediated dysregulation of myelopoiesis in mice lacking I kappa B alpha. Immunity 22, 479–491. https://doi.org/10.1016/j.immuni.2005.02.009.

Saliba, J., Saint-Martin, C., Di Stefano, A., Lenglet, G., Marty, C., Keren, B., Pasquier, F., Valle, V.D., Secardin, L., Leroy, G., Mahfoudhi, E., Grosjean, S., Droin, N., Diop, M., Dessen, P., Charrier, S., Palazzo, A., Merlevede, J., Meniane, J.-C., Delaunay-Darivon, C., Fuseau, P., Isnard, F., Casadevall, N., Solary, E., Debili, N., Bernard, O.A., Raslova, H., Najman, A., Vainchenker, W., Bellanné-Chantelot, C., Plo, I., 2015. Germline duplication of ATG2B and GSKIP predisposes to familial myeloid malignancies. Nat. Genet. 47, 1131–1140. https://doi.org/10.1038/ng.3380.

Sangkhae, V., Etheridge, S.L., Kaushansky, K., Hitchcock, I.S., 2014. The thrombopoietin receptor, MPL, is critical for development of a JAK2V617F-induced myeloproliferative neoplasm. Blood 124, 3956–3963. https://doi.org/10.1182/blood-2014-07-587238.

Sano, S., Oshima, K., Wang, Y., Katanasaka, Y., Sano, M., Walsh, K., 2018. CRISPR-mediated gene editing to assess the roles of Tet2 and Dnmt3a in clonal hematopoiesis and cardiovascular disease. Circ. Res. 123, 335–341. https://doi.org/10.1161/CIRCRESAHA.118.313225.

Santos, F.P.S., Tam, C.S., Kantarjian, H., Cortes, J., Thomas, D., Pollock, R., Verstovsek, S., 2014. Splenectomy in patients with myeloproliferative neoplasms: efficacy, complications and impact on survival and transformation. Leuk. Lymphoma 55, 121–127. https://doi.org/10.3109/10428194.2013.794269.

Sasaki, M., Knobbe, C.B., Munger, J.C., Lind, E.F., Brenner, D., Brüstle, A., Harris, I.S., Holmes, R., Wakeham, A., Haight, J., You-Ten, A., Li, W.Y., Schalm, S., Su, S.M., Virtanen, C., Reifenberger, G., Ohashi, P.S., Barber, D.L., Figueroa, M.E., Melnick, A., Zúñiga-Pflücker, J.-C., Mak, T.W., 2012. IDH1(R132H) mutation increases murine haematopoietic progenitors and alters epigenetics. Nature 488, 656–659. https://doi.org/10.1038/nature11323.

Sashida, G., Wang, C., Tomioka, T., Oshima, M., Aoyama, K., Kanai, A., Mochizuki-Kashio, M., Harada, H., Shimoda, K., Iwama, A., 2016. The loss of Ezh2 drives the pathogenesis of myelofibrosis and sensitizes tumor-initiating cells to bromodomain inhibition. J. Exp. Med. 213, 1459–1477. https://doi.org/10.1084/jem.20151121.

Schneider, R.K., Mullally, A., Dugourd, A., Peisker, F., Hoogenboezem, R., Van Strien, P.M.H., Bindels, E.M., Heckl, D., Büsche, G., Fleck, D., Müller-Newen, G., Wongboonsin, J., Ventura Ferreira, M., Puelles, V.G., Saez-Rodriguez, J., Ebert, B.L., Humphreys, B.D., Kramann, R., 2017. Gli1 + mesenchymal stromal cells are a key driver of bone marrow fibrosis and an important cellular therapeutic target. Cell Stem Cell 20, 785–800.e8. https://doi.org/10.1016/j.stem.2017.03.008.

Scott, L.M., Tong, W., Levine, R.L., Scott, M.A., Beer, P.A., Stratton, M.R., Futreal, P.A., Erber, W.N., McMullin, M.F., Harrison, C.N., Warren, A.J., Gilliland, D.G., Lodish, H.F., Green, A.R., 2007. JAK2 exon 12 mutations in polycythemia vera and idiopathic erythrocytosis. N. Engl. J. Med. 356, 459–468. https://doi.org/10.1056/NEJMoa065202.

Shide, K., Shimoda, H.K., Kumano, T., Karube, K., Kameda, T., Takenaka, K., Oku, S., Abe, H., Katayose, K.S., Kubuki, Y., Kusumoto, K., Hasuike, S., Tahara, Y., Nagata, K., Matsuda, T., Ohshima, K., Harada, M., Shimoda, K., 2008. Development of ET, primary myelofibrosis and PV in mice expressing JAK2 V617F. Leukemia 22, 87–95. https://doi.org/10.1038/sj.leu.2405043.

Shide, K., Kameda, T., Shimoda, H., Yamaji, T., Abe, H., Kamiunten, A., Sekine, M., Hidaka, T., Katayose, K., Kubuki, Y., Yamamoto, S., Miike, T., Iwakiri, H., Hasuike, S., Nagata, K., Marutsuka, K., Iwama, A., Matsuda, T., Kitanaka, A., Shimoda, K., 2012. TET2 is essential for survival and hematopoietic stem cell homeostasis. Leukemia 26, 2216–2223. https://doi.org/10.1038/leu.2012.94.

Shide, K., Kameda, T., Yamaji, T., Sekine, M., Inada, N., Kamiunten, A., Akizuki, K., Nakamura, K., Hidaka, T., Kubuki, Y., Shimoda, H., Kitanaka, A., Honda, A., Sawaguchi, A., Abe, H., Miike, T., Iwakiri, H., Tahara, Y., Sueta, M., Hasuike, S., Yamamoto, S., Nagata, K., Shimoda, K., 2017. Calreticulin mutant mice develop essential thrombocythemia that is ameliorated by the JAK inhibitor ruxolitinib. Leukemia 31, 1136–1144. https://doi.org/10.1038/leu.2016.308.

Shide, K., Kameda, T., Kamiunten, A., Oji, A., Ozono, Y., Sekine, M., Honda, A., Kitanaka, A., Akizuki, K., Tahira, Y., Nakamura, K., Hidaka, T., Kubuki, Y., Abe, H., Miike, T., Iwakiri, H., Tahara, Y., Sueta, M., Hasuike, S., Yamamoto, S., Nagata, K., Ikawa, M., Shimoda, K., 2019. Mice with Calr mutations homologous to human CALR mutations only exhibit mild thrombocytosis. Blood Cancer J. 9, 42. https://doi.org/10.1038/s41408-019-0202-z.

Shih, A.H., Abdel-Wahab, O., Patel, J.P., Levine, R.L., 2012. The role of mutations in epigenetic regulators in myeloid malignancies. Nat. Rev. Cancer 12, 599–612. https://doi.org/10.1038/nrc3343.

Shimizu, T., Kubovcakova, L., Nienhold, R., Zmajkovic, J., Meyer, S.C., Hao-Shen, H., Geier, F., Dirnhofer, S., Guglielmelli, P., Vannucchi, A.M., Feenstra, J.D.M., Kralovics, R., Orkin, S.H., Skoda, R.C., 2016. Loss of Ezh2 synergizes with JAK2-V617F in initiating myeloproliferative neoplasms and promoting myelofibrosis. J. Exp. Med. 213, 1479–1496. https://doi.org/10.1084/jem.20151136.

Skoda, R.C., 2010. JAK2 impairs stem cell function? Blood 116, 1392–1393. https://doi.org/10.1182/blood-2010-06-287318.

Sykes, S.M., Scadden, D.T., 2013. Modeling human hematopoietic stem cell biology in the mouse. Semin. Hematol. 50, 92–100. https://doi.org/10.1053/j.seminhematol.2013.03.029.

Tefferi, A., 2016. Somatic JAK2 mutations and their tumor phenotypes. Blood 128, 748–749. https://doi.org/10.1182/blood-2016-06-722645.

Tefferi, A., Wassie, E.A., Guglielmelli, P., Gangat, N., Belachew, A.A., Lasho, T.L., Finke, C., Ketterling, R.P., Hanson, C.A., Pardanani, A., Wolanskyj, A.P., Maffioli, M., Casalone, R., Pacilli, A., Vannucchi, A.M., Passamonti, F., 2014. Type 1 versus type 2 calreticulin mutations in essential thrombocythemia: a collaborative study of 1027 patients. Am. J. Hematol. 89, E121–E124. https://doi.org/10.1002/ajh.23743.

Teofili, L., Martini, M., Cenci, T., Petrucci, G., Torti, L., Storti, S., Guidi, F., Leone, G., Larocca, L.M., 2007. Different STAT-3 and STAT-5 phosphorylation discriminates among Ph-negative chronic myeloproliferative diseases and is independent of the V617F JAK-2 mutation. Blood 110, 354–359. https://doi.org/10.1182/blood-2007-01-069237.

Tiedt, R., Hao-Shen, H., Sobas, M.A., Looser, R., Dirnhofer, S., Schwaller, J., Skoda, R.C., 2008. Ratio of mutant JAK2-V617F to wild-type Jak2 determines the MPD phenotypes in transgenic mice. Blood 111, 3931–3940. https://doi.org/10.1182/blood-2007-08-107748.

Toki, T., Katsuoka, F., Kanezaki, R., Xu, G., Kurotaki, H., Sun, J., Kamio, T., Watanabe, S., Tandai, S., Terui, K., Yagihashi, S., Komatsu, N., Igarashi, K., Yamamoto, M., Ito, E., 2005. Transgenic expression of BACH1 transcription factor results in megakaryocytic impairment. Blood 105, 3100–3108. https://doi.org/10.1182/blood-2004-07-2826.

Toppaldoddi, K.R., da Costa Cacemiro, M., Bluteau, O., Panneau-Schmaltz, B., Pioch, A., Muller, D., Villeval, J.-L., Raslova, H., Constantinescu, S.N., Plo, I., Vainchenker, W., Marty, C., 2019. Rare type 1-like and type 2-like calreticulin mutants induce similar myeloproliferative neoplasms as prevalent type 1 and 2 mutants in mice. Oncogene 38, 1651–1660. https://doi.org/10.1038/s41388-018-0538-z.

Tsuruta-Kishino, T., Koya, J., Kataoka, K., Narukawa, K., Sumitomo, Y., Kobayashi, H., Sato, T., Kurokawa, M., 2017. Loss of p53 induces leukemic transformation in a murine model of Jak2 V617F-driven polycythemia vera. Oncogene 36, 3300–3311. https://doi.org/10.1038/onc.2016.478.

Tyner, J.W., Bumm, T.G., Deininger, J., Wood, L., Aichberger, K.J., Loriaux, M.M., Druker, B.J., Burns, C.J., Fantino, E., Deininger, M.W., 2010. CYT387, a novel JAK2 inhibitor, induces hematologic responses and normalizes inflammatory cytokines in murine myeloproliferative neoplasms. Blood 115, 5232–5240. https://doi.org/10.1182/blood-2009-05-223727.

Ueda, K., Ikeda, K., Ikezoe, T., Harada-Shirado, K., Ogawa, K., Hashimoto, Y., Sano, T., Ohkawara, H., Kimura, S., Shichishima-Nakamura, A., Nakamura, Y., Shikama, Y., Mori, T., Mason, P.J., Bessler, M., Morishita, S., Komatsu, N., Shide, K., Shimoda, K., Koide, S., Aoyama, K., Oshima, M., Iwama, A., Takeishi, Y., 2017. Hmga2 collaborates with JAK2V617F in the development of myeloproliferative neoplasms. Blood Adv. 1, 1001–1015. https://doi.org/10.1182/bloodadvances.2017004457.

Ulich, T.R., del Castillo, J., Senaldi, G., Kinstler, O., Yin, S., Kaufman, S., Tarpley, J., Choi, E., Kirley, T., Hunt, P., Sheridan, W.P., 1996. Systemic hematologic effects of PEG-rHuMGDF-induced megakaryocyte hyperplasia in mice. Blood 87, 5006–5015.

Uni, M., Masamoto, Y., Sato, T., Kamikubo, Y., Arai, S., Hara, E., Kurokawa, M., 2019. Modeling ASXL1 mutation revealed impaired hematopoiesis caused by derepression of p16Ink4a through aberrant PRC1-mediated histone modification. Leukemia 33, 191–204. https://doi.org/10.1038/s41375-018-0198-6.

Vannucchi, A.M., Bianchi, L., Cellai, C., Paoletti, F., Rana, R.A., Lorenzini, R., Migliaccio, G., Migliaccio, A.R., 2002. Development of myelofibrosis in mice genetically impaired for GATA-1 expression (GATA-1(low) mice). Blood 100, 1123–1132. https://doi.org/10.1182/blood-2002-06-1913.

Vannucchi, A.M., Lasho, T.L., Guglielmelli, P., Biamonte, F., Pardanani, A., Pereira, A., Finke, C., Score, J., Gangat, N., Mannarelli, C., Ketterling, R.P., Rotunno, G., Knudson, R.A., Susini, M.C., Laborde, R.R., Spolverini, A., Pancrazzi, A., Pieri, L., Manfredini, R., Tagliafico, E., Zini, R., Jones, A., Zoi, K., Reiter, A., Duncombe, A., Pietra, D., Rumi, E., Cervantes, F., Barosi, G., Cazzola, M., Cross, N.C.P., Tefferi, A., 2013. Mutations and prognosis in primary myelofibrosis. Leukemia 27, 1861–1869. https://doi.org/10.1038/leu.2013.119.

Verstovsek, S., Manshouri, T., Pilling, D., Bueso-Ramos, C.E., Newberry, K.J., Prijic, S., Knez, L., Bozinovic, K., Harris, D.M., Spaeth, E.L., Post, S.M., Multani, A.S., Rampal, R.K., Ahn, J., Levine, R.L., Creighton, C.J., Kantarjian, H.M., Estrov, Z., 2016. Role of neoplastic monocyte-derived fibrocytes in primary myelofibrosis. J. Exp. Med. 213, 1723–1740. https://doi.org/10.1084/jem.20160283.

Villeval, J.L., Metcalf, D., Johnson, G.R., 1992. Fatal polycythemia induced in mice by dysregulated erythropoietin production by hematopoietic cells. Leukemia 6, 107–115.

Villeval, J.L., Cohen-Solal, K., Tulliez, M., Giraudier, S., Guichard, J., Burstein, S.A., Cramer, E.M., Vainchenker, W., Wendling, F., 1997. High thrombopoietin production by hematopoietic cells induces a fatal myeloproliferative syndrome in mice. Blood 90, 4369–4383.

Walkley, C.R., Olsen, G.H., Dworkin, S., Fabb, S.A., Swann, J., McArthur, G.A., Westmoreland, S.V., Chambon, P., Scadden, D.T., Purton, L.E., 2007. A microenvironment-induced myeloproliferative syndrome caused by retinoic acid receptor gamma deficiency. Cell 129, 1097–1110. https://doi.org/10.1016/j.cell.2007.05.014.

Walz, C., Ahmed, W., Lazarides, K., Betancur, M., Patel, N., Hennighausen, L., Zaleskas, V.M., Van Etten, R.A., 2012. Essential role for Stat5a/b in myeloproliferative neoplasms induced by BCR-ABL1 and JAK2(V617F) in mice. Blood 119, 3550–3560. https://doi.org/10.1182/blood-2011-12-397554.

Wang, J., Li, Z., He, Y., Pan, F., Chen, S., Rhodes, S., Nguyen, L., Yuan, J., Jiang, L., Yang, X., Weeks, O., Liu, Z., Zhou, J., Ni, H., Cai, C.-L., Xu, M., Yang, F.-C., 2014. Loss of Asxl1 leads to myelodysplastic syndrome-like disease in mice. Blood 123, 541–553. https://doi.org/10.1182/blood-2013-05-500272.

Wendling, F., Maraskovsky, E., Debili, N., Florindo, C., Teepe, M., Titeux, M., Methia, N., Breton-Gorius, J., Cosman, D., Vainchenker, W., 1994. cMpl ligand is a humoral regulator of megakaryocytopoiesis. Nature 369, 571–574. https://doi.org/10.1038/369571a0.

Wernig, G., Mercher, T., Okabe, R., Levine, R.L., Lee, B.H., Gilliland, D.G., 2006. Expression of Jak2V617F causes a polycythemia vera-like disease with associated myelofibrosis in a murine bone marrow transplant model. Blood 107, 4274–4281. https://doi.org/10.1182/blood-2005-12-4824.

Wernig, G., Kharas, M.G., Okabe, R., Moore, S.A., Leeman, D.S., Cullen, D.E., Gozo, M., McDowell, E.P., Levine, R.L., Doukas, J., Mak, C.C., Noronha, G., Martin, M., Ko, Y.D., Lee, B.H., Soll, R.M., Tefferi, A., Hood, J.D., Gilliland, D.G., 2008. Efficacy of TG101348, a selective JAK2 inhibitor, in treatment of a murine model of JAK2V617F-induced polycythemia vera. Cancer Cell 13, 311–320. https://doi.org/10.1016/j.ccr.2008.02.009.

Wolach, O., Sellar, R.S., Martinod, K., Cherpokova, D., McConkey, M., Chappell, R.J., Silver, A.J., Adams, D., Castellano, C.A., Schneider, R.K., Padera, R.F., DeAngelo, D.J., Wadleigh, M., Steensma, D.P., Galinsky, I., Stone, R.M., Genovese, G., McCarroll, S.A., Iliadou, B., Hultman, C., Neuberg, D., Mullally, A., Wagner, D.D., Ebert, B.L., 2018. Increased neutrophil extracellular trap formation promotes thrombosis in myeloproliferative neoplasms. Sci. Transl. Med. 10 (436), eaan8292. https://doi.org/10.1126/scitranslmed.aan8292.

Xie, M., Lu, C., Wang, J., McLellan, M.D., Johnson, K.J., Wendl, M.C., McMichael, J.F., Schmidt, H.K., Yellapantula, V., Miller, C.A., Ozenberger, B.A., Welch, J.S., Link, D.C., Walter, M.J., Mardis, E.R., Dipersio, J.F., Chen, F., Wilson, R.K., Ley, T.J., Ding, L., 2014. Age-related mutations associated with clonal hematopoietic expansion and malignancies. Nat. Med. 20, 1472–1478. https://doi.org/10.1038/nm.3733.

Xing, S., Wanting, T.H., Zhao, W., Ma, J., Wang, S., Xu, X., Li, Q., Fu, X., Xu, M., Zhao, Z.J., 2008. Transgenic expression of JAK2V617F causes myeloproliferative disorders in mice. Blood 111, 5109–5117. https://doi.org/10.1182/blood-2007-05-091579.

Yamazaki, S., Ema, H., Karlsson, G., Yamaguchi, T., Miyoshi, H., Shioda, S., Taketo, M.M., Karlsson, S., Iwama, A., Nakauchi, H., 2011. Nonmyelinating Schwann cells maintain hematopoietic stem cell hibernation in the bone marrow niche. Cell 147, 1146–1158. https://doi.org/10.1016/j.cell.2011.09.053.

Yan, X.Q., Lacey, D., Fletcher, F., Hartley, C., McElroy, P., Sun, Y., Xia, M., Mu, S., Saris, C., Hill, D., Hawley, R.G., McNiece, I.K., 1995. Chronic exposure to retroviral vector encoded MGDF (mpl-ligand) induces lineage-specific growth and differentiation of megakaryocytes in mice. Blood 86, 4025–4033.

Yan, X.Q., Lacey, D., Hill, D., Chen, Y., Fletcher, F., Hawley, R.G., McNiece, I.K., 1996. A model of myelofibrosis and osteosclerosis in mice induced by overexpressing thrombopoietin (mpl ligand): reversal of disease by bone marrow transplantation. Blood 88, 402–409.

Yan, D., Hutchison, R.E., Mohi, G., 2012. Critical requirement for Stat5 in a mouse model of polycythemia vera. Blood 119, 3539–3549. https://doi.org/10.1182/blood-2011-03-345215.

Yan, D., Jobe, F., Hutchison, R.E., Mohi, G., 2015. Deletion of Stat3 enhances myeloid cell expansion and increases the severity of myeloproliferative neoplasms in Jak2V617F knock-in mice. Leukemia 29, 2050–2061. https://doi.org/10.1038/leu.2015.116.

Yang, Y., Akada, H., Nath, D., Hutchison, R.E., Mohi, G., 2016. Loss of Ezh2 cooperates with Jak2V617F in the development of myelofibrosis in a mouse model of myeloproliferative neoplasm. Blood 127, 3410–3423. https://doi.org/10.1182/blood-2015-11-679431.

Yang, H., Kurtenbach, S., Guo, Y., Lohse, I., Durante, M.A., Li, J., Li, Z., Al-Ali, H., Li, L., Chen, Z., Field, M.G., Zhang, P., Chen, S., Yamamoto, S., Li, Z., Zhou, Y., Nimer, S.D., Harbour, J.W., Wahlestedt, C., Xu, M., Yang, F.-C., 2018. Gain of function of ASXL1 truncating protein in the pathogenesis of myeloid malignancies. Blood 131, 328–341. https://doi.org/10.1182/blood-2017-06-789669.

Yue, L., Bartenstein, M., Zhao, W., Ho, W.T., Han, Y., Murdun, C., Mailloux, A.W., Zhang, L., Wang, X., Budhathoki, A., Pradhan, K., Rapaport, F., Wang, H., Shao, Z., Ren, X., Steidl, U., Levine, R.L., Zhao, Z.J., Verma, A., Epling-Burnette, P.K., 2017. Efficacy of ALK5 inhibition in myelofibrosis. JCI Insight 2. https://doi.org/10.1172/jci.insight.90932, e90932.

Zaleskas, V.M., Krause, D.S., Lazarides, K., Patel, N., Hu, Y., Li, S., Van Etten, R.A., 2006. Molecular pathogenesis and therapy of polycythemia induced in mice by JAK2 V617F. PLoS One 1, e18. https://doi.org/10.1371/journal.pone.0000018.

Zhou, W., Toombs, C.F., Zou, T., Guo, J., Robinson, M.O., 1997. Transgenic mice overexpressing human c-mpl ligand exhibit chronic thrombocytosis and display enhanced recovery from 5-fluorouracil or antiplatelet serum treatment. Blood 89, 1551–1559.

Zingariello, M., Martelli, F., Ciaffoni, F., Masiello, F., Ghinassi, B., D'Amore, E., Massa, M., Barosi, G., Sancillo, L., Li, X., Goldberg, J.D., Rana, R.A., Migliaccio, A.R., 2013. Characterization of the TGF-β1 signaling abnormalities in the Gata1low mouse model of myelofibrosis. Blood 121, 3345–3363. https://doi.org/10.1182/blood-2012-06-439661.

CHAPTER SIX

Lab tests for MPN

Alice Moncada[a] and Alessandro Pancrazzi[a,*]
[a]Laboratory Medicine Department, Molecular and Clinical Pathology Sector, Azienda USL Toscana Sudest, Ospedale San Donato, Arezzo, Italy
*Corresponding author: e-mail address: alessandro.pancrazzi@uslsudest.toscana.it

Contents

Abstract

Molecular laboratory investigations for myeloproliferative neoplasm (MPN) can ideally be divided into two distincts groups, those for the detection of the BCR-ABL rearrangement (suspect of chronic myeloid leukemia) and those for the variants determination of the driver genes of the negative Philadelphia forms (MPN Ph neg).

The BCR-ABL detection is based on RT-Polymerase Chain Reaction techniques and more recently on droplet digital PCR (ddPCR). For this type of analysis, combined

International Review of Cell and Molecular Biology, Volume 366
ISSN 1937-6448
https://doi.org/10.1016/bs.ircmb.2021.02.010

with chromosome banding analysis (CBA) and Fluorescent in situ hybridization (FISH), it is essential to quantify BCR-ABL mutated copies by standard curve method.

The investigation on driver genes for MPN Ph neg forms includes activity for erythroid forms such as Polycythemia Vera (test JAK2V617F and JAK2 exon 12), for non-erythroid forms such as essential thrombocythemia and myelofibrosis (test JAK2V617F, CALR exon 9, MPL exon 10), for "atypical" ones such as mastocytosis (cKIT D816V test) and for hypereosinophilic syndrome (FIP1L1-PDGFRalpha test). It's crucial to assign prognosis value through calculating allelic burden of JAK2 V617F variant and determining CALR esone 9 variants (type1/1like, type2/2like and atypical ones).

A fundamental innovation for investigating triple negative cases for JAK2, CALR, MPL and for providing prognostic score is the use of Next Generation Sequencing panels containing high molecular risk genes as ASXL1, EZH2, TET2, IDH1/IDH2, SRSF2. This technique allows to detect additional or subclonal mutations which are usually acquired in varying sized sub-clones of hematopoietic progenitors. These additional variants have a prognostic significance and should be indagated to exclude false negative cases.

1. Philadelphia chromosome (Ph) positive myeloproliferative neoplasms (Ph+MPNs)

1.1 Introduction

Chronic myeloid leukemia (CML) is a clonal myeloproliferative disease characterized by a reciprocal translocation between chromosomes 9 and 22t(9;22)(q34;q11) with the juxtaposition of two genes, ABL and BCR, which recombine to form a chimeric BCR–ABL fusion gene. This specific type of rearrangement is pathognomonic of all cases of CML and is identified by the presence of the Philadelphia chromosome (Ph) in the leukemic cells. The resulting BCR-ABL gene codes for an abnormal tyrosine kinase enzyme that is responsible for white blood cell proliferation. It also is a therapeutic target for tyrosine kinase inhibitor (TKI) drugs. Patients newly diagnosed with CML are routinely treated with tyrosine kinase inhibitors; however, the clinical course of the disease can vary, and this variance may be associated with genetic heterogeneity (Daley et al., 1990; Faderl et al., 1999; Nowell and Hungerford, 1960).

The Ph chromosome became of great historical interest because it was the first cytogenetic marker found that recurrently characterized a clinically specific group of human neoplasms (Apperley, 2015; Molica et al., 2017). The laboratory should be able to detect chromosomal rearrangements (translocations, inversions, duplications, deletions) because this event is crucial in establishing a disease diagnosis, evaluating the prognosis and making therapeutic decisions.

In CML patients, cytogenetics techniques completely revolutionized the history of diagnosis and therapy by permitting the identification of the pathognomonic chromosome translocation of the disease (Deininger et al., 2000). Furthermore, despite the increasing refinement of the molecular techniques [the qualitative and quantitative reverse transcription Polimerase Chain Reaction (rt-PCR) and, more recently, droplet digital PCR (ddPCR)], use of the cytogenetic techniques including banding cytogenetics and FISH remains the gold standard for the detection of the Ph chromosome and for monitoring treatment response in CML patients. Indeed, during therapy, the results of banding cytogenetics may detect additional chromosomal abnormalities, in Ph-negative or -positive clones, which may represent a possible cause of progression to advanced phase of disease (Molica et al., 2017).

European Leukemia Net (ELN) and the National Comprehensive Cancer Network (NCCN), have drawn up guidelines for diagnosis, treatment, and management of CML (Baccarani et al., 2013; Pallera et al., 2016). These evidence-based recommendations suggest that response monitoring in CML must be performed at specific time points using different tests, such as banding cytogenetics on bone marrow samples, FISH on peripheral blood and real-time quantitative PCR (RQ-PCR).

1.2 Banding cytogenetics

The cytogenetic analysis of bone marrow or peripheral blood cells continues to preserve a crucial role in CML scenario. Diagnostic assays are based on the standard banding cytogenetics or "chromosome banding analysis." Bone marrow is the tissue of choice when using banding cytogenetics. The t(9;22)(q34;q11) may be clearly identified when the proliferation process of the leukemic cells is arrested at metaphase. Indeed, cell synchronization agents are added to the culture, which can typically increase the quality and yield of the analyzable cells. After treating with trypsin and staining with Giemsa, the identification of the t(9;22) requires the analysis of at least 20 G-banded metaphases from cultured leukemic cell preparations (Morris and Fitzgerald, 1985; Webber and Garson, 1983). The Ph chromosome is usually present in 100% of metaphases, often as the sole abnormality, but addictional cytogenetic alterations in Ph positive cells can be found in 5% of patients at diagnosis (El-Zimaity et al., 2004; Fabarius and Leitner, 2011). Karyotypes were usually reported and described according to the International System for Human Cytogenetic Nomenclature (McGowan-Jordan et al., 2016).

Conventional metaphase cytogenetic examination is still the gold standard of early TKI response, and cytogenetic analysis has the high prognostic value, high specificity and allows detection of additional cytogenetic abnormalities. Disadvantages of this technique include low sensitivity 5–10%, longer turnaround time and higher cost due to its laborious nature (Molica et al., 2017). Other disadvantages can be caused by the inconsistency of cell samples derived from bone marrow aspiration that prevent a correct analysis, due to the lack of metaphases that can be analyzed.

1.3 Fluorescent in situ hybridization (FISH)

Over the last three decades, in most diagnostic laboratories, conventional karyotyping, in conjunction with molecular cytogenetic analysis, is routinely performed to detect BCR-ABL positive patients. Molecular cytogenetic assays mainly include the Fluorescent in situ hybridization (FISH) technique, a sensitive and quantitative method to detect specific chromosomal aberrations. FISH technique implies the use of several labeled fluorescent DNA probes that specifically bind to regions in BCR or ABL (Douet-Guilbert et al., 2004; Shaver and Tommerup, 2009; Swansbury, 2012). This test allows a large number of cells to be analyzed (for standardization there should be at least 200 nuclei) and for this reason, it is characterized by higher sensitivity (1–10%) than banding cytogenetics. Furthermore, metaphase and interphase (nondividing) cells can be used, and assay that can be performed on both peripheral blood and bone marrow cells (Molica et al., 2017). The greatest limitation of this technique in CML is its inability to identify additional chromosomal alterations: FISH should be performed in combination with banding cytogenetics at the time of diagnosis or to monitor clonal evolution (Egan and Radich, 2016).

1.4 Reverse transcriptase-polymerase chain reaction (RT-PCR)

Reverse transcriptase-polymerase chain reaction (RT-PCR) amplifies the region around the splice junction between BCR and ABL genes. The qualitative PCR testing is useful for diagnosis of CML allowing detection of the presence of the BCR-ABL transcript, the quantitative test is important to assess the amount of BCR-ABL transcripts and so is gold standard for monitoring residual disease; It is highly sensitive in detecting minimal residual disease (Jabbour et al., 2008; Kantarjian et al., 2008).

A major benefit of quantitative PCR is that peripheral blood can be used rather than bone marrow, making monitoring less invasive; RT-PCR testing uses mRNA extracted from either peripheral blood or bone marrow. Disadvantages include the inability to detect rare variants, lower specificity due to RNA cross-contamination, and false positivity. Poor-quality RNA or failure of the PCR reaction may generate false-negative results; false-positive results can be due to carry over contamination. A 0.5–1 log difference in some samples can occur depending on testing procedures, sample handling and laboratory experience (Schoch et al., 2002).

Quantitative RT-PCR is specifically designed to quantify BCR-ABL mRNA for monitoring disease progression. Commercially available assays use real time RT-PCR. Quantitative PCR is generally able to detect 1 CML cell among approximately 100,000 normal cells (Guo et al., 2002). The quantity of BCR-ABL mRNA is standardized against a reference housekeeping gene such as ABL. A common measure of relative CML response is the magnitude of reduction in BCR-ABL transcripts from a standardized baseline value, and the major molecular response (MMR) is defined as a 3-log or greater reduction in the BCR-ABL/control gene ratio (Rudzki et al., 2004). The major problem with quantitative PCR for BCR-ABL is a lack of standardization. Different laboratories use different control reference genes and have different BCR-ABL/control gene ratios.

In the last 15 years a great standardization work has been carried out throughout Europe, there are many networks like in Italy in which every laboratory is linked to specific conversion factor to calculating exact copy number expressed in IS (i.e., international unit of measurement). This approach, obtained through validating studies and by interlaboratory quality proficiency test, allows to obtain max concordance in final quantification of BCR-ABL in spite of the different material and methods used (Foroni et al., 2011).

Break-points in ABL occur in the first and second introns. Break-points in BCR occur in one of three regions: major break-point cluster region (M-bcr) involving exons b2 or b3, minor break-point region (m-bcr) involving exon e1 or micro break-point region (u-bcr) involving exon e19. Because of these different BCR break-points distinct fusion mRNAs of variable length including p190, p210 or p230 are generated. Most CML patients have p210 mRNA. High levels of p190 mRNA are primarily detected in acute lymphoblastic leukemia or the blast phase of CML. Low levels of p190 mRNA have been recently reported in chronic-phase CML. Historically, diagnosis and monitoring of CML required cytogenetic studies to detect the Ph chromosome. Newer and more sensitive techniques are now available for diagnosis and post therapy monitoring. A brief overview of the various techniques available for detection of these fusion mRNAs follows.

Recently, several droplet digital PCR (ddPCR) assays are described to detect BCR-ABL (Chung et al., 2020; Maier et al., 2019). ddPCR technique allows to obtain high performance in terms of sensitivity without use of standards and not needs to perform calculating alghoritm to quantify copy number (Cross, 2009). The critical point for this method remains probe and oligo design which should assure a complete coverage of the molecular target. This method provided promising results in comparative studies (Alikian et al., 2017; Franke et al., 2020; Yan et al., 2019).

1.4.1 European LeukemiaNet recommendations 2020

"Cytogenetics should be performed by CBA (Chromosome Binding Analysis) of Giemsa-stained metaphases from bone marrow cells. A qualitative reverse transcriptase PCR on peripheral blood cells is mandatory to identify the type of BCR-ABL1 transcripts that can be appropriately followed when assessing response to TKI therapy. About 2–4% of patients harbor atypical BCR-ABL1 transcripts lacking ABL1 exon2 (e13a3 or e14a3) or resulting from atypical BCR breakpoints (e.g., e1a2, e6a2, e8a2, or e19a2) that may yield a false negative PCR using routine primer/probe sets in qualitative or quantitative reverse transcriptase PCR protocols. If not tested at diagnosis, a false impression may be given that a patient is in complete molecular response after TKI treatment. A quantitative PCR is not mandatory at diagnosis. If a molecular assay detects BCR-ABL1, but the Ph-chromosome cannot be identified by cytogenetics, a FISH test is required *(Hochhaus et al., 2020)*".

1.5 Additional molecular investigations

Additional investigation in addition to BCR-ABL test should be considered, according to the ELN criteria, in case of failure or suboptimal response to imatinib (TKI, tyrosine kinase inhibitor) first-line. These additional molecular tests must be conducted to detect point mutations in the kinase domain (KD) of Bcr-Abl were identified that could impair or even totally abrogate imatinib binding. The recommended methodology is bidirectional Sanger sequencing, but can be performed a screening by other techniques like denaturing-high performance liquid chromatography. The positive results are considered usefull data for therapeutic change and for TKI types selection (Soverini et al., 2011).

2. Philadelphia chromosome (Ph)-negative myeloproliferative neoplasms (MPNs)

2.1 Introduction

The molecular diagnostic for MPNs imply a breadthy and ordinated exploration of the various molecular targets. The diagnostic algorithm is based on relative incidence of the driver genes variants (JAK2, CALR and MPL). The frequency of JAK2-V617F is around 95% in polycythemia vera (PV) and between 50% and 60% in essential thrombocythemia (ET) and primary myelofibrosis (PMF). Other somatic mutations in exon 12 have been found in PV (about 1%) (Baxter et al., 2005; James et al., 2005; Kralovics et al., 2005; Levine et al., 2005). Somatic mutations in exon 9 of calreticulin (CALR) occur in 20–35% of patients with ET or PMF (Nangalia et al., 2013; Park et al., 2015). Genetic variants can be found in in the thrombopoietin (Tpo) receptor MPL, these are germline mutations in rare cases of familial thrombocytosis (MPL-S505N) (Chung et al., 2020) or somatic mutations that occur in 3–8% of patients with PMF or ET (MPL-W515) (Cross, 2009; Franke et al., 2020). More broadly, for these disease forms, there is a rare instability of the whole exon 10 sequence of MPL gene. Have been described a set of of mutations in genes affecting epigenetic regulation (prevalent in specific exons or distribuited in whole Coding DNA Sequence) such as ASXL1 (exon 12), DNMT3A (exons 7–23), TET2 (CDS), IDH1/2 (exon 4), and EZH2 (CDS), often but not always co-occur with phenotypic driver mutations and convey important prognostic information in MPN. Other frequently co-occurring mutations affect genes involved in splicing, including SF3B1 (exons 12–16), U2AF1 (exons 2–8), ZRSR2 (CDS), and SRSF2 (exon 1) (Bain et al., 2008; Deininger et al., 2000; Lundberg et al., 2014; Tefferi et al., 2014a; Vannucchi et al., 2013). Considering all these evidences, the laboratory should respond using a dual diagnostic approach. A first level, dedicated to the promptly screening driver mutations, is useful for diagnostic setting and a second one is necessary for triple negative cases, to provide prognostic precious data and for the purpose of the correct therapeutic strategy in particular cases like young patients with MF.

For the rare eosinophilic forms, the laboratory should consider a valid molecular approach to investigate PDGFRA genic rearrangiaments and PCM1-JAK2 fusion products (Arber et al., 2016; Sadrzadeh et al., 2011). In the scenario of the mastocytosis is necessary to respond at clinical request searching c-KIT D816V variant. This mutation is frequent in about 100% of these forms (Reiter and Gotlib, 2017; Semenza, 2009).

Rare JAK2 V617F negative cases of acquired erythrocytosis have been described. Also in pre era NGS some specialized laboratories, through a laboriousus Sanger sequencing, investigated a set of genes which are involved in oxygen sensing pathway such as prolyl hydroxylase domain protein 2 (PHD2) (exons 2 and 3), von Hippel-Lindau tumor suppressor (VHL) (exons 1,2,3) protein, α-subunit of the hypoxia-inducible factor (HIF-α) (exon 12) protein and in erythropoietin activity such as EPO receptor (exons 5,6,7) (Arber et al., 2016; Schofield and Ratcliffe, 2005; Vočanec et al., 2019; Wang et al., 1995). Even if these are rare variants, several NGS panels cover these genic regions.

2.2 Ph-MPNs: Molecular analysis of the driver genes

Molecular diagnosis of driver mutations includes searching genetic variants in the JAK2 CALR and MPL genes. The diagnostic value of testing JAK2, CALR and MPL mutations is well established and reinforced by the 2016 revision of the WHO classification of myeloid neoplasms (Guglielmelli et al., 2017; *Hematology Am. Soc. Hematol. Educ. Program*, 2016). WHO criteria include JAK2 mutations (V617F and those in exon 12) as major diagnostic criteria in PV and JAK2V617F, CALR and MPL mutations in ET and PMF (Gong et al., 2013).

2.3 JAK2 (Janus kinase 2)

2.3.1 JAK2V617F

A lot of different approaches for the detection of the JAK2V617F mutation have been described. In the diagnostic routine and research activity have been used for each type of technique different assays that use various instruments, primers and/or probes sequence and detection methods. The techniques described can be divided into two main categories: assays designed to specifically target the c.1849G>T (V617F) mutation and those that analyze wide regions of exon 14 JAK2 (e.g., direct Sanger sequencing, HRM analysis). Although there are no Food and Drug Administration (FDA) approved test, a number of methods have been developed for detecting JAK2V617F mutation: qualitative end point PCR (epPCR) or real-time PCR (rt-PCR), allele-specific PCR/amplification refractory mutation system (ARMS/PCR), High Resolution Melting curve analysis (HMR), Pyrosequencing, PCR–restriction fragment length polymorphism (RFLP), Direct Sanger sequencing and recently, digital PCR (ddPCR) and, Next Generation Sequencing (NGS) technique. The techniques differ for a set of performance data, such as sensitivity (ranging from 0.01–20%), the potential for high-throughput workflow and for equipment required (Lippert et al., 2009). An overview of currently available assays is presented in Table 1.

The most widely used method for the detection of V617F is pragmatically based on DNA allele-specific PCR. Quantitative PCR (qPCR)

Table 1 Main methods in detecting JAK2 mutations.

Method	Sensitivity (%)	Advantage	Disadvantage
qPCR or PCR Real Time (AS)	0.008–1	High sensitivity; quantitative	Detects only target mutations
ddPCR		Standard curve is not necessary	
ARMS/PCR (AS)	0.1–5	High sensitivity; simple to perform	Detects only target mutations; not quantitative
High Resolution-Melting curve analysis	5–10	Simple to perform; semiquantitative; low cost	Moderate to low sensitivity; poor reproducibility in low positive samples; needs confirmation
Pyrosequencing	5–10	Simultaneous detection of other mutations; Simple to perform; quantitative; low cost	Relatively low sensitivity; Pyrosequencing equipment required, expensive
PCR-RFLP	1–10	No special equipment requie; Low cost	Relatively low sensitivity; requires post-PCR manipulation; unreliable in low positive samples; not quantitative
Direct Sanger sequencing	10–20	Simultaneous detection of several mutations; bidirectional confirmation	Low sensitivity; Labor intensive, sequencing equipment required, not quantitative
Next Generation Sequencing	1–2	Simultaneous detection of several mutations; sensitivity comparable to qPCR	Oversized to detect single variant; expensive; time consuming versus qPCR

methods should be preferred over qualitative ones for their greater reproducibility, sensitivity and accuracy. As a matter of fact, in the last years many quantitative PCR (qPCR) methods have been published with sensitivity from 0.01% to 1%, and specificity of 100% (Merker et al., 2010; Vannucchi et al., 2007a).

The qPCR allows to quantify the mutated JAK2 allele percentage, this is an evaluation of disease burden. Using standard curves for both the wild-type and mutant forms, it is possible to perform this calculation; however, there is no effective global harmonization for choising reference material.

WHO guidelines talk about "presence of JAK2V617F" and not impose quantitative determination of the allele burden, so the clinical utility of quantification of V617F has not yet been established. The cause of this generic recommendation must be exclusively accepted in a diagnostic setting, at the same time, several studies demonstrated evident prognostic value for JAK2V617F allelic burden. Proportionally linked to relatively high JAK2V617F allelic burden (or to burden progression) have been identified main following parameters: thrombotic events in PV and ET (Silver et al., 2011; Vannucchi et al., 2007b), fibrosis in PV (Passamonti et al., 2010), increase in risk of post-PV MF (Gong et al., 2013), global prognosis in MF cases (Gong et al., 2013). A limitation of allele-specific qPCR is that, in many cases, signal is generated in the mutant PCR reaction at high Ct (Cycle threshold) counts even for known negative samples. In this situation, it is necessary for the laboratory to establish the minimal amount of signal needed to confidently call a sample positive (Jovanovic et al., 2013). A huge global study that has tried to define gold standard qPCR assay among main several published methods is joint European Leukemia Net/MPN&MPNr-EuroNet (COST action BM0902) study (Li et al., 2008). This study measured analytical performance of nine several assays (six of these published) (Ishii et al., 2006; Kröger et al., 2007; Larsen et al., 2007; Lippert et al., 2006; Nussenzveig et al., 2007; WHO, 2016), valuated different normalization approaches for housekeeping signal quantifing many DNA mix from cell lines transfected for JAK2V617F gene. The goal of this interlaboratories activity has been identifying performanter assay for tracking of minimal residual disease in JAK2-V617F-associated MPNs. The results of this study can be summaryze in these main points:

- assays working in separate determination of wild type (wt) and mutant genotipe are more performant than other that are designed for one only mix where wt/V617F primers or probes compete for target matching;

- the accuracyer assays allow to quantify relatively low JAK2V617F allelic burden (about 0.08%), this sensitivity is particulary useful for minimal residual disease but currently is not request for diagnostyc setting.

The first step to managing these severals qPCR methods in routine activity and to try identifying acceptability analytical range has been defining reference material standard. In this field has been decisive National Institute for Biological Standards and Control (NIBSC) study (WHO, 2017) accepted from WHO and documented in the Sixty-seventh report 2017 (Bench et al., 2013).

In the 2013 the United Kingdom National External Quality Assessment Service (UKNEQAS) suggested for acceptability of JAK2V617F determinations a sensitivity generic range about 1–3% (Waterhouse et al., 2016). The NIBSC experience allowed to obtain reference material standard (cell lines synthezed in vitro and mixed for different JAK2V617F allelic burden) but also to define LOD threshold at least to 0.1% JAK2V617F and to demonstrating high sensitivity and accuracy for ddPCR methods (about 0.03%).

Several quantitative methods have been developed in the years, most of them based on real-time allele-specific PCR, with sensitivity ranging from 0.05% to 1%, and specificity of 100%. The ddPCR has also been proposed with the aim to achieve an absolute quantification of the target variant without a standard curve (Abdelhamid et al., 2013; Fontanelli et al., 2015; Link-Lenczowska et al., 2018). Recently, the widespread use of NGS has showed the possibility to detect JAK2 V617F mutation with comparable analytical performances to qPCR (Er et al., 2009) but in a first level investigation setting, targeting the alone 617 codon, is not recomended for high costs and time consuming to responding to the clinical request (Lundberg et al., 2014; Park et al., 2015; Tefferi et al., 2014a; Vannucchi et al., 2013). In the global literature many others methods to detect JAK2V617F mutation like ARMS/PCR, HRM, Pyrosequencing, PCR-RFLP and Direct Sanger sequencing have been described. Part of these assays seem able to provide comparable perfomance in terms of accuracy (HRM, Pyrosequencing PCR-RFLP, Sanger) and sensitivity (ARMS/PCR) but considering necessary analytical characteristics for this specific diagnostic investigation (execution short time, high global performance and low costs) are not used in routine analysis but only in research field (Didone et al., 2015; Scott et al., 2007; Zapparoli et al., 2013).

2.3.2 JAK2 exon 12

In the routine activity of the laboratory, specialized in Ph-MPNs molecular diagnostic, there is a sporadic necessity to investigate on JAK2 exon 12 genotype. This analysis is necessary in case of JAK2 V617F negative rare PV forms or in idiopathic erythrocytosis patients with low serum erythropoietin levels. More than 10 different sequence variations have been described between codons 536 and 544 (Albiero et al., 2008; Bernardi et al., 2009; Carillo et al., 2011; Guo et al., 2002; Kouroupi et al., 2008; Ma et al., 2009; Martínez-Avilés et al., 2007; Pardanani et al., 2007; Percy et al., 2007; Pietra et al., 2008; Williams et al., 2007). Due these two aspects, rarity and heterogeneity of these mutations described in this region, the routine analysis of the JAK2 exon 12 sequence is usually performed by a method based on Sanger sequencing of exon 12.

Nested High-Resolution Melting (HRM) curve analysis has been proposed as highly sensitive screening method eventually followed by Sanger sequencing for the precise characterization of the variant (Laughlin et al., 2010). This last approach could represent a paradox situation due to the more high analytical sensitivity of the screening method (HRM) versus to the confirmation one (Sanger). Due the exon 12 mutations are frequently presenting in only a small fraction of the granulocytes in some patients (Geay et al., 2020) and considering the complicated heterogeneity of the mutations and low allelic load, is recommended to perform NGS confirmation in doubt cases (Lasho et al., 2018).

2.4 CALR (calreticulin)

CALR assessment enters the 2016 WHO classification specifically for JAK2 negative cases of ET and PMF. CALR exon 9 mutations are predominantly heterozygous ($\leq$50% allelic burden) with the majority of cases (80–90%) carrying a 52-bp deletion: (CALRdel52/type I; c.1092_1143del; L367 fs*46)or a 5-bp insertion CALRins5/type II; c1154_1155insTTGTC; K385 fs*47). More than 50 different other variations have been identified and part of these are considered "type 1 like| or "type 2 like" or "atypicals" (Jones et al., 2015). Several methods have been proposed to detect CALR variants, including Sanger Sequencing, fragment length analysis, qPCR, HRM, ddPCR, pyrosequencing and NGS (Tefferi and Pardanani, 2014). Due to the heterogeneity of CALR insertions and deletions detected, conventional Sanger sequencing or fragment length analysis are the main methods of testing. These assays for the screening of CALR must have a sensitivity level of at least 10%. The HRM analysis, due to its simplicity, rapidity and low cost, is an approach frequently used for the screening. Therefore, the HRM is the most appropriate screening method. Several methods based on HRM have been reported with a sensitivity of 3–5% mutation load and a turnaround time of 1 h for the experimental procedure. However, these assays need to be paired to Sanger sequencing to provide an accurate genotyping of the different *indels* in positive samples (Chi et al., 2015; Jones et al., 2015; Maier et al., 2015; Mansier et al., 2016a; Mehrotra et al., 2015; Murugesan et al., 2016).

Although NGS has been considered the most accurate method (sensitivity less than 1%), it is still a costly and time-consuming test, and detection of variants may require great bioinformatic efforts. Use of NGS as a screening test for CALR mutations is disproportionate at present time.

As an alternative diagnostic approach for CALR exon 9 mutations the combined use of a standardized PCR-HRM analysis method or fragment analysis method together with a Taqman Real Time quantitative PCR (RT-qPCR) assay was proposed (Pietra et al., 2016).

Finally, due the prognostic value of distinct variants types (Bilbao-Sieyro et al., 2016; Guglielmelli et al., 2015; Tefferi et al., 2014b), the direct sequencing must be considered appropriate to solve clinical request in case of doubt results or positive profiles generated from aspecific methods like HRM (Pietra et al., 2011).

2.5 MPL (Proto-oncogene, thrombopoietin receptor)

MPL exon 10 variants represent the third group of driver mutations in terms of frequency in MPNs because they have been reported in 3–5% of ET and 6–10% of PMF (Passamonti et al., 2016). The screening of the MPL mutations must be performed in patients with ET or PMF without JAK2 and CALR mutations. To date, as explained for JAK2 gene analysis, different methods have been applied for this investigation, which can be grouped into two main categories: the targeted identification of a specific mutation or assays that scan the entire exon 10 (Beer et al., 2008; Ghaderi et al., 2010; Ivanova et al., 2011; Pancrazzi et al., 2008; Zhuge et al., 2010). Among the methods included in the first category, greater sensitivity was reported for allele specific AS-qPCR and bead-based assay with LNA-probes and amplification refractory mutation system-PCR (Furtado et al., 2013) (at least 0.1–0.5%). In 2013, a multiplexed AS-PCR assay that detects the four most frequent MPL exon 10 mutations (W515L, W515K, W515A and S505N) with 100% specificity and sensitivity of 2.5% has been developed (Boyd et al., 2010). However, several assays are required to detect all possibly occurring mutations. Second category includes High-Resolution Melting (HRM) curve analysis, sequencing, pyrosequencing and NGS. In particolar, HMR allows the detection of all W515, S505 and unknown exon 10 mutations in one reaction with sensitivities of 2–5% (Mansier et al., 2016b; Schnittger et al., 2009). The most appropriate method forthe screening of MPL exon 10 mutations should have sufficient analytical sensitivity to detect mutations that occur at low allelic levels. Differently from CALR exon 9 specific investigation, in setting analysis of exon 10 MPL is not strictly necessary classify somatic mutations for valuating prognositc significance but is appropriate exclude eventual not damaging polymorphisms that could affecting HRM profile and generating false positive results. For this fact should be confirmed every screening result by direct sequencing. The use of multiplex PCR causes a lack of exonic global coverage, therefore should be executed a design of appropriate specific primers for HRM application in the screening setting. To this aim, HRM analysis followed by Sanger sequencing is the elective approach to perform.

2.5.1 Algorithm analysis for uncertain results on driver genes: Searching for double mutated cases

The main aim for first level examination is to characterize most of patients with MPN suspect. Even if, in this context, the laboratory challenge is represented by triple negative cases (about 10%), should be reduced false negative cases. It is primary to estabilish analitycal positivity threshold for identifing of uncertain results that could be uncorrectly considered like negative ones. The double execution of JAK2V617F and JAK2 exon 12 in clinical phenotype of PV and triple one of JAK2 V617F, CALR, MPL investigations in suspect cases of PMF or TE should be performed in laboratory standard scheme. This approach is particularly important taken in count that about 15% of JAK2V617F cases with very low allelic burden (<5%) show CALR variants at relatively high-level burdens (Usseglio et al., 2017). This last fact allows to give a clearer information to clinician than a doubt JAK2V617F case. Moreover, in JAK2V617F patients, other coexisting mutation such MPL are described but these cases seem to show prevalently at low molecular burden levels (Vainchenker et al., 2011).

2.6 Molecular analysis for additional or subclonal mutations detection

In the last years, the study approach of MPNs has become particularly complex due discovery several other mutations which are usually acquired in varying sized sub-clones of hematopoietic progenitors. These additional or "subclonal" mutations may be present in addition to or in independent lineage from mutant clone (Ortmann et al., 2015).

There are three main reasons to consider this addictional mutations research like second level setting in the laboratory approach. First of these is linked to specific diagnostic value, second to the relatively low frequency, the third to effective costs to perform distinct validated techniques or NGS panels.

As regards specific diagnostic value, altough the presence of these subcolnal or additional mutations is factual evidence of the clonality, is in the other part clearly evident that these ones are commonly present in myelodysplastic syndromes and other myeloid malignancies, including acute leukemia. Due these evidences they do not have a specific diagnostic value, as opposed to the driver gene mutations cases, these other variants are undoubtly associated with primary myelofibrosis (PMF) patients, characterized by reduced survival or increased risk of evolution to acute leukemia, thereby acquire a very specific prognostic significance. For this reason, additional or subclonal mutations could be detected after diagnosis to explore disease progression (Nangalia et al., 2015, 2015; Gong et al., 2013).

The frequency of these rare variants, considering necessity to reduce TAT (Tourn Around Time) of the diagnostic tests and to enhance the efficacy of the analysis, places this laboratory research in second level setting. Approximately from 1 to max value of 30% of patients with chronic MPN show mutations in genes such as DNMT3A, TET2, IDH1 and IDH2 (epigenetic regulation function), in others like ASXL1 and EZH2 (maintenance of chromatin structure), in SF3B1, U2AF1 and SRSF2 (splicing activity) or in oncogenes/tumor suppressor genes, such as MYB, RUNX1, TP53 (Busque et al., 2012; Tenedini et al., 2014).

The multitarget analysis requested for second level exams concerns mainly NGS method, for this reason many studies are focused on the design of specific panels (Aguilera-Diaz et al., 2020; Kuo and Dong, 2015). Although, in the last decades, the widespread use of NGS technique determined a significative exam cost decrement, the frequent use of this technology is relatively expensive comparing to reagents cost for conventional method (RT-PCR, HRM, Sanger, etc.) (Kedes and Campany, 2011). Moreover, this intensive use determines a significative investment in terms of the specialized personal such as bioinformatics and genetists. For all these considerations, these molecular investigations on MPN patients should be conducted by NGS technique but exclusively in in second level context. At the core of debate in multitarget analysis there is panels design that affect real efficacy of laboratory response. The making of panel should consider several molecular targets that are mainly involved in clinical valuation of High Molecular Risk (HMR) status such as following genes: ASXL1, EZH2, TET2, IDH1/IDH2, SRSF2 (Guglielmelli et al., 2016) (Table 2). This apparently short set of genes justify a large molecular exploration cause these exhibit large coding sequences (for example ASXL1, TET2) whose sequencing may be much time consuming and expensive if Sanger sequencing is employed. In addition to genetic instability of these known targets should be considered the rare presence of atypical mutations of the driver genes such as JAK2 (JAK2 V626F and JAK2 F556V) and MPL (MPL S204P and MPL Y591N).

Many genetic panels are commercially available and can be used in the diagnostic routine, but to date there is no specific recommendation can be made at this regard; choice of the different genetic panels, number of gene analyzed, interpretation of identified pathogenetic variants depends mainly on specific request of clinician, individual (and familiar) clinical hystory of patient and laboratory facilities.

Recent studies based on NGS techniques show relatively impact in whole gene JAK2 sequencing. In case of triple negative patients could be usefull include in NGS panel whole gene JAK2 coverage. Germline variants as JAK2 V625F and F556V, found in low percentage of the triple negative cases, are definied like gain-of-function mutations. This fact suggest that triple-negative cases of ET and PMF could do not represent a homogenous disease entity. Further studies are necessary to propose a laboratory investigation on germinal DNA for exploring side of myeloproliferative neoplasms hereditarity (Milosevic Feenstra et al., 2016).

Regarding multi-genic panels must be reported important NGS approach to detect genic alterations that are typical signs of progression in Acute Myeloid Leukemia (AML). Post-MPN AML usually occurs years after the initial MPN diagnosis with an average age of onset between 64 and 68 years (Heaney and Soriano, 2013). The standard protocols for complete molecular investigations in suspect cases of AML includes cytogenetic analysis and molecular investigations to detect mutations in single

Table 2 Minimum recommended list of molecular targets (in addiction to driver genes) and region instability of each one (based on literature data) to take in count for design of the NGS panels. LAMs, secondary acute myeloid leukemia; OS, overall survival; LFS, leukemia-free survival.

List of the distinted genes for noted activity/effect	Chromosomal Locus	Target region	Mutations frequency described in literature related to clinical phenotype			Mutation type	Clinical correlation	References
			PV	TE	PMF			
Signaling								
CBL	11q23	***exons 8–9***	rare	rare	5–10	Loss of function	LAMs	Vannucchi et al. (2013)
SH2B3	12q24	***exons 2–4***	1–2	–	rare	Loss of function	Erythrocytosis, LAMs	Pardanani et al. (2010)
Epigenetics								
TET2	4q24	***CDS***	CDS	10–20	5	Loss of function	LAMs (26%)	Vannucchi et al. (2013)
ASXL1	20q11	***exon 12***	exon12	2–5	2–5	Loss of function	<OS, <LFS, LAMs (19%)	Lundberg et al. (2014)
EZH2	7q35	***CDS***	CDS	1–3	rare	Loss of function	<OS, <LFS	Lundberg et al. (2014)
Splicing								
SRSF2	17q25	***exon 1***	rare	rare	10–20	Loss of function	<OS, <LFS, LAMs	Lundberg et al. (2014)
SF3B1	2q33	***exon 12–16***	rare	rare	5–10	Loss of function	Ring sideroblasts, >OS	Tefferi et al. (2014a)

Continued

Table 2 Minimum recommended list of molecular targets (in addiction to driver genes) and region instability of each one (based on literature data) to take in count for design of the NGS panels. LAMs, secondary acute myeloid leukemia; OS, overall survival; LFS, leukemia-free survival.—cont'd

List of the distinted genes for noted activity/effect	Chromosomal Locus	Target region	Mutations frequency described in literature related to clinical phenotype			Mutation type	Clinical correlation	References
			PV	TE	PMF			
U2AF1	21q22	***exon 2–8***	rare	rare	5–15	Loss of function	<OS	Bartels et al. (2015)
Leukemia progression								
IDH1	2q33	***exon 4***	rare	rare	2–5	Neo-enzyme	<OS, <LFS, LAMs (8%)	Lundberg et al., 2014
IDH2	15q26	***exon 4***	rare	rare	2–5	Neo-enzyme	<OS, <LFS, LAMs (18%)	Lundberg et al. (2014)
TP53	17p13.1	***CDS***	rare	rare	rare	Loss of function	<OS, <LFS, LAMs (20%)	Lundberg et al. (2014)
DNMT3A	2p23	***exons 7–23***	5–10	1–5	5–10	Loss of function	<OS, LAMs	Park et al. (2020)
IKZF1	7p21	***CDS***	–	–	rare	Delezione	LAMs (21%)	Milosevic et al. (2012)
RUNX1	21q22	***CDS***	–	–	rare	Loss of function	LAMs (37%)	
NRAS	1p13	***CDS***	rare	rare	rare	Function Acquisition	LAMs (7–13%)	
KRAS	12p12	***CDS***	rare	rare	rare	Function Acquisition	LAMs (5–10%)	

LAMs, secondary acute myeloid leukemia; OS, overall survival; LFS, leukemia-free survival

genes such as FLT3, NPM1, CEBPA, KIT, IDH1 and IDH2. The presence of genetic variants in these targets is important to define risk assessment, prognosis and treatment decisions. Recent advances in genetic technologies push to use multi-gene panel diagnostics wich detects simultaneously genic variants and chromosomal abnormalities. The molecular testing should be perform at diagnosis and also after a relapse because it is possible an acquisition of genetic abnormalities after the completion of first-line treatment (Kayser and Levis, 2019; Roloff and Griffiths, 2018; Shen et al., 2011).

3. Source of sample and DNA template for genetic testing in MPNs

Reported cellular type and fraction of samples used for the analysis of MPN-associated mutations vary widely. Initial testing of MNPs mutations is most performed on peripheral blood samples.

3.1 Philadelphia chromosome (Ph)+positive myeloproliferative neoplasms (Ph+MPNs)

The most commonly analized and used sample for detection and monitoring BCR-ABL transcript levels is pherypheral blood (PB). The mononuclear cell separation normally is avoided due lymphoprep method could remove the granulocytes, which are affected cells, and could enrich the sample for lymphocytes (most of which are BCR-ABL negative). The blood sample is treated with ammonium chloride for lysing of red cells and subsequently is washed with phosphate buffered saline (Foroni et al., 2011). Bone marrow (BM) aspirate sample could be used for molecular applications in diagnosis phase. In monitoring context, in doubt cases of disease relapse, is debated necessity to use bone marrow sample to deeply explore minimal BCR-ABL copies (Jiang et al., 2012; Stock et al., 2006). However, pheripheral blood is confirmed analysis standard sample.

3.2 Philadelphia chromosome (Ph)-negative myeloproliferative neoplasms (MPNs)

The PB is the preferred type of specimen, this choice is sometime motivated for certain aviability versus BM specimen which are not always collected in PV cases. The aviability and minor invasive effect in sample collecting are clear reasons to generate general recomandation of PB use in routine activity of the MPN molecular diagnostic.

The comparing between PB in toto and purified granulocytes (GN) use was performed prevalently to verify concordance on the quantitative molecular data such as JAK2V617F analysis. Taking in count that no major differences in the sensitivity of mutation detection between PB and GN have been reported, the indiscriminate use of both ones is accepted in diagnosis context.

This consideration could be not valid if we consider the monitoring of minimal residua disease. The studies on comparing JAK2V617F allelic burden quantification show about 15% of calculating discrepancy (Bench et al., 2013; Hermouet et al., 2007; Park et al., 2015), indicating more high sensitivity for analysis on GN sample. This difference is not important for diagnostic context but could be significatively inficiant to detect low allelic burden post transplant.

As regards the sample processing finalyze to eosinophils analysis (in case of searching genic rearrangement such as FIP1L1-PDGFRa in eosinophilic disorders) may be evaluated additional purification methods based on double Histopaque gradient and negative eluition anti-CD16.

Due to evidence of somatic variants presence in cDNA obtained by retrotrascription of platelets RNA, could be helpfull gene expression analysis in triple negative patients. In one-third of these patient's somatic mutations are described (Angona et al., 2016) but is unclear if this data are generated cause low sensitivity of routine analysis. For these reason is auspicable further studies to define correct algorithm analysis on triple negative patients.

NGS analysis of MPNs pushed to identify rare germline variants. For these cases are used buccal swabs and these samples often contains cells contamination from blood (Gilles et al., 2017). In genetic specialized centres are perfomed analysis on DNA extracted from fibroblasts coltures obtained from bone marrow (Specimen Considerations For Individuals With Hematologic Disease, 2018). This approach is clearly complex and is not a practical solution for routine analyisis.

3.3 Laboratory investigations for rare forms

There are a set of exams which are considered desuete analysis for routine activity of specialized laboratory but are in fact requested in differential diagnosis context. These test are used to investigate rare PV or triple negative TE cases.

3.3.1 Endogenous erythroid colonies (EECs) as diagnostic marker

Endogenous erythroid colonies (EECs) test, even if is not a closely molecular investigation, has been an instrument of the molecular laboratory dedicated to MPNs research. This analysis, in part obsolete and fallen into disuse, remains in rare cases the last chance to detect a diagnostic minor criterion. Progenitor cells from myeloproliferative disorders patients show a characteristic and specific abnormality when grown to hematopoietic colonies in vitro. Unlike healthy progenitors, MPNs precursors are potentially able to generate erythroid colonies in a semi-solid medium (methylcellulose) in the absence of erythropoietin (EPO). These colonies have been termed "Endogenous Erythroid Colonies (EECs)" and was considered for 30 years as diagnostic marker of PV, although it was also observed in some ET and PMF patients (Lemoine et al., 1986; Prchal and Axelrad, 1974; Shih et al., 1997, 1998).

For a long time, EEC colture was rarely performed due execution complexity, difficulty standardization and cost. However, in the late 1990, value of EEC test in PV cases has been demonstrated through standardized media use. This fact has led the scientific community to consider EEC as rilevant diagnostic criterion in the WHO classification, and this diagnostic tool remained a minor criterion in the WHO diagnostic classification until 2008. The development of the molecular investigation and the discovery of the driver genes mutations in the 2005–2015 period, made the EEC test questionable since only JAK2 mutations (either V617F or exon 12) are associated to EEC formation. Patogenetic variants in CALR and MPL genes do not induce formation of EEC even though the formation of EEC is functionally due to an activation of BCL-xL/STAT5. Because of its cost, the development of JAK2V617F test standardization and the shift from functional to molecular tests, EEC colture as a diagnostic marker disappeared from the criteria in the WHO 2016 classification. However, today EEC formation still remains an important tool to properly study rare clinical situations which may not be easy to distinguish from PV cases such as high affinity hemoglobinopathies which require specialized analysis to make the diagnosis. These clonogenic assays also may prove useful to study the clonal architecture of hematopoietic progenitors that has been shown to directly impact the phenotype and prognosis of PV (Ortmann et al., 2015). If will increase standardization of ECC test, the role of cell culture (in addition to molecular landscape) could be revisionated in the next few years for accurate prognostication of patients.

3.3.2 *Clonality assays: HUMARA (human androgen receptor) and TCA (transcriptional-based clonality assay)*

These investigations are aimed to provide clonality demonstration through X-chromosome inactivation (XCI) analysis. All these methods are particularly complex and laborosious. HUMARA test needs granulocytes and CD3 cells purification, the revelation is generally performedby capillary electrophoresis of digestion product amplified by specific PCR (Mitterbauer et al., 1999). TCA test is performed through ligase detection reaction and ligase chain reaction (LDR/LCR) on cDNA which is extracted from swabs sample and urinary epithelial cells (Prchal and Guan, 1993). Even if these laboratory applications are dated solutions, in recently years have been revalutated for triple negative cases. TCA test seems to have relatively high efficacy to identify clonality status in triple negative cases of ET and PMF (respective incidence about 82% and 75%) (Swierczek et al., 2015).

3.4 Laboratory investigations for systemic mastocytosis (SM) and hypereosinophilic syndrome

The somatic substitution of valine for aspartic acid at codon 816 (D816V) in exon 17 of KIT gene is an important event and constitute diagnostic value (Pardanani, 2005; Valent et al., 2005). KIT mutations play a critical role in the pathogenesis of SM and have been postulated as diagnostic markers and potential therapeutic targets (Arock et al., 2015). Multilineage involvement by KIT D816V is associated with disease progression and advanced disease (Garcia-Montero et al., 2006). Several assays for detection of the *KIT* D816V mutation have been reported. Based on literature-data, the most sensitive assays are RT-PCR plus restriction fragment length polymorphism (RFLP), PNA-mediated PCR, and Allele Specific Oligonucleotide-qPCR on DNA or RNA/cDNA (Arock et al., 2015) (Table 3). These techniques are fast, reliable, sensitive, and cost-effective methods for routine screening of *KIT* D816V in all SM subtypes. When applied on BM cells, all these assays have demonstrated their ability to detect the *KIT* D816V mutation in >80–90% of SM patients, a sensitivity considered sufficient in daily practice (Kristensen et al., 2014). Quantification of the *KIT* mutational load with ASO-qPCR may be rapidly adopted as gold standard for assessing the disease burden in advanced SM, thereby improving prognostication and monitoring of the disease course prior, during and after cytoreductive therapies, including KIT-targeting drugs (Arock et al., 2015). In a small percentage of cases (<5–10%) no mutation is detectable in codon 816 of *KIT*, therefore it is recommended in these few cases analyze the entire exon

Table 3 Performance characteristics of the main KIT D816V detection methods.

Technique	RT-PCR plus restriction fragment length polymorphism (RFLP)	Nested RT-PCR followed by D-HPLC of PCR amplicons	PNA-mediated PCR	ASO-qPCR on DNA or RNA/cDNA
Advantages	– Simple – Fast – Reliable – Cost-saving	– Detects different *KIT* mutations at position 816	– Allows detection of *KIT* mutations at position 816 or at adjacent positions – Recommended for formalin-fixed paraffin-embedded tissues	– Simple – Fast – Cost-saving – Highly sensitive test -quantitative: allows the quantification of the *KIT* D816V allele burden in blood or BM at diagnosis or in the follow up
Disadvantages	– Detects only *KIT* D816V mutant – Not quantitative	– Relatively low sensitivity – Not quantitative – Time-consuming – Needs special facilities (HPLC)	– Not quantitative – Intermediate sensitivity	– Detects only *KIT* D816V mutant – Needs standardization, validation and harmonization
Sensitivity	~0.05%	0.5–1.0%	~0.1%	~0.01%

17 with Sanger sequencing analysis. It has also been shown that additional genetic lesions such as mutations in *TET2*, *SRSF2*, *ASXL1*, *SRSF2* or *RUNX1* are frequently identified in patients with advanced SM. The sequencing of the whole KIT coding sequence by NGS may be underestimated, the sensitivity, using myeloid gene mutation panels for detection of KIT mutations, is only 5–10% and may fail to detect low Variant Allele Frequency (VAF) (Shomali and Gotlib, 2018).

The MPN specialized laboratory sporadically receives samples on which are requested molecular investigations to characterize hypereosinophilic status. The identification of the FIP1L1-PDGFRA fusion gene (FIP gene) provides a molecular explanation for the pathogenesis of approximately half of the patients with the hypereosinophilic syndrome (HES). This novel fusion gene results from a cryptic 4q12 interstitial deletion involving an 800 kb region between FIP1L1 and PDGFRA (Vandenberghe et al., 2004). Deletion and rearrangement remain undetectable by standard cytogenetics but can be detected by FISH with specific probes that hybridize to the region between the FIP1L1 and PDGFRA genes (Gotlib and Cools, 2008). Detection was performed by a two-color FISH assay on peripheral blood.

If PB screening is negative, the next step is to perform BM examination. A more sensitive way to detect the presence of the FIP1L1–PDGFRA fusion gene in the blood of eosinophilia patients is the use of (nested) reverse transcription (RT)-PCR (La Starza et al., 2005). However, due to low-level expression of the fusion gene and heterogeneity in the FIP1L1 break points, the fusion is difficult to detect. Even if nested RT-PCR or quantitative RT-PCR is most choiced technique for monitoring response to therapy (Gotlib and Cools, 2008) and FISH analysis has been ridimensionated for lower sensitivity and time consuming, the dual methodological approach remains widespread overall in diagnostic context.

4. Conclusions

The Ph + MPNs may be considered an example of molecular investigation importance for oncohematologic context. BCR-ABL detection and quantification is a real disease data. The widespread use of heterogeneous laboratory methods like citogenetic tests and molecular biology assays shows how is important to provide accurate results and to use correct instrument to respond at clinicians request. The FISH data for BCR-ABL rearrangiaments, which are concretly significant for diagnosis, should be accompanished to RT-PCR quantification for minimal residual disease monitoring.

The standardization actions of these last methods (use of correct calculating procedure and reference materials) is the core of good laboratory practice.

The discovery of recently identified molecular features produced new perspectives regarding diagnosis and prognosis of Ph-MPNs. The identification of multiple molecular abnormalities opened new horizons expecially for prognosis and disease monitoring. Genetic testing for Ph- MPNs are much more complex than initially hypothesized. The Ph-MPNs are characterized by mutations defined as "driver or phenotypic mutations" and "additional or subclonal mutations". Taking in count that driver or phenotypic mutations determine the clinical phenotype of these diseases and mutated espression of these genes (JAK2, CALR, MPL) in cell lines causes cytokine independent or hypersensitive growth, the investigations on these targets should be considered a laboratory priority. The second level step should be performed to detect additional or subclonal mutations which are usually acquired in varying sized sub-clones of hematopoietic progenitors. The detection of these additional variants should be considered for potential disease progression (prognostic value), in this context, to avoide laboratory response delay and incomplete analysis, revelation of this large number of targets should be approached through NGS methods.

References

Abdelhamid, E., Figeac, M., Renneville, A., et al., 2013. Quantification of JAK2V617F mutation by next-generation sequencing technology. Am. J. Hematol. 88 (6), 536–537. https://doi.org/10.1002/ajh.23446.

Aguilera-Diaz, A., Vazquez, I., Ariceta, B., Mañú, A., Blasco-Iturri, Z., Palomino-Echeverría, S., et al., 2020. Assessment of the clinical utility of four NGS panels in myeloid malignancies. Suggestions for NGS panel choice or design. PLoS One 15 (1), e0227986. https://doi.org/10.1371/journal.pone.0227986.

Albiero, E., Madeo, D., Ruggeri, M., Bernardi, M., Giorgetti, A., Rodeghiero, F., 2008. Loss of the JAK2 intramolecular auto-inhibition mechanism is predicted by structural modelling of a novel exon 12 insertion mutation in a case of idiopathic erythrocytosis. Br. J. Haematol. 142 (6), 986–990. https://doi.org/10.1111/j.1365-2141.2008.07180.x.

Alikian, M., Whale, A.S., Akiki, S., et al., 2017. RT-qPCR and RT-digital PCR: a comparison of different platforms for the evaluation of residual disease in chronic myeloid Leukemia. Clin. Chem. 63 (2), 525–531. https://doi.org/10.1373/clinchem.2016.262824.

Angona, A., Fernández-Rodríguez, C., Alvarez-Larrán, A., et al., 2016. Molecular characterisation of triple negative essential thrombocythaemia patients by platelet analysis and targeted sequencing. Blood Cancer J. 6 (8), e463. Published 2016 Aug 26 https://doi.org/10.1038/bcj.2016.75.

Anon., 2016. Update from the latest WHO classification of MPNs: a user's manual. Hematology Am. Soc. Hematol. Educ. Program 2016 (1), 534–542. https://doi.org/10.1182/asheducation-2016.1.534.

Apperley, J.F., 2015. Chronic myeloid leukaemia. Lancet 385, 1447–1459.

Arber, D.A., Orazi, A., Hasserjian, R., et al., 2016. The 2016 revision to the World Health Organization classification of myeloid neoplasms and acute leukemia. Blood 127 (20), 2391–2405. https://doi.org/10.1182/blood-2016-03-643544.
Arock, M., Sotlar, K., Akin, C., et al., 2015. KIT mutation analysis in mast cell neoplasms: recommendations of the European Competence Network on Mastocytosis. Leukemia 29 (6), 1223–1232.
Baccarani, M., Deininger, M.W., Rosti, G., et al., 2013. European LeukemiaNet recommendations for the management of chronic myeloid leukemia: 2013. Blood 122 (6), 872–884.
Bain, B.J., Gilliland, D.G., Horny, H.-P., et al., 2008. Myeloid and lymphoid neoplasms with eosinophilia and abnormalities of PDGFRA, PDGFRB, or FGFR1. In: Swerdlow, S., Harris, N.L., Stein, H., Jaffe, E.S., Theile, J., Vardiman, J.W. (Eds.), World Health Organization Classification of Tumours. Pathology and Genetics of Tumours of Haematopoietic and Lymphoid Tissues. IARC Press, Lyon, France, pp. 68–73.
Bartels, S., Lehmann, U., Büsche, G., et al., 2015. SRSF2 and U2AF1 mutations in primary myelofibrosis are associated with JAK2 and MPL but not calreticulin mutation and may independently reoccur after allogeneic stem cell transplantation. Leukemia 29 (1), 253–255. https://doi.org/10.1038/leu.2014.277.
Baxter, E.J., Scott, L.M., Campbell, P.J., et al., 2005. Acquired mutation of thetyrosine kinase JAK2 in human myeloproliferative disorders. Lancet 365, 1054–1061.
Beer, P.A., Campbell, P.J., Scott, L.M., et al., 2008. MPL mutations in myeloproliferative disorders: analysis of the PT-1 cohort. Blood 112, 141–149.
Bench, A.J., White, H.E., Foroni, L., et al., 2013. Molecular diagnosis of the myeloproliferative neoplasms: UK guidelines for the detection of JAK2 V617F and other relevant mutations. Br. J. Haematol. 160 (1), 25–34. https://doi.org/10.1111/bjh.12075.
Bernardi, M., Ruggeri, M., Albiero, E., Madeo, D., Rodeghiero, F., 2009. Isolated erythrocytosis in V617F negative patients with JAK2 exon 12 mutations: report of a new mutation. Am. J. Hematol. 84 (4), 258–260. https://doi.org/10.1002/ajh.21357.
Bilbao-Sieyro, C., Florido, Y., Gómez-Casares, M.T., 2016. CALR mutation characterization in myeloproliferative neoplasms. Oncotarget 7 (33), 52614–52617. https://doi.org/10.18632/oncotarget.10376.
Boyd, E.M., Bench, A.J., Goday-Fernández, A., et al., 2010. Clinical utility of routine MPL exon 10 analysis in the diagnosis of essential thrombocythaemia and primary myelofibrosis. Br. J. Haematol. 149, 250–257.
Branford, S., Rudzki, Z., Grigg, A., et al., 2004. BCR-ABL levels continue to decrease up to 42 months after commencement of standard dose Imatinib in patients with newly diagnosed chronic phase CML who achieve a major molecular response. Blood 104 (11), 274. https://doi.org/10.1182/blood.V104.11.274.274.
Busque, L., Patel, J.P., Figueroa, M.E., et al., 2012. Recurrent somatic TET2 mutations in normal elderly individuals with clonal hematopoiesis. Nat. Genet. 44, 1179–1181.
Carillo, S., Henry, L., Lippert, E., et al., 2011. Nested high-resolution melting curve analysis a highly sensitive, reliable, and simple method for detection of JAK2 exon 12 mutations–clinical relevance in the monitoring of polycythemia. J. Mol. Diagn. 13, 263–270. https://doi.org/10.1016/j.jmoldx.2010.12.002.
Chi, J., Manoloukos, M., Pierides, C., Nicolaidou, V., et al., 2015. Calreticulin mutations in myeloproliferative neoplasms and new methodology for their detection and monitoring. Ann. Hematol. 94, 399–408.
Chung, H.J., Hur, M., Yoon, S., et al., 2020. Performance evaluation of the QXDx BCR-ABL %IS droplet digital PCR assay. Ann. Lab. Med. 40 (1), 72–75. https://doi.org/10.3343/alm.2020.40.1.72.

Cross, N.C., 2009. Standardisation of molecular monitoring for chronic myeloid leukaemia. Best Pract. Res. Clin. Haematol. 22 (3), 355–365. https://doi.org/10.1016/j.beha.2009.04.001.

Daley, G.Q., Van Etten, R.A., Baltimore, D., 1990. Induction of chronic myelogenousleukemia in mice by the P210bcr/abl gene of the Philadelphia chromosome. Science 247, 824–830.

Deininger, M.W., Goldman, J.M., Melo, J.V., 2000. The molecular biology of chronic myeloid leukemia. Blood 96, 3343–3356.

Didone, A., Nardinelli, L., Marchiani, M., et al., 2015. Comparative study of different methodologies to detect the JAK2 V617F mutation in chronic BCR-ABL1 negative myeloproliferative neoplasms. Pract. Lab. Med. 4, 30–37. Published 2015 Dec 14 https://doi.org/10.1016/j.plabm.2015.12.004.

Douet-Guilbert, N., Morel, F., Le Charpentier, T., et al., 2004. Interphase FISH for follow-up of Philadelphia chromosome positive chronic myeloid leukemia treatment. Anticancer Res 24, 2535–2540.

Egan, D., Radich, J., 2016. Monitoring disease burden in chronic myeloid leukemia: past, present, and future. Am. J. Hematol. 91 (7), 742–746.

El-Zimaity, M.M., Kantarjian, H., Talpaz, M., et al., 2004. Results of imatinibmesylate therapy in chronic myelogenousleukemia with variant Philadelphia chromosome. Br. J. Haematol. 125, 187–195.

Er, T.K., Lin, S.F., Chang, J.G., Hsieh, L.L., Lin, S.K., Wang, L.H., Lin, C.W., Chang, C.S., Liu, T.C., 2009. Detection of the JAK2 V617F missense mutation by high resolution melting analysis and its validation. Clin. Chim. Acta 408 (1–2), 39–44. https://doi.org/10.1016/j.cca.2009.07.002. Epub 2009 Jul 10. PubMed PMID 19595684.

Fabarius, A., Leitner, A., 2011. HochhausA et al. impact of additional cytogenetic aberrations at diagnosis on prognosis of CML: long-term observation of 1151 patients from the randomized CML Study IV. Blood 118 (26), 6760–6768.

Faderl, S., Talpaz, M., Estrov, Z., et al., 1999. Chronic myelogenousleukemia: biology and therapy. Ann. Intern. Med. 131, 207–219.

Fontanelli, G., Barate, C., Ciabatti, E., et al., 2015. Real-time PCR and droplet digital PCR: two techniques for detection of the JAK2(V617F) mutation in Philadelphia-negative chronic myeloproliferative neoplasms. Int. J. Lab. Hematol. 37, 766–773. https://doi.org/10.1111/ijlh.12404.

Foroni, L., Wilson, G., Gerrard, G., et al., 2011. Guidelines for the measurement of BCR-ABL1 transcripts in chronic myeloid leukaemia. Br. J. Haematol. 153 (2), 179–190. https://doi.org/10.1111/j.1365-2141.2011.08603.x.

Franke, G.N., Maier, J., Wildenberger, K., et al., 2020. Comparison of real-time quantitative PCR and digital droplet PCR for BCR-ABL1 monitoring in patients with chronic myeloid leukemia. J. Mol. Diagn. 22 (1), 81–89. https://doi.org/10.1016/j.jmoldx.2019.08.007.

Furtado, L.V., Weigelin, H.C., Elenitoba-Johnson, K.S., Betz, B.L., 2013. Detection of MPL mutations by a novel allele-specific PCR-based strategy. J. Mol. Diagn. 15, 810–818. https://doi.org/10.1016/j.jmoldx.2013.07.006.

Garcia-Montero, A.C., Jara-Acevedo, M., Teodosio, C., et al., 2006. KIT mutation in mast cells and other bone marrow hematopoietic cell lineages in sistemi mast cell disorders: a prospective study of the Spanish Network on Mastocytosis (REMA) in a series of 113 patients. Blood 108 (7), 2366–2372.

Geay, A., Aral, B., Bourgeois, V., et al., 2020. Diagnosis of exon 12-positive polycythemia vera rescued by NGS. Clin. Case Rep. 8 (5), 790–792. Published 2020 Mar 21 https://doi.org/10.1002/ccr3.2720.

Ghaderi, M., Strömberg, O., Porwit, A., 2010. Rapid real-time PCR assay for detection of MPL W515L mutation in patients with chronic myeloproliferative disorders. Int. J. Lab. Hematol. 32, 122–126.

Gilles, S.R., Baughn, L.B., Schomaker, M.L., Courville, E.L., Nelson, A.C., Sachs, Z., 2017. Buccal epithelial cells display somatic, bone marrow-derived CALR mutation. Blood Adv. 1 (25), 2302–2306. Published 2017 Nov 16 https://doi.org/10.1182/bloodadvances.2017012229.

Gong, J.Z., Cook, J.R., Greiner, T.C., et al., 2013. Laboratory practice guidelines for detecting and reporting JAK2 and MPL mutations in myeloproliferative neoplasms: a report of the Association for Molecular Pathology. J. Mol. Diagn. 15 (6), 733–744. https://doi.org/10.1016/j.jmoldx.2013.07.002.

Gotlib, J., Cools, J., 2008. Five years since the discovery of FIP1L1–PDGFRA: what we have learned about the fusion and other molecularly defined eosinophilias. Leukemia 22, 1999–2010. https://doi.org/10.1038/leu.2008.287.

Guglielmelli, P., Rotunno, G., Fanelli, T., et al., 2015. Validation of the differential prognostic impact of type 1/type 1-like versus type 2/type 2-like CALR mutations in myelofibrosis. Blood Cancer J. 5 (10), e360. Published 2015 Oct 16 https://doi.org/10.1038/bcj.2015.90.

Guglielmelli, P., Rotunno, G., Pacilli, A., et al., 2016. Prognostic impact of bone marrow fibrosis in primary myelofibrosis. A study of the AGIMM group on 490 patients. Am. J. Hematol. 91 (9), 918–922. https://doi.org/10.1002/ajh.24442.

Guglielmelli, P., Pietra, D., Pane, F., Pancrazzi, A., et al., 2017. Recommendations for molecular testing in classical Ph1-neg myeloproliferative disorders-A consensus project of the Italian Society of Hematology. Leuk. Res. 58, 63–72. https://doi.org/10.1016/j.leukres.2017.04.006.

Guo, J.Q., Lin, H., Kantarjian, H., et al., 2002. Comparison of competitive-nested PCR and real-time PCR in detecting BCR-ABL fusion transcripts in chronic myeloid leukemia patients. Leukemia 16 (12), 2447–2453. https://doi.org/10.1038/sj.leu.2402730.

Heaney, M.L., Soriano, G., 2013. Acute myeloid leukemia following a myeloproliferative neoplasm: clinical characteristics, genetic features and effects of therapy. Curr. Hematol. Malig. Rep. 8 (2), 116–122. https://doi.org/10.1007/s11899-013-0154-5. PMID 23572311.

Hermouet, S., Dobo, I., Lippert, E., et al., 2007. Comparison of whole blood vs purified blood granulocytes for the detection and quantitation of JAK2(V617F). Leukemia 21 (5), 1128–1130. Epub 2007 Feb 15. PubMed PMID 17301814.

Hochhaus, A., Baccarani, M., Silver, R.T., et al., 2020. European LeukemiaNet 2020 recommendations for treating chronic myeloid leukemia. Leukemia 34, 966–984. https://doi.org/10.1038/s41375-020-0776-2.

Ishii, T., Bruno, E., Hoffman, R., Xu, M., 2006. Involvement of various hematopoietic-cell lineages by the JAK2V617F mutation in polycythemia vera. Blood 108 (9), 3128–3134. https://doi.org/10.1182/blood-2006-04-017392.

Ivanova, M.I., Shivarov, V.S., Hadjiev, E.A., Naumova, E.J., 2011. Novel multiplex beadbased assay with LNA-modified probes for detection of MPL exon 10 mutations. Leuk. Res. 35, 1120–1123.

Jabbour, E., Cortes, J.E., Kantarjian, H.M., 2008. Molecular monitoring in chronic myeloid leukemia: response to tyrosine kinase inhibitors and prognostic implications. Cancer 112 (10), 2112–2118.

James, C., Ugo, V., Le Couedic, J.P., et al., 2005. A unique clonal JAK2 mutationleading to constitutive signalling causespolycythaemia vera. Nature 434, 1144–1148.

Jiang, Q., Zhao, X.Y., Qin, Y.Z., et al., 2012. The differences and correlations of BCR-ABL transcripts between peripheral blood and bone marrow assays are associated with the molecular responses in the bone marrow for chronic myelogenous leukemia. Am. J. Hematol. 87 (12), 1065–1069. https://doi.org/10.1002/ajh.23321.

Jones, A.V., Ward, D., Lyon, M., et al., 2015. Evaluation of methods to detect CALR mutations in myeloproliferative neoplasms. Leuk. Res. 39 (1), 82–87. https://doi.org/10.1016/j.leukres.2014.11.019.

Jovanovic, J.V., Ivey, A., Vannucchi, A.M., et al., 2013. Establishing optimal quantitative-polymerase chain reaction assays for routine diagnosis and tracking of minimal residual disease in JAK2-V617F-associated myeloproliferative neoplasms: a joint European LeukemiaNet/MPN&MPNr-EuroNet (COST action BM0902) study. Leukemia 27 (10), 2032–2039. https://doi.org/10.1038/leu.2013.219.

Kantarjian, H., Schiffer, C., Jones, D., Cortes, J., 2008. Monitoring the response and course of chronic myeloid leukemia in the modern era of BCR-ABL tyrosine kinase inhibitors: practical advice on the use and interpretation of monitoring methods. Blood 111 (4), 1774–1780.

Kayser, S., Levis, M.J., 2019. Clinical implications of molecular markers in acute myeloid leukemia. Eur. J. Haematol. 102, 20–35. https://doi.org/10.1111/ejh.13172.

Kedes, L., Campany, G., 2011. The new date, new format, new goals and new sponsor of the archon genomics X PRIZE competition. Nat. Genet. 43 (11), 1055–1058. https://doi.org/10.1038/ng.988.

Kouroupi, E., Zoi, K., Parquet, N., et al., 2008. Mutations in exon 12 of JAK2 are mainly found in JAK2 V617F-negative polycythaemia vera patients. Br. J. Haematol. 142 (4), 676–679. https://doi.org/10.1111/j.1365-2141.2008.07223.x.

Kralovics, R., Passamonti, F., Buser, A.S., et al., 2005. A gain-of-function mutationofJAK2 in myeloproliferative disorders. N. Engl. J. Med. 352, 1779–1790.

Kristensen, T., Vestergaard, H., Bindslev-Jensen, C., et al., 2014. Sensitive KIT D816V mutation analysis of blood as a diagnostic test in mastocytosis. Am. J. Hematol. 89 (5), 493–498.

Kröger, N., Badbaran, A., Holler, E., Hahn, J., Kobbe, G., Bornhäuser, M., et al., 2007. Monitoring of the JAK2-V617F mutation by highly sensitive quantitative real-time PCR after allogeneic stem cell transplantation in patients with myelofibrosis. Blood 109, 1316–1321.

Kuo, F.C., Dong, F., 2015. Next-generation sequencing-based panel testing for myeloid neoplasms. Curr. Hematol. Malig. Rep. 10 (2), 104–111.

La Starza, R., Specchia, G., Cuneo, A., et al., 2005. The hypereosinophilic syndrome: fluorescence in situ hybridization detects the del(4)(q12)-FIP1L1/PDGFRA but not genomic rearrangements of other tyrosine kinases. Haematologica 90 (5), 596–601.

Larsen, T.S., Christensen, J.H., Hasselbalch, H.C., Pallisgaard, N., 2007. The JAK2 V617F mutation involves B- and T-lymphocyte lineages in a subgroup of patients with Philadelphia-chromosome negative chronic yeloproliferative disorders. Br. J. Haematol. 136, 745–751.

Lasho, T.L., Finke, C.M., Tischer, A., Pardanani, A., Tefferi, A., 2018. Mayo CALR mutation type classification guide using alpha helix propensity. Am. J. Hematol. 93 (5), E128–E129. https://doi.org/10.1002/ajh.25065.

Laughlin, T.S., Moliterno, A.R., Stein, B.L., Rothberg, P.G., 2010. Detection of exon 12 mutations in the JAK2 gene: enhanced analytical sensitivity using clamped PCR and nucleotide sequencing [published correction appears in J Mol Diagn. 2010 Sep;12(5):730]. J. Mol. Diagn. 12 (3), 278–282. https://doi.org/10.2353/jmoldx.2010.090177.

Lemoine, F., Najman, A., Baillou, C., et al., 1986. A prospective study of the value of bone marrow erythroid progenitor cultures in polycythemia. Blood 68, 996–1002.

Levine, R.L., Wadleigh, M., Cools, J., et al., 2005. Activating mutation in thetyrosine kinase JAK2 in polycythemia vera, essential thrombocythemia,and myeloid metaplasia with myelofibrosis. Cancer Cell 7, 387–397.

Li, S., Kralovics, R., De Libero, G., Theocharides, A., Gisslinger, H., Skoda, R.C., 2008. Clonal heterogeneity in polycythemia vera patients with JAK2 exon12 and

JAK2-V617F mutations. Blood 111 (7), 3863–3866. https://doi.org/10.1182/blood-2007-09-111971.

Link-Lenczowska, D., Pallisgaard, N., Cordua, S., Zawada, M., Czekalska, S., Krochmalczyk, D., et al., 2018. A comparison of qPCR and ddPCR used for quantification of the JAK2 V617F allele burden in Ph negative MPNs. Ann. Hematol. 97, 2299–2308. https://doi.org/10.1007/s00277-018-3451-1.

Lippert, E., Boissinot, M., Kralovics, R., et al., 2006. The JAK2-V617F mutation is frequently present at diagnosis in patients with essential thrombocythemia and polycythemia vera. Blood 108 (6), 1865–1867. https://doi.org/10.1182/blood-2006-01-013540.

Lippert, E., Girodon, F., Hammond, E., et al., 2009. Concordance of assays designed for the quantification of JAK2V617F: a multicenter study. Haematologica 94 (1), 38–45. https://doi.org/10.3324/haematol.13486.

Lundberg, P., Karow, A., Nienhold, R., et al., 2014. Clonal evolution and clinical correlates of somatic mutations in myeloproliferative neoplasms. Blood 123 (14), 2220–2228. https://doi.org/10.1182/blood-2013-11-537167.

Ma, W., Kantarjian, H., Zhang, X., et al., 2009. Mutation profile of JAK2 transcripts in patients with chronic myeloproliferative neoplasias. J. Mol. Diagn. 11 (1), 49–53. https://doi.org/10.2353/jmoldx.2009.080114.

Maier, C.L., Fisher, K.E., Jones, H.H., et al., 2015. Development and validation of CALR mutation testing for clinical diagnosis. Am. J. Clin. Pathol. 144, 738–745. https://doi.org/10.1309/AJCPXPA83MVCTSOQ.

Maier, J., Lange, T., Cross, M., Wildenberger, K., Niederwieser, D., Franke, G.N., 2019. Optimized digital droplet PCR for BCR-ABL. J. Mol. Diagn. 21 (1), 27–37. https://doi.org/10.1016/j.jmoldx.2018.08.012.

Mansier, O., Migeon, M., Saint-Lezer, A., James, C., Verger, E., Robin, M., et al., 2016a. Quantification of the mutant CALR allelic burden by digital PCR: application to minimal residual disease evaluation after bone marrow transplantation. J. Mol. Diagn. 18, 68–74. https://doi.org/10.1016/j.jmoldx.2015.07.007.

Mansier, O., Migeon, M., Etienne, G., Bidet, A., Lippert, E., 2016b. JAK2V617F and CALR double mutations are more frequently encountered in patients with low JAK2V617F allelic burdens. Leuk. Lymphoma 57 (8), 1949–1951. https://doi.org/10.3109/10428194.2015.1116122.

Martínez-Avilés, L., Besses, C., Alvarez-Larrán, A., Cervantes, F., Hernández-Boluda, J.C., Bellosillo, B., 2007. JAK2 exon 12 mutations in polycythemia vera or idiopathic erythrocytosis. Haematologica 92 (12), 1717–1718. https://doi.org/10.3324/haematol.12011.

McGowan-Jordan, J., Simons, A., Schmid, M., International Standing Committee on Human Cytogenomic Nomenclature, ISCN : an international system for human cytogenomic nomenclature (2016), Basel ; New York : Karger, [2016]. NLM ID: 101678834 [Book].

Mehrotra, M., Luthra, R., Singh, R.R., et al., 2015. Clinical validation of a multipurpose assay for detection and genotyping of CALR mutations in myeloproliferative neoplasms. Am. J. Clin. Pathol. 144, 746–755. https://doi.org/10.1309/AJCP5LA2LDDNQNNC.

Merker, J.D., Jones, C.D., Oh, S.T., et al., 2010. Design and evaluation of a real-time PCR assay for quantification of JAK2 V617F and wild-type JAK2 transcript levels in the clinical laboratory. J. Mol. Diagn. 12 (1), 58–64. https://doi.org/10.2353/jmoldx.2010.090068.

Milosevic Feenstra, J.D., Nivarthi, H., Gisslinger, H., et al., 2016. Whole-exome sequencing identifies novel MPL and JAK2 mutations in triple-negative myeloproliferative neoplasms. Blood 127 (3), 325–332. https://doi.org/10.1182/blood-2015-07-661835.

Milosevic, J.D., Puda, A., Malcovati, L., et al., 2012. Clinical significance of genetic aberrations in secondary acute myeloid leukemia. Am. J. Hematol. 87 (11), 1010–1016. https://doi.org/10.1002/ajh.23309.

Mitterbauer, G., Winkler, K., Gisslinger, H., Geissler, K., Lechner, K., Mannhalter, C., 1999. Clonality analysis using X-chromosome inactivation at the human androgen receptor gene (Humara). Evaluation of large cohorts of patients with chronic myeloproliferative diseases, secondary neutrophilia, and reactive thrombocytosis. Am. J. Clin. Pathol. 112 (1), 93–100. https://doi.org/10.1093/ajcp/112.1.93.

Molica, M., Massaro, F., Breccia, M., 2017. Diagnostic and prognostic cytogenetics of chronic myeloid leukaemia: an update. Expert Rev. Mol. Diagn. 17 (11), 1001–1008. https://doi.org/10.1080/14737159.2017.1383156.

Morris, C.M., Fitzgerald, P.H., 1985. An evaluation of high-resolution chromosome banding of hematologic cells by methotrexate synchronization and thymidine release. Cancer Genet. Cytogenet. 14, 275–284.

Murugesan, G., Guenther-Johnson, J., Mularo, F., et al., 2016. Validation of a molecular diagnostic assay for CALR exon 9 indels in myeloproliferative neoplasms: identification of coexisting JAK2 and CALR mutations and a novel 9 bp deletion in CALR. Int. J. Lab. Hematol. 38, 284–297. https://doi.org/10.1111/ijlh.12484.

Nangalia, J., Massie, C.E., Baxter, E.J., et al., 2013. Somatic CALR mutations inmyeloproliferative neoplasms with nonmutated JAK2. N. Engl. J. Med. 369, 2391–2405.

Nangalia, J., Nice, F.L., Wedge, D.C.D., et al., 2015. DNMT3A mutations occur early or late in patients with myeloproliferative neoplasms and mutation order influences phenotype. Haematologica 100, e438–e442.

Nowell, P.C., Hungerford, D.A., 1960. Chromosome studies on normal and leukemic human leukocytes. J. Natl. Cancer Inst. 25, 85–109.

Nussenzveig, R.H., Swierczek, S.I., Jelinek, J., Gaikwad, A., Liu, E., Verstovsek, S., et al., 2007. Polycythemia vera is not initiated by JAK2V617F mutation. Exp. Hematol. 35, 32–38.

Ortmann, C.A., Kent, D.G., Nangalia, J., et al., 2015. Effect of mutation order on myeloproliferative neoplasms. N. Engl. J. Med. 372, 601–612.

Pallera, A., Altman, J.K., Berman, E., et al., 2016. NCCN guidelines insights: chronic myeloid leukemia, version 1.2017. J. Natl. Compr. Canc. Netw. 14 (12), 1505–1512.

Pancrazzi, A., Guglielmelli, P., Ponziani, V., et al., 2008. A sensitivedetection method for MPLW515L or MPLW515K mutation inchronic myeloproliferative disorders with locked nucleic acidmodifiedprobes and real-time polymerase chain reaction. J. Mol. Diagn. 10, 435–441.

Pardanani, A., 2005. Systemic mastocytosis: bone marrow pathology, classification, and current therapies. Acta Haematol. 114, 41–51.

Pardanani, A., Lasho, T.L., Finke, C., Hanson, C.A., Tefferi, A., 2007. Prevalence and clinicopathologic correlates of JAK2 exon 12 mutations in JAK2V617F-negative polycythemia vera. Leukemia 21 (9), 1960–1963. https://doi.org/10.1038/sj.leu.2404810.

Pardanani, A., Lasho, T., Finke, C., Oh, S.T., Gotlib, J., Tefferi, A., 2010. LNK mutation studies in blast-phase myeloproliferative neoplasms, and in chronic-phase disease with TET2, IDH, JAK2 or MPL mutations. Leukemia 24 (10), 1713–1718. https://doi.org/10.1038/leu.2010.163.

Park, S.H., Kim, S.Y., Lee, S.M., et al., 2015. Incidence, clinical features, and prognostic impact of CALR exon 9 mutations in essential thrombocythemia and primary myelofibrosis: an experience of a single tertiary hospital in Korea. Ann. Lab. Med. 35 (2), 233–237. https://doi.org/10.3343/alm.2015.35.2.233.

Park, D.J., Kwon, A., Cho, B.S., et al., 2020. Characteristics of DNMT3A mutations in acute myeloid leukemia. Blood Res. 55 (1), 17–26. https://doi.org/10.5045/br.2020.55.1.17.

Passamonti, F., Rumi, E., Pietra, D., et al., 2010. A prospective study of 338 patients with polycythemia vera: the impact of JAK2 (V617F) allele burden and leukocytosis on fibrotic or leukemic disease transformation and vascular complications. Leukemia 24 (9), 1574–1579. https://doi.org/10.1038/leu.2010.148.

Passamonti, F., Mora, B., Maffioli, M., 2016. New molecular genetics in the diagnosis and treatment of myeloproliferative neoplasms. Curr. Opin. Hematol. 23 (2), 137–143. https://doi.org/10.1097/MOH.0000000000000218. PMID 26825696.

Percy, M.J., Scott, L.M., Erber, W.N., et al., 2007. The frequency of JAK2 exon 12 mutations in idiopathic erythrocytosis patients with low serum erythropoietin levels. Haematologica 92 (12), 1607–1614. https://doi.org/10.3324/haematol.11643.

Pietra, D., Li, S., Brisci, A., et al., 2008. Somatic mutations of JAK2 exon 12 in patients with JAK2 (V617F)-negative myeloproliferative disorders. Blood 111 (3), 1686–1689. https://doi.org/10.1182/blood-2007-07-101576.

Pietra, D., Brisci, A., Rumi, E., et al., 2011. Deep sequencing reveals double mutations in cis of MPL exon 10 in myeloproliferative neoplasms. Haematologica 96, 607–611.

Pietra, D., Rumi, E., Ferretti, V.V., et al., 2016. Differential clinical effects of different mutation subtypes in CALR-mutant myeloproliferative neoplasms. Leukemia 30 (2), 431–438. https://doi.org/10.1038/leu.2015.277.

Prchal, J.F., Axelrad, A.A., 1974. Bone-marrow responses in polycythemia vera. N. Engl. J. Med. 290, 1382.

Prchal, J.T., Guan, Y.L., 1993. A novel clonality assay based on transcriptional analysis of the active X chromosome. Stem Cells 11, 62–65.

Reiter, A., Gotlib, J., 2017. Myeloid neoplasms with eosinophilia. Blood 129 (6), 704–714. https://doi.org/10.1182/blood-2016-10-695973.

Roloff, G.W., Griffiths, E.A., 2018. When to obtain genomic data in acute myeloid leukemia (AML) and which mutations matter. Blood Adv. 2 (21), 3070–3080. https://doi.org/10.1182/bloodadvances.2018020206.

Sadrzadeh, H., Abdel-Wahab, O., Fathi, A.T., 2011. Molecular alterations underlying eosinophilic and mast cell malignancies. Discov. Med. 12 (67), 481–493.

Schnittger, S., Bacher, U., Haferlach, C., et al., 2009. Characterization of 35 new cases with four different MPLW515 mutations and essential thrombocytosis or primary myelofibrosis. Haematologica 94, 141–144.

Schoch, C., Schnittger, S., Bursch, S., et al., 2002. Comparison of chromosome banding analysis, interphase-and hypermetaphase-FISH, qualitative and quantitative PCR for diagnosis and for follow-up in chronic myeloid leukemia: a study on 350 cases. Leukemia 16 (1), 53.

Schofield, C.J., Ratcliffe, P.J., 2005. Signalling hypoxia by HIF hydroxylases. Biochem. Biophys. Res. Commun. 338 (1), 617–626.

Scott, L.M., Tong, W., Levine, R.L., et al., 2007. JAK2 exon 12 mutations in polycythemia vera and idiopathic erythrocytosis. N. Engl. J. Med. 356 (5), 459–468. https://doi.org/10.1056/NEJMoa065202.

Semenza, G.L., 2009. Involvement of oxygen-sensing pathways in physiologic and pathologic erythropoiesis. Blood 114 (10), 2015–2019.

Shaver, L.G., Tommerup, N., 2009. An International System of Human Cytogenetic Nomenclature. Karger: ISCN.

Shen, Y., Zhu, Y.M., Fan, X., et al., 2011. Gene mutation patterns and their prognostic impact in a cohort of 1185 patients with acute myeloid leukemia. Blood 118 (20), 5593–5603. https://doi.org/10.1182/blood-2011-03-343988.

Shih, L.Y., Lee, C.T., Ou, Y.C., 1997. Prediction of clinical course in patients with idiopathic erythrocytosis by endogenous erythroid colony assay but not by serum erythropoietin levels. Exp. Hematol. 25, 288–292.

Shih, L.Y., Lee, C.T., See, L.C., et al., 1998. In vitro culture growth of erythroid progenitors and serum erythropoietin assay in the differential diagnosis of polycythaemia. Eur. J. Clin. Invest. 28, 569–576.

Shomali, W., Gotlib, J., 2018. The new tool "KIT" in advanced systemic mastocytosis. Hematology Am. Soc. Hematol. Educ. Program 2018 (1), 127–136. https://doi.org/10.1182/asheducation-2018.1.127.

Silver, R.T., Vandris, K., Wang, Y.L., et al., 2011. JAK2(V617F) allele burden in polycythemia vera correlates with grade of myelofibrosis, but is not substantially affected by therapy. Leuk. Res. 35 (2), 177–182. https://doi.org/10.1016/j.leukres.2010.06.017.

Soverini, S., Hochhaus, A., Nicolini, F.E., Gruber, F., Lange, T., Saglio, G., Pane, F., Müller, M.C., Ernst, T., Rosti, G., Porkka, K., Baccarani, M., Cross, N.C., Martinelli, G., 2011. BCR-ABL kinase domain mutation analysis in chronic myeloid leukemia patients treated with tyrosine kinase inhibitors: recommendations from an expert panel on behalf of European LeukemiaNet. Blood 118 (5), 1208–1215. https://doi.org/10.1182/blood-2010-12-326405. Epub 2011 May 11. PMID 21562040.

Specimen Considerations For Individuals With Hematologic Disease, 2018. https://www.genedx.com/wp-content/uploads/2018/08/Specimen-Considerations-for-Individuals-with-Hematologic-Disease.pdf.

Stock, W., Yu, D., Karrison, T., et al., 2006. Quantitative real-time RT-PCR monitoring of BCR-ABL in chronic myelogenous leukemia shows lack of agreement in blood and bone marrow samples. Int. J. Oncol. 28 (5), 1099–1103.

Swansbury, J., 2012. Cancer Cytogenetics: Methods and Protocols (Methods in Molecular Biology). 220 Humana Press, Totowa, New Jersey.

Swierczek, S., Lima, L.T., Tashi, T., Kim, S.J., Gregg, X.T., Prchal, J.T., 2015. Presence of polyclonal hematopoiesis in females with Ph-negative myeloproliferative neoplasms. Leukemia 29 (12), 2432–2434. https://doi.org/10.1038/leu.2015.249.

Tefferi, A., Pardanani, A., 2014. Genetics: CALR mutations and a new diagnostic algorithm for MPN. Nat. Rev. Clin. Oncol. 11, 125–126. https://doi.org/10.1038/nrclinonc.2014.16.

Tefferi, A., Guglielmelli, P., Lasho, T.L., et al., 2014a. CALR and ASXL1 mutations-based molecular prognostication in primary myelofibrosis: an international study of 570 patients. Leukemia 28 (7), 1494–1500. https://doi.org/10.1038/leu.2014.57.

Tefferi, A., Lasho, T.L., Tischer, A., et al., 2014b. The prognostic advantage of calreticulin mutations in myelofibrosis might be confined to type 1 or type 1-like CALR variants. Blood 124 (15), 2465–2466. https://doi.org/10.1182/blood-2014-07-588426.

Tenedini, E., Bernardis, I., Artusi, V., et al., 2014. Targeted cancer exome sequencing reveals recurrent mutation in myeloproliferative neoplasms. Leukemia 28, 1052–1059.

Usseglio, F., Beaufils, N., Calleja, A., Raynaud, S., Gabert, J., 2017. Detection of CALR and MPL mutations in low allelic burden JAK2 V617F essential Thrombocythemia. J. Mol. Diagn. 19 (1), 92–98. https://doi.org/10.1016/j.jmoldx.2016.08.006.

Vainchenker, W., Delhommeau, F., Constantinescu, S.N., et al., 2011. New mutations and pathogenesis of myeloproliferative neoplasms. Blood 118, 1723–1735.

Valent, P., Akin, C., Sperr, W.R., et al., 2005. Mastocytosis: pathology, genetics, and current options for therapy. Leuk. Lymphoma 46, 35–48.

Vandenberghe, P., Wlodarska, I., Michaux, L., et al., 2004. Clinical and molecular features of FIP1L1-PDFGRA (+) chronic eosinophilic leukemias. Leukemia, 734–742. https://doi.org/10.1038/sj.leu.2403313.

Vannucchi, A.M., Antonioli, E., Guglielmelli, P., et al., 2007a. Prospective identification of high-risk polycythemia vera patients based on JAK2(V617F) allele burden. Leukemia 21 (9), 1952–1959. https://doi.org/10.1038/sj.leu.2404854.

Vannucchi, A.M., Antonioli, E., Guglielmelli, P., et al., 2007b. Clinical profile of homozygous JAK2 617V>F mutation in patients with polycythemia vera or essential thrombocythemia. Blood 110 (3), 840–846. https://doi.org/10.1182/blood-2006-12-064287.

Vannucchi, A.M., Lasho, T.L., Guglielmelli, P., et al., 2013. Mutations and prognosis in primary myelofibrosis. Leukemia 27 (9), 1861–1869. https://doi.org/10.1038/leu.2013.119.

Vočanec, D., Prijatelj, T., Debeljak, N., Kunej, T., 2019. Genetic variants of erythropoietin (EPO) and EPO receptor genes in familial erythrocytosis. Int. J. Lab. Hematol. 41 (2), 162–167. https://doi.org/10.1111/ijlh.12949.

Wang, G.L., Jiang, B.H., Rue, E.A., Semenza, G.L., 1995. Hypoxia-inducible factor 1 is a basic-helix-loop-helix-PAS heterodimer regulated by cellular O2 tension. Proc. Natl. Acad. Sci. U. S. A. 92 (12), 5510–5514.

Waterhouse, M., Follo, M., Pfeifer, D., et al., 2016. Sensitive and accurate quantification of JAK2 V617F mutation in chronic myeloproliferative neoplasms by droplet digital PCR. Ann. Hematol. 95, 739–744. https://doi.org/10.1007/s00277-016-2623-0.

Webber, L.M., Garson, O.M., 1983. Fluorodeoxyuridine synchronization of bone marrow cultures. Cancer Genet. Cytogenet. 8, 123–132.

WHO, 2016, WHO/BS/2016.2293 Collaborative study to evaluate the proposed WHO 1st International Reference Panel for Genomic JAK2 V617F : http://www.who.int/biologicals/ ECBS_2016_BS2293_JAK2_WHO_reference_panel. pdf?ua=1.

WHO, 2017. WHO Expert Committee on Biological Standardization, Sixty-Seventh Report(WHO Technical Report Series; No. 1004). ISBN 978-92-4-121013-3ISBN (PDF) 978-92-4-069645-7ISSN 0512-3054, World Health Organization.

Williams, D.M., Kim, A.H., Rogers, O., Spivak, J.L., Moliterno, A.R., 2007. Phenotypic variations and new mutations in JAK2 V617F-negative polycythemia vera, erythrocytosis, and idiopathic myelofibrosis. Exp. Hematol. 35 (11), 1641–1646. https://doi.org/10.1016/j.exphem.2007.08.010.

Yan, Z., Sun, Q., Zhang, H., et al., 2019. Advantages of digital PCR in the detection of low abundance BCR-ABL1 gene in patients with chronic myeloid leukemia. Oncol. Lett. 18 (5), 5139–5144. https://doi.org/10.3892/ol.2019.10861.

Zapparoli, G.V., Jorissen, R.N., Hewitt, C.A., et al., 2013. Quantitative threefold allele-specific PCR (QuanTAS-PCR) for highly sensitive JAK2V617F mutant allele detection. BMC Cancer 13, 206. https://doi.org/10.1186/1471-2407-13-206.

Zhuge, J., Zhang, W., Zhang, W., Xu, M., Hoffman, R., 2010. Sensitive detection of MPLW515L/K mutations by amplification refractory mutation system (ARMS)-PCR. Clin. Chim. Acta 411, 122–123.

FSC
www.fsc.org
MIX
Papier | Fördert gute Waldnutzung
FSC® C083411

Kolibri 360 GmbH
Ferdinand-Jühlke-Straße 7
99095 Erfurt, Deutschland
produktsicherheit@kolibri360.de